Cell Proliferation & Apoptosis

Cell Proliferation & Apoptosis

D. Hughes, PhD
Clinical Scientist, Nuffield Department of Surgery, University of Oxford

H. Mehmet, PhD
Weston Senior Lecturer in Neurobiology, Imperial College of Science, Technology and Medicine, University of London

A CIP catalogue record for this book is available from the British Library.

ISBN 1 85996 193 2

BIOS Scientific Publishers Ltd
9 Newtec Place, Magdalen Road, Oxford OX4 1RE, UK
Tel. +44 (0)1865 726286. Fax +44 (0)1865 246823
World Wide Web home page: http://www.bios.co.uk/

Production Editor: Phil Dines
Typeset by Charon Tec Pvt. Ltd, Chennai, India
Printed by The Cromwell Press, Trowbridge, UK

Contents

4 Cell Cycle Proteins 77

Nicholas C. Lea and N. Shaun B. Thomas

**Colour plates can be found between pages 46 and 47, 110 and 111,
and 206 and 207.**

Contributors

Colin Adrain, Molecular Cell Biology Laboratory, Department of Genetics, The Smurfit Institute, Trinity College, Dublin 2, Ireland

Eric Baehrecke, Center for Agricultural Biotechnology, University of Maryland, Biotechnology Institute, College Park, MD 20742, USA

Helen M. Beere, La Jolla Institute for Allergy and Immunology, 10355 Science Center Drive, San Diego, CA 92121, USA

Marco Corazzari, Department of Experimental Medicine, University of Rome, 'Tor Vergata', via Tor Vergata, 00133 Rome, Italy

Emma M. Creagh, Molecular Cell Biology Laboratory, Department of Genetics, The Smurfit Institute, Trinity College, Dublin 2, Ireland

Vincenzo de Laurenzi, Department of Experimental Medicine, University of Rome, 'Tor Vergata', via Tor Vergata, 00133 Rome, Italy

Katharina D'Herde, Dept Anatomy and Embryology, University of Ghent, Godshuizenlaan, 4, B-9000 Ghent, Belgium

Ewald Dumont, Department Biochemistry, Cardiovascular Research Institute Maastricht, University Maastricht, PO Box 616, 6200 MD Maastricht, the Netherlands

Douglas R. Green, Adjunct Professor of Biology, La Jolla Institute for Allergy and Immunology, 10355 Science Center Drive, La Jolla, CA92121, USA

Leo Hofstra, Department Biochemistry, Cardiovascular Research Institute Maastricht, University Maastricht, PO Box 616, 6200 MD Maastricht, the Netherlands

David Hughes, Nuffield Department of Surgery, John Radcliffe Hospital, Headington, Oxford, OX3 9DU, UK

Heidi Kenis, Department Biochemistry, Cardiovascular Research Institute Maastricht, University Maastricht, PO Box 616, 6200 MD Maastricht, the Netherlands

Nicholas C. Lea, Guy's, King's, St Thomas' School of Medicine and Dentistry, Department of Haematological Medicine, Leukaemia Sciences, King's Denmark Hill Campus, The Rayne Institute, 123 Cold Harbour Lane, London SE5 9NU, UK

Seamus J. Martin, Smurfit Professor of Medical Genetics, Molecular Cell Biology Laboratory, Department of Genetics, The Smurfit Institute, Trinity College, Dublin 2, Ireland

Christopher P Maske, Sir William Dunn School of Pathology, South Parks Road, Oxford OX1 3RE, UK

Huseyin Mehmet, Weston Laboratory, Division of Paediatrics, Obstetrics and Gynaecology, Imperial College of Science, Technology and Medicine, Hammersmith Hospital, Du Cane Road, London W12 0NN, UK

Sylvie Mussche, Department of Human Anatomy, Embryology, Histology and Medical Physics, Faculty of Medicine and Health Sciences, University of Ghent, Godshuizenlaan, 4, B-9000 Ghent, Belgium

Andrew Oberst, Department of Experimental Medicine, University of Rome, 'Tor Vergata', via Tor Vergata, 00133 Rome, Italy

Marco Ranalli, Department of Experimental Medicine, University of Rome, 'Tor Vergata', via Tor Vergata, 00133 Rome, Italy

Chris P. Reutelingsperger, Department Biochemistry, Cardiovascular Research Institute Maastricht, University Maastricht, PO Box 616, 6200 MD Maastricht, the Netherlands

Jean-Ehrland Ricci, La Jolla Institute for Allergy and Immunology, 10355 Science Center Drive, San Diego, CA 92121, USA

Karin Roberg, Division of Pathology 2, Faculty of Health Science, Linköping University, S-581 85 Linköping, Sweden

Predrag Slijepcevic, Department of Biological Sciences, Brunel University, Uxbridge, UB8 3PH, UK

N. Shaun B. Thomas, Guy's, King's, St Thomas' School of Medicine and Dentistry, Department of Haematological Medicine, Leukaemia Sciences, King's Denmark Hill Campus, The Rayne Institute, 123 Cold Harbour Lane, London SE5 9NU, UK

Antonio Torres-Montaner, Servicio de Anatomía Patológica, Hospital Universitario de Puerto Real, C/Nacional IV, km 665, 11510 Puerto Real, Cádiz, Spain

Joseph A. Trapani, Cancer cell death laboratory, Peter MacCallum Cancer Institute, St. Andrews Place, East Melbourne, Victoria 3002, Australia

Hugo van Genderen, Department Biochemistry, Cardiovascular Research Institute Maastricht, University Maastricht, PO Box 616, 6200 MD Maastricht, the Netherlands

Waander van Heerde, Department Biochemistry, Cardiovascular Research Institute Maastricht, University Maastricht, PO Box 616, 6200 MD Maastricht, the Netherlands

David J. Vaux, Sir William Dunn School of Pathology, South Parks Road, Oxford OX1 3RE, UK

Nigel J. Waterhouse, Cancer cell death laboratory, Peter MacCallum Cancer Institute, St. Andrews Place, East Melbourne, Victoria 3002, Australia

Ian Zachary, Rayne Institute, Department of Medicine, University College London, Gower Street, London WC1E 6BT, UK

Abbreviations

$\Delta\psi_m$	mitochondrial transmembrane potential	**DAPI**	4',6'-diamidino-2-phenylindole
2D	two-dimensional	**DCFDA**	2',7',-dichlorodihydrofluorescein diacetate
3D	three-dimensional	**DEAD-box**	DEAD represents the one letter code
7-AAD	7-aminoactinomycin D	**proteins**	for the tetrapeptide Asp-Glu-Ala-Asp
ADK	adenylate kinase	**DED**	death effector domain
AFC	7-amino-4-fluoromethyl coumarin	**DFC**	dense fibrillar component
AgNOR	argyrophilic nucleolar organiser region	**DFF**	DNA fragmentation factor
AIF	apoptosis-inducing factor	**DIABLO**	direct IAP binding protein with low PI
ALLN	N-acetyl-leu-leu-norleucinal	**DIFP**	di-isopropyl fluoro phosphate
ALT	alternative lengthening of telomeres	**Dig-dUTP**	digoxygenin-11-dUTP
AMC	7-amino-4-methyl coumarin	**DISC**	death-inducing signalling complex
ANT	adenine nucleotide transporter	**DKC**	dyskeratosis congenita
AO	acridine orange	**DMEM**	Dulbecco's modified Eagle's medium
APAF-1	apoptosis protein-activating factor-1	**DMSO**	dimethylsulphoxide
APT	aminophospholipid translocase	**DNA**	deoxyribonucleic acid
AT	ataxia telangiectasia	**DNA-PK**	DNA-dependent protein kinase
ATM	ataxia telangiectasia gene	**DNAse**	deoxyribonuclease
ATP	adenosine triphosphate	**DPX**	Distrene Plasticizer Xylene
BCA	bicinchoninic acid	**DSB**	double strand break
BrdU	5-bromo-2'-deoxyuridine	**DTT**	dithiothreitol
BSA	bovine serum albumin	**ECL**	enhanced chemiluminescence
CAD	caspase-activated DNAse	**EDTA**	ethylenediaminetetraacetic acid
CARD	caspase-associated recruitment domain	**EGTA**	ethylene glycol-bis-(β-amino ethyl ether) N,N,N',N',-tetra acetic acid
CBs	coiled bodies	**ELISA**	enzyme-linked immuno-sorbent assay
CCCP	carbamoyl cyanide n-chlorophenylhydrazone	**EM**	electron microscope
CCD	charge-coupled device	**EndoG**	endonuclease G
CDKs	cyclin-dependent kinases	**ER**	endoplasmic reticulum
CEB	cell extract buffer	**ESMS**	electrospray mass spectrometry
CFSE	CFDA-SE – carboxyfluorescein diacetate, succinimidyl ester	**FACS**	fluorescence-activated cell sorting
CHAPS	3-[(3-chloramidopropyl)dimethyl-ammonio]-1-propanesulphonate	**FADD**	Fas-associated death domain
CML	chronic myeloid leukaemia	**FC**	fibrillar centre
con A	concanavalin A	**FCCP**	carbonyl cyanide p-(trifluoromethoxy) phenylhydrazone
CRP	C-reactive protein		
CSK	cytoskeleton		
CSLM	confocal scanning laser microscopy		
CV	coefficient of variation		

FCM	flow cytometry
FCS	fetal calf serum
FDA	fluorescein-diacetate
FISH	fluorescence *in situ* hybridisation
FIGE	field inversion gel electrophoresis
FITC	fluorescein isothiocyanate
FLIM	fluorescence lifetime imaging
FLIP	fluorescence loss in photobleaching
FPLC	fast performance liquid chromatography
FRAP	fluorescence recovery after photobleaching
FRET	fluorescence resonance energy transfer
FSC	forward scatter
FSH	follicle stimulating hormone
G1-pm	G1 post-mitosis
G1-ps	G1 pre-S phase
GC	granular component
GFP	green fluorescent protein
GlcNAc	N-acetylglucosamine
HBSS	Hank's balanced salt solution
HDAC	histone deacetylase
Hepes	N-[2-hydroxyethyl]piperazine-N′-[2-ethanesulphonic acid]
HERDS	heterogeneous ectopic RNP-derived structures
HIV-1	human immunodeficiency virus-1
HPLC	high performance liquid chromatography
HR	homologous recombination
HRP	horseradish peroxidase
HSP	heat shock protein
HSV	Herpes simplex virus
IAP	inhibitor of apoptosis proteins
ICAD	inhibitor of caspase-activated deoxyribonuclease
IEF	isoelectric focusing
IMS	industrial methylated spirit
IR	ionising radiation
Its	interstitial telomeric sequences
ITT	*in vitro* transcription and translation
LAP	lamina-associated polypeptides
LBR	lamin B receptor
LM	light microscope
LMP	low melting point
MALDI-MS	matrix-assisted laser desorption ionisation mass spectrometry
MC 540	fluorescent merocyanine 540
MCM	mini chromosome maintenance protein
MOPS	3-(N-morpholino) propanesulphonic acid
mRNA	messenger RNA
MTS	3-(4,5-dimethylthiazol-2-yl)-5-(3-carboxymethoxyphenyl)-2-(4-sulphophenyl)-2H-tetrazolium, inner salt
MTT	methylthiazoletetrazolium
NBS	Nijmegen breakage syndrome
NE	nuclear envelope
NEM	N-ethyl maleimide
NFDM	non-fat dried milk
NHEJ	non-homologous end joining
NMP	nuclear matrix protein
NOR	nucleolar organiser region
NPC	nuclear pore complex
Nrap	nucleolar RNA-associated protein
PAGE	polyacrylamide gel electrophoresis
PARP	poly ADP-ribose polymerase
PBS	phosphate-buffered saline
PCD	programmed cell death
PCNA	proliferating cell nuclear antigen
PCR	polymerase chain reaction
PEG	polyethylene glycol
PFA	paraformaldehyde
PFGE	pulse-filed gel electrophoresis
PHA	phytohaemagglutinin
PI	propidium iodide
PIPES	1,4-piperazinediethanesulphonic acid
PLSCR	phospholipid scramblase
PML	polymorphonuclear leucocyte
PMSF	phenylmethylsulphonyl fluoride
pNA	p-nitroaniline
PNA	peptide nucleic acid
PNBs	pre-nucleolar bodies
PNC	peri-nucleolar compartment
PNGase	N-linked-glycopeptide-[N-acetyl β-D-glucosaminyl]-L-asparagine amidohydrolase
pRB	retinoblastoma protein
pre-rRNAs	rRNA precursor molecules
PS	phosphatidylserine
PSP	paraspeckle protein
pot	protection of telomeres
PT	permeability transition
PTB	polypyrimidine tract binding protein
PTD	protein transduction domain

Q-FISH	quantitative fluorescence *in situ* hybridisation		**TBF**	TATA-binding protein
RANKL	receptor activator of nuclear factor-kappa B ligand		**TBP**	telomere binding protein
RB	retinoblastoma		**TBS**	Tris-buffered saline
rDNA	ribosomal DNA		**TBST**	Tris-buffered saline – Tween
RNA	ribonucleic acid		**TCA**	trichloroacetic acid
RNA pol-I	RNA polymerase-1		**TdT**	terminal deoxynucleotidyl transferase
RNAse	ribonuclease		**TE**	Tris-EDTA buffer
RNP	ribonucleoprotein		**TF**	telomeric fusion
ROS	reactive oxygen species		**TMPD**	N,N,N,N′-tetramethyl-p-phenylenediamine dihydrochloride
rRNA	ribosomal RNA		**TMRE**	tetramethylrhodamine ethyl ester
SAP	serum amyloid protein		**TNF**	tumour necrosis factor
SNB	Sam68 nuclear body		**TPCK**	N-tosyl-L-phenylalamine chloromethyl ketone
scid	severe combined immuno-deficiency		**TPP**	tetraphenylphosphonium
SDS	sodium dodecyl sulphate		**TRADD**	TNF receptor associated death domain
SDS-PAGE	sodium dodecyl sulphate polyacrylamide gel electrophoresis		**TRAIL**	tumour necrosis factor-related apoptosis-inducing ligand
SM	sphingomyelin		**TRAP**	telomere repeat amplification protocol
SMAC	second mitochondria-derived activator of caspase (also called DIABLO)		**TRF**	telomere repeat factor
			TRFs	terminal restriction fragments
SnRNP	small nuclear ribonucleo-protein		**TUNEL**	terminal deoxynucleotidyl transferase-mediated dUTP nick end-labelling
SPECT	single photon emission computed tomography		**UBF**	upstream binding factor
SRF	serum response factor		**VDAC**	voltage-dependent anion channel
SSBR	single strand break repair		**VSMC**	vascular smooth muscle cell
SSC	saline sodium citrate		**WGA**	wheat germ agglutinin
TBE	Tris-borate EDTA buffer		**XIAP**	X-linked inhibitor of apoptosis
			YFP	yellow fluorescent protein

Preface

Cell Proliferation and Apoptosis is an essential and comprehensive practical guide to an important and expanding area. The chapters have been contributed by a panel of internationally recognised authors. The book provides a detailed practical description of cell proliferation and apoptosis detection methods – the novel approach of combining both of these areas allows important comparisons to be made. In order to help the reader make the correct choice of procedure, the wide range of laboratory protocols is supported by full appraisals of current underlying theory.

Following a general introduction, the first three specialist chapters are devoted to the study of dividing cells covering the topics of cell number determination, the characterisation of nuclear changes in dividing cells, and the study of cell cycle proteins. The protocols include all aspects of tissue handling from collection, storage, fixation and processing through to locating and quantifying cells in different stages of the cell cycle.

The central portion of the book contains two chapters describing the study of telomeres and the nucleolus in cell division. The final six chapters deal with different aspects of the study of apoptosis, including characterising membrane alterations and morphological changes in dying cells, caspase detection and analysis, and identification of mitochondrial and nuclear changes during apoptosis. The wide ranging contributions to this book are drawn together in a closing chapter on flow cytometry applications for studying cell death, with special reference being made to the combined study of cell proliferation and cell death in the same experiments.

David Hughes
Huseyin Mehmet

Introduction to Cell Proliferation and Cell Death

David Hughes and Huseyin Mehmet

Contents

The cell cycle and cell death lie at the heart of cell biology. This is reflected in the vast array of published procedures that have developed for studying these fields of intense interest. Until now, researchers have frequently had to search the literature for suitable methodologies and the innumerable associated variations and updates. This book has been designed to bring under one cover a comprehensive range of complementary cell proliferation and cell death procedures. The protocol content of each chapter is supported by sufficient theoretical background to enable the reader to identify the most appropriate procedures knowledgeably. Although the protocols have been grouped under particular cell proliferation and apoptosis headings, many of the procedures may be used in combination in multiple labelling experiments; this is particularly so in the case of flow cytometry studies.

1.1 The structure and function of the mammalian cell and nucleus

Cytoplasmic and nuclear organelles are all affected by cell division, whether actively or passively. It is also well established that organelles are intricately involved in cell death, whether it be the part played by lysosomes in necrosis or the central roles of particularly the cell membrane, the mitochondria and the nucleus in the apoptotic cascade. Although nuclear DNA damage and ligation of plasma membrane death receptors have long been recognised as initial triggers of apoptosis, evidence is

Cell Proliferation and Apoptosis, David Hughes and Huseyin Mehmet (Eds)
© 2003 BIOS Scientific Publishers Ltd, Oxford

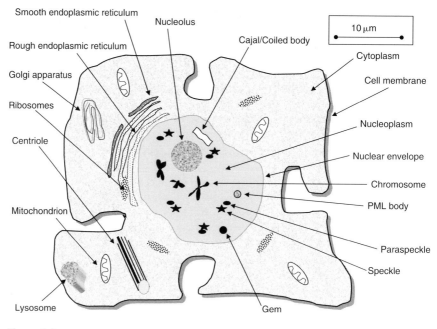

Smooth endoplasmic reticulum
Nucleolus
Rough endoplasmic reticulum
Cajal/Coiled body
Golgi apparatus
Cytoplasm
Ribosomes
Cell membrane
Centriole
Nucleoplasm
Nuclear envelope
Mitochondrion
Chromosome
PML body
Paraspeckle
Speckle
Lysosome
Gem
10 μm

Figure 1.1

A schematic drawing of a generalised animal cell showing the main cytoplasmic
and nuclear ultrastructural features. (Abbreviation: PML body, promyelocyte
leukaemia body.)

accumulating to suggest that other organelles, including endoplasmic reticulum,
lysosomes and Golgi apparatus, are also major points for pro-apoptotic signalling
and damage sensing [1]. Cell division and cell death are realms in which molecular,
biochemical and cell biological methods are beginning to shed light on how the
nucleus functions in relation to cell structure in general. An understanding of basic
cell structure and function, and of the nucleus in particular, is essential to any study
of cell proliferation and cell death (*Figure 1.1*) [2].

The cell nucleus was first described in 1802 by Franz Bauer and – mainly due to
advances in microscopy, biochemistry and molecular biology technology – this most
prominent of intracellular structures became notable over the next 150 years for its
intriguing behaviour during cell division and reproduction. Since the 1980s, studies
have clearly illustrated the relationship between the highly dynamic features of
nuclear compartmental architecture and function. Although the nucleus is microscopically seen to have distinctive sub-structures, these structures are notable for
their lack of delineating membranes. However, there are reasons why they should be
considered as separate microanatomical entities, namely because they contain characteristic groups of proteins, they can be morphologically characterised by light and
electron microscopy, and they can often be isolated and studied biochemically. In
the last decade, the concept of the nuclear matrix has emerged, this being the structure remaining after extraction of membranes, nucleic acids and histones. The
nuclear matrix comprises nuclear lamina (composed of lamins), the nucleolar
matrix (proteins involved in ribosomal RNA (rRNA) processing) and a fibrillogranu-
lar network (the nuclear matrix proteins). There is still much controversy surrounding the existence and functional processes of nuclear compartments and how they
might be spatially organised by the presence of the nuclear matrix.

While the nucleolus and splicing factor compartments remain the best identified of the nuclear structures, the function of other entities such as the Cajal bodies ('nuclear accessory bodies') and small dot-like nuclear bodies (including promyelocyte leukaemia bodies) remains to be established [3]. Chromatin, messenger RNA (mRNA) and nuclear proteins are highly dynamic molecules within the nuclear matrix and apart from being the targets of influences originating from the cytoplasm they exhibit a self-organisational ability that plays a substantial role in the maintenance and operation of the nuclear environment. Pre-nucleolar bodies are small, fibrogranular bodies appearing in the nucleus during telophase to eventually fuse at the nucleolar organiser region (NOR) to form the telophase nucleolus. Coiled bodies are small (0.5–1.0 μm diameter) round bodies composed of coiled fibrils whose numbers increase along with increasing cell activity. Paraspeckles are found in the interchromatin space, generally at 10–20 copies per cell, and they often co-localise with splicing speckles. Paraspeckles represent novel nuclear domains that are often in close proximity to splicing factor compartments known as speckles. Paraspeckle domains are distinct from other nuclear sub-organelles, such as Cajal bodies. In transcriptionally active cells, paraspeckle protein 1 (PSP1) is enriched in paraspeckles and cycles between paraspeckles and nucleoli [4]. Nuclear transfer experiments, microinjection assays, monitoring interspecies heterokaryons and genetic observations have all provided evidence over the last 50 years, but especially in the last decade or so, to show that proteins originally thought to exist only in the nucleus or the cytoplasm do in fact shuttle between the two. Such proteins play important roles both as carriers of cargo in transit between the nucleus and the cytoplasm and in relaying information between the two major cellular compartments. They include transport receptors and adaptors, steroid hormone receptors, transcription factors, cell cycle regulators and RNA binding proteins. The dynamic nature of the nucleus can be studied at the chromatin, RNA, protein and compartmental level [5]. New microscopy methods including video time-lapse photography and fluorescence recovery after photobleaching (FRAP) are being used to demonstrate that many nuclear proteins are mobile within the nucleus. It is against this background that the methodologies described in this book are so relevant to the modern cell biologist.

1.2 Cell proliferation and cell death – essential and interconnected cellular processes

In multicellular organisms, cell proliferation and death must be regulated to maintain tissue homeostasis. Many observations suggest that this regulation may be achieved, in part, by coupling the process of cell cycle progression and programmed cell death by using and controlling a shared set of factors [6, 7]. An argument in favour of a link between the cell cycle and apoptosis arises from the accumulated evidence that manipulation of the cell cycle may either prevent or induce an apoptotic response. This linkage has been recognised for tumour suppressor genes such as p53 and RB, the dominant oncogene, c-myc, and several cyclin-dependent kinases (CDKs) and their regulators. These proteins that function in proliferative pathways may also act to sensitise cells to apoptosis. Indeed, unregulated cell proliferation can result in pathologic conditions including neoplasias if it is not countered by the appropriate cell death. Translating the knowledge gained by studying the connection between cell death and cell proliferation may aid in identifying novel therapies to circumvent disease progression or improve clinical outcome [8]. Genes involved in cell proliferation (e.g. c-myc, p53 and cyclin D1) have also been found to play a role in apoptosis. Green and Evan [9] have proposed that deregulation of proliferation, together with a reduction in apoptosis, creates a platform that is both necessary and can be sufficient for cancer. The secondary traits of diverse neoplasms are a consequence of cell proliferation, tissue expansion and other outcomes of this platform.

1.3   The laboratory study of dividing and dying cells

An ever-expanding array of cell proliferation and cell death procedures is available to the cell biologist, each with its own bewildering number of modifications. It is now commonplace to combine molecular analysis with high-resolution microscopy to relate structural and functional information. Until recent times, such methods were applied to *in vitro* investigations, but it is now possible to identify sites of activity in living cells with the potential for making movies of incorporation and progression of labels through the cell cycle and cell death pathways. Many of these procedures are still only accessible to the researcher by means of time-consuming literature surveys; this book for the first time brings a comprehensive, well-integrated selection of such methodologies under one cover. Cell population kinetics can be studied in a wide range of fields. *In vitro* methods may be applied to cells that have been sampled and stored for examination in the laboratory or that have been harvested from tissue culture studies. Samples may come from investigations in such fields as oncology, tissue repair, the immunobiology of cell fate and free radical-induced injury following ischaemia/reperfusion. Many different types of sample may be encountered, including solid tissues prepared as frozen, paraffin or resin sections as well as free cell suspensions and exfoliates, aspirates, cultures and digests (e.g. pancreatic islets) prepared as cytospins, smears or imprints. Techniques commonly used to study cell kinetics can be divided into microscopical and non-microscopical. In general, microscopical techniques are employed to determine the identity, location and molecular activity of proliferating and dying cells in a tissue. Nowadays, it is possible to study cell growth and death using an array of different types of instrumentation at the light microscope and electron microscope. Non-microscopical procedures are used to provide quantitative information on cell kinetics. Widely used non-microscopical procedures include flow cytometry and biochemical techniques such as enzyme-linked immunosorbent assay (ELISA) and liquid scintillation counting of radionucleotide-labelled replicating DNA. More recently, the introduction of quantitative microscopy technology based on the confocal microscope and the laser scanning microscope has bridged the gap between the various divisions by providing simultaneous locational and numerical information. Modifications in the approach to sample preparation and storage may have a drastic effect on the success of any subsequent assay procedure. Depending on the sample source and the technique under study, different approaches will be required for sample storage. For short-term storage of tissues and cell suspensions, special fixatives, transport media, preservatives or cooling systems may be required. For long-term storage of samples with a view to subsequent investigation, attention must be paid to drying, freezing, wrapping and desiccation. Delay in correct processing and poor preservation procedure will lead to inconsistency or even failure in results. All of the protocols included in this book pay special attention to sample handling along with details of appropriate controls to check the specificity of the procedure.

1.4   Studying cell proliferation

Cell proliferation is one of the most fundamental, highly organised and complex of biological processes. Tissues grow primarily by increasing the number of cells, although this is accompanied by parallel increases in cell size and the amount of intercellular substance. Cell division is a fundamental process that is required throughout the life of all eukaryotes. Proliferation from progenitor cells occurs in embryonic, fetal, neonatal, juvenile and mature tissues. Some cells, such as neurones, do not replicate once fully differentiated. Other cells, such as those in the kidney and liver, have the ability to replicate in response to damage. The time taken for one cell cycle to take place varies between different tissues and species but even rapidly dividing cells, such as those involved in epithelial layer repair, will take at least 24 hours to divide once. Although it has been known for many years that cells have the ability to grow and replicate, the actual mechanisms involved have only been discovered relatively recently and there is still a great deal to be explained. The first part of this book will cover the current ideas about cell proliferation and its regulation with a description of a concise range of laboratory methods for assessing cell proliferation in various tissue preparations (*Figure 1.2*).

There is always a fraction of the cell population of any tissue that is actively cycling – this is the proliferative or growth fraction. Likewise, there are always cells that are either resting or can no longer divide. Consequently, it is possible to define proliferative or growth fraction as the ratio of cycling to cycling plus non-cycling cells. Proliferative activity is dependent on the speed of the cycle and the proportion of cells committed to the cycle (the growth fraction). The fraction of cells that are actively proliferating can be measured *in vivo* and *in vitro* using techniques described in Chapter 2.

Cell division involves both the nucleus (mitosis) and the cytoplasm (cytokinesis) to give rise to two new cells. The process can be viewed as an irreversible reiteration of a precisely timed sequence of two main sets of events. One is mitosis (M phase) in which actual cell division takes place and comprises four sub-phases, namely,

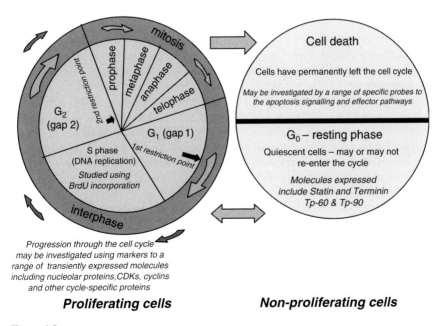

Figure 1.2

A schematic representation of the cell cycle showing the temporal expression of some proliferation-related markers. (Abbreviation: BrdU, 5-bromo-2'-deoxyuridine.)

prophase, metaphase, anaphase and telophase. Between actual cell divisions is interphase, when the nuclear material is duplicated. Interphase, during which DNA synthesis occurs, comprises 3 sub-phases – G_1, S and G_2. These phases are not distinctive parts of the cell cycle but just stages with no definite start or finish points for the cell to pass through when cycling. Numerous external signals can initiate cell division, including growth factors, hormones, viral agents and artificial stimulants. A quiescent phase (G_0) also exists when viable cells are not actually cycling. Methods for manipulating the cell cycle and studying nuclear changes during cell division are described in Chapter 3.

The prime purpose of the cell cycle is to ensure that the cell accurately completes DNA replication before cell division commences. Cell cycle controls ensure that all stages are carried out in the correct sequence. All stages must be complete before proceeding to the next. Damaged cells are not allowed to proliferate unchecked. Assessment of proliferation has become particularly popular in the study of checkpoint failure, tumours and their recurrence, metastatic potential and the growth of metastases [10]. This is mainly due to the extensive range of laboratory methodology available, including incorporation techniques.

Molecular control of cellular proliferation is remarkably conserved through evolution. There are three main groups of proteins that regulate the cell cycle: CDKs, cyclins and CDK inhibitors. Cascades of regulatory genes have a role in controlling cell division. Other molecules have been described where expression is indicative of cell cycle status. Statin is a non-proliferating, cell-specific nuclear protein of mass 57 kDa whose presence can be used to distinguish between growing and non-growing cells. Terminin, which can be used to distinguish between temporarily and permanently growth-arrested cells, is a cytoplasmic protein which has three forms, Tp-90, Tp-60 and Tp-30. Tp-90 (the 90 kDa form) is only present in growing and quiescent non-growing cells, Tp-60 is found in senescent cells, while Tp-30 is found in cells committed to apoptotic death [11]. Methods for studying the proteins involved in the control of the cell cycle are described in Chapter 4.

All cellular DNA is duplicated prior to mitosis. This duplication is controlled by the enzyme DNA polymerase. In 'mortal' cells, the ends of the chromosomes – the so-called telomeric regions – do not duplicate and these regions consequently get shorter with every cell division. Telomeres in vertebrates are composed of sequences of six nucleotides (TTAGGG) and they have a protective function keeping the chromosome ends intact. Telomeres maintain chromosome stability and when they are lost the chromosomes rearrange, after which the cells arrest and eventually die. In 'immortal' cells, such as those encountered in malignancy, the telomere length is maintained by the action of the enzyme telomerase – such a phenomenon is central to our ever-developing understanding of germ cell division, haematopoiesis, epithelial renewal, longevity, ageing and cancer [12]. With the aid of modern fluorescence *in situ* hybridisation (FISH) and image analysis techniques, it has been possible to unravel much of the mystery behind the maintenance of chromosome stability during cell division, and techniques applicable for studying the role of telomeres in cell division are described in Chapter 5.

Cell proliferation under normal physiological conditions is accompanied by increased ribosome biogenesis. The level of transcriptional activity is related to the amount of rDNA transcription machinery present in NORs. The strict positive correlation between cell proliferation and rDNA transcription activity underlies the use of the argyrophilic nucleolar organiser region (AgNOR) method for evaluating proliferation activity and prognosis in malignant neoplasms. There is now evidence that ribosome biogenesis antagonises cell cycle progress until the cell has grown to an adequate size, preventing cell cycle progression until the growth response is complete [13]. It is also becoming evident that the nucleolus together with spindle bodies plays a key role in the inactivation of mitotic CDKs and control of the end of the cell's cycle [14]. Methods for studying AgNORs and the protein basis to ribosomal biogenesis and nucleolar activity in dividing cells are described in Chapter 6.

1.5 Studying cell death

Cells that die accidentally as a result of injury degenerate by an uncontrolled process. This is called necrosis and is characterised by swelling and bursting of the cells together with the induction of an inflammatory response in the surrounding tissue. Apoptosis, on the other hand, is a genetically regulated process that was first described over 30 years ago (*Figure 1.3*) [15, 16]. Examples of apoptosis are seen throughout the entire animal and plant kingdoms. In necrosis, nuclear DNA remains mainly intact or, in advanced degeneration, becomes disrupted into fragments of different lengths. A seminal feature of apoptosis, on the other hand, is internucleosomal fragmentation of nuclear DNA that results in the production of oligonucleosomes of different but distinct lengths. Terminal differentiation (denucleation), a specialised type of apoptosis, is a process in which cells lose their nuclei but remain functional. It is a unique type of cell death and a good example are fibres in the terminally differentiating adult lens – they undergo denucleation but remain viable as anucleated fibre cells [17]. Terminally differentiated cells may be directed to undergo apoptosis when about to enter an aberrant cell cycle.

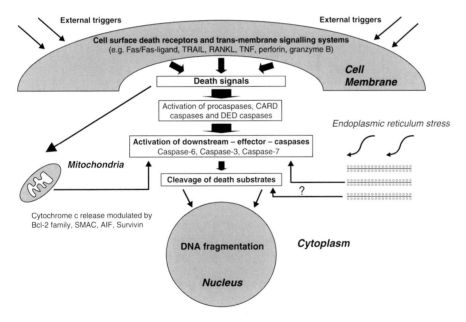

Figure 1.3

A brief overview of the stages of apoptosis. Stimulation leads to various pathways of intracellular response. Our knowledge of how molecules interact in these pathways is becoming increasingly more complex. (Abbreviations: TRAIL, tumour necrosis factor-related apoptosis-inducing ligand; RANKL, receptor activator of nuclear factor-kappa B ligand; TNF, tumour necrosis factor; CARD, caspase-associated recruitment domain; DED, death effector domain; SMAC, second mitochondrial-derived activator of caspase; AIF, apoptosis-inducing factor.)

Stimulation is the first stage in the apoptosis pathway and this may or may not be receptor-mediated. For apoptosis to commence the cell must first encounter a signal that activates the genetic control machinery. The mechanisms that are involved in initiating the intracellular cascade are still only partially understood but the composition of the environment surrounding the cell is certainly involved in starting and controlling the pace of the process. Various external triggers include heat shock, infection and drugs and toxins such as staurosporine, camptothecin, cycloheximide, dexamethasone and etoposide. Cell surface death factors include tumour necrosis factor-related apoptosis-inducing ligand (TRAIL), receptor activator of nuclear factor-kappa B ligand (RANKL), Fas and members of the tumour necrosis factor (TNF) superfamily.

Stimulation of cell surface receptors is followed by an intracellular response that includes signal transduction, activation of transcription factors and induction of apoptosis-related genes. The intracellular response culminates in cell dismantling and DNA degradation processes characteristic of apoptosis. Expression of phagocytic recognition molecules at the surface of apoptotic cells may lead at any stage to their removal from the tissue. The second half of this book will cover the range of procedures now available for studying these processes in detail.

One of the earliest changes in apoptosis is the movement of negatively charged phospholipid molecules (phosphatidylserine) from the inside of the cell membrane to the outside. Annexin-V labelling of phosphatidylserine exposure is an established procedure for identifying cells at an earlier stage in apoptosis [18]. It is probable that phagocytes recognise and eliminate such cells at an early stage and the involvement of phosphatidylserine exposure and class B scavenger receptor CD14 mediation in apoptotic cell removal has been well described [19]. Failure of phagocyte recognition leads to secondary necrosis when the apoptotic cell is broken down releasing its intracellular contents and assuming necrotic morphology. Defects in the phagocytic elimination of apoptotic cells may be implicated in the persistence of inflammatory disease as these cells eventually become necrotic with release of intracellular contents. Methods for identifying membrane alterations in dying cells are described in Chapter 7. This is followed in Chapter 8 by a description of procedures suitable for identifying the series of morphological changes that occur in cells as a result of the cascade of intracellular events.

Caspases are central to the apoptosis process [20]. They are cysteine proteases that cleave aspartic acid residues. At least 14 members of the family are known with many being involved in inflammation. Caspase enzymes act in a cascade fashion with caspases -8, -9 and -10 (long domain) being initiators of and are nearer the beginning of the cascade while caspases -3, -6 and -7 (short domain) are apoptosis effector molecules. Methods for caspase detection and analysis are described in Chapter 9.

The Bcl-2 family of proteins regulates the integrity of the outer mitochondrial membrane. The family is comprised of two opposing groups, namely pro-apoptotic and the anti-apoptotic. Pro-apoptotic members induce the release of cytochrome c from mitochondria and include Bax, BAK and BIM/BOD. Anti-apoptotic members prevent the release of cytochrome c from mitochondria and include Bcl-2, Bcl-x_L, Bcl-w, Mcl-1, A1-BFL-1 and BOO/DIVA. The balance between Bcl-2 and Bax forms an apoptotic rheostat, which seems to determine sensitivity to apoptosis [21].

Mitochondria have long been associated with cell injury, but their central role in the initiation of cell death is only now being elucidated. As well as involvement in adenosine triphosphate (ATP) synthesis, modulation of cell redox status, osmotic regulation, cytosolic calcium homeostasis and cell signalling, mitochondria are centrally involved in apoptosis [22]. Physiological dysfunction and structural alteration of mitochondria can both lead to cell death. Dissipation of mitochondrial membrane potential and release of cytochrome c from mitochondria appear to be key events during apoptosis. Second mitochondrial-derived activator of caspase (SMAC, or DIABLO (direct IAP binding protein with low PI)), the mammalian functional homologue of Drosophila Hid, Grim and Reaper, regulates apoptosis by neutralising inhibitor of apoptosis proteins and its release appears to be regulated by anti-apoptotic members of the Bcl-2 family [23]. Methods for identifying mitochondrial changes in apoptotic cells are described in Chapter 10. This is followed in Chapter 11 with a description of methods applicable to the identification of changes within the nucleus during apoptosis.

Programmed cell death is a highly regulated but double-edged sword – it is crucial to developmental processes in multi-cellular organisms but loss of regulatory control is implicated in a growing number of diseases, including inflammation, malignancy, autoimmunity and neurodegeneration. Apoptosis-related genes have been identified in most tissues and successful characterisation of such genes presents potential targets for specific pharmacological intervention in a wide range of diseases. Many of today's medical conditions can be attributed directly or indirectly to defects in the regulation of apoptosis that result in either a cell accumulation, in which cell eradication or cell turnover is impaired, or cell loss, in which the cell-suicide programme is inappropriately triggered. Identification of the genes and gene products that are responsible for apoptosis, together with emerging information about the mechanisms of action and structures of apoptotic regulatory and effector proteins, has laid a foundation for the discovery of drugs, some of which are now undergoing evaluation in human clinical trials [24]. It is now increasingly accepted that part of the efficacy of conventional chemotherapeutic drugs is due to their ability to induce apoptosis with the aim of providing death signals and abrogating survival signals. There is a wide range of novel approaches to the induction of apoptosis by down-regulating survival signalling along with many alternative strategies aimed at targeting particular molecular abnormalities of neoplastic cells as a means of inducing apoptosis [25].

1.6 New molecules – where cell proliferation and cell death meet?

Over recent years many new cell cycle and apoptosis-related molecules have been discovered, often with blurred roles. Survivin is a bifunctional molecule that has a role both as an inhibitor in apoptosis and an orchestrator of cell division; the resulting cross-talk between the processes promotes a balance between proliferation and death with limits on the growth and survival of cells suffering oncogenic mutations.

Use of affinity probes to localise proteasomal components – activators and inhibitors – reveals fundamental roles in normal homeostasis and disease. It has become increasingly clear that intracellular proteolysis via the ubiquitin/proteasome pathway has a fundamental role in cell cycle regulation, protein degradation in mitosis and apoptosis, as well as in other fundamental physiological processes including antigen processing, signal transduction, transcription and the flux of substrates through metabolic pathways.

The calcium-dependent thiol proteases, calpains, are widely expressed with ubiquitous and tissue-specific isoforms [26]. Calpains have been implicated in basic cellular processes including cell proliferation, apoptosis, differentiation, cytoskeletal rearrangements and cell migration.

Ras genes are ubiquitously expressed in tissues but mutations in each Ras gene are frequently found in different tumours, suggesting their involvement in the development of specific neoplasia. These mutations lead to a constitutively active and potentially oncogenic protein that could cause a deregulation of cell cycle. Recent observations have begun to clarify the complex relationship between Ras activation, apoptosis and cellular proliferation. Macaluso *et al.* [27] summarise the current knowledge of the structural and functional characteristics of Ras proteins and their links with cell cycle, apoptosis and cancer.

In addition to the massive array of molecules now known to be involved in cell division and cell death, another ever-increasing list has developed of molecules involved in related processes. Hyaluronan, a high molecular weight glycosaminogly-can, is an extracellular matrix component of tissues that facilitates cell locomotion which plays a major role in wound healing, cell proliferation, mitotic cell rounding, de-adhesion and migration. Furthermore, recent studies have shown that hyaluro-nan is involved in the intracellular regulation of the cell cycle and gene transcription; the immunocytochemical techniques described in this book may be applied to the identification of such molecules and their receptors in tissue preparations [28].

The wide-ranging contributions to this book are drawn together in Chapter 12 with a description of flow cytometry applications for studying cell death. Advances in the design of the flow cytometer, especially with the development of multi-channel fluorescence technology, has meant that many different combinations of the cell proliferation and cell death assay repertoire can be performed concomitantly.

1.7 References

1. Ferri, K.F. and Kroemer, G. (2001) Organelle-specific initiation of cell death pathways. *Nat. Cell Biol.* **3**: E255–E263.
2. Cook, P.R. (2001) *Nuclear Structure and Function.* J. Wiley and Sons, New York.
3. Dundr, M. and Misteli, T. (2001) Functional architecture in the cell nucleus. *Biochem. J.* **356**: 297–310.
4. Fox, A.H., Lam, Y.W., Leung, A.K., Lyon, C.E., Andersen, J., Mann, M. and Lamond, A.I. (2002) Paraspeckles: a novel nuclear domain. *Curr. Biol.* **12**: 13–25.
5. Phair, R.D. and Misteli, T. (2000) High mobility of proteins in the mammalian nucleus. *Nature* **404**: 604–609.
6. Evan, G.I. and Vousden, K.H. (2001) Proliferation, cell cycle and apoptosis in cancer. *Nature* **411**: 342–348.
7. Guo, M. and Hay, B.A. (1999) Cell proliferation and apoptosis. *Curr. Opin. Cell Biol.* **11**: 745–752.
8. Pucci, B., Kasten, M. and Giordano, A. (2000) Cell cycle and apoptosis. *Neoplasia* **2**: 291–299.
9. Green, D.R. and Evan, G.I. (2002) A matter of life and death. *Cancer Cell* **1**: 19–30.
10. Van Diest, P.J., Brugal, G., and Baak, J.P.A. (1998) Proliferation markers in tumours: interpretation and clinical value. *J. Clin. Pathol.* **51**: 716–724.
11. Miller, M.M., Teng, C.J., Mitmaker, B. and Wang, E. (1995) Characterization of the tissue regression process in the uterus of older mice as apoptotic by the presence of Tp30, an isoform of terminin. *Eur. J. Histochem.* **39**: 91–100.
12. Hayflick, L. (1998) A brief history of the mortality and immortality of cultured cells. *Keio J. Med.* **47**: 174–182.
13. Thomas, G. (2000) An encore for ribosome biogenesis in the control of cell proliferation. *Nat. Cell Biol.* **2**: E71–E72.
14. Cerutti, L. and Simanis V. (2000) Controlling the end of the cell cycle. *Curr. Opin. Genet. Dev.* **10**: 65–69.
15. Kerr, J.F., Wyllie, A.H. and Currie, R. (1972) Apoptosis: a basic biological phenomenon with wide-ranging implications in tissue kinetics. *Br. J. Cancer* **26**: 239–257.

16. Lockshin, R.A. and Zakeri, Z. (2001) Timeline: Programmed cell death and apoptosis: origins of the theory. *Nat. Rev. Mol. Cell Biol.* **2**: 545–550.

17. Gagna, C.E., Kuo, H-R., Florea, E., Shami, W., Taormina, R., Vaswani ,N., Gupta, M., Vijh, R. and Lambert, W.C. (2001) Comparison of apoptosis and terminal differentiation: the mammalian aging process. *J. Histochem. Cytochem.* **49**: 929–930.

18. van Engeland, M., Nieland, L.J., Ramaekers, F.C., Schutte, B. and Reutelingsperger, C.P. (1998) Annexin V-affinity assay: a review on an apoptosis detection system based on phosphatidylserine exposure. *Cytometry* **31**: 1–9.

19. Fadok, V.A. and Chimini, G. (2001) The phagocytosis of apoptotic cells. *Semin. Immunol.* **13**: 365–372.

20. Mehmet, H. (2000) Caspases find a new place to hide. *Nature* **403(6765)**: 29–30.

21. Van Der Vliet, H.J., Wever, P.C., Van Diepen, F.N., Yong, S.L. and Ten Berge, I.J. (1997) Quantification of Bax/Bcl-2 ratios in peripheral blood lymphocytes, monocytes and granulocytes and their relation to susceptibility to anti-Fas (anti-CD95)-induced apoptosis. *Clin. Exp. Immunol.* **110**: 324–328.

22. Green, D.R. and Reed, J.C. (1998) Mitochondria and apoptosis. *Science* **281**: 1309–1312.

23. Shi, Y. (2002) Mechanisms of caspase activation and inhibition during apoptosis. *Mol. Cell* **9**: 459–470.

24. Reed, J.C. (2002) Apoptosis-based therapies. *Nat. Rev. Drug Discov.* **1**: 111–121.

25. Makin, G. (2002) Targeting apoptosis in cancer chemotherapy. *Expert Opin. Ther. Targets* **6**: 73–84.

26. Perrin, B.J. and Huttenlocher, A. (2002) Calpain. *Int. J. Biochem. Cell Biol.* **34**: 722–725.

27. Macaluso, M., Russo, G., Cinti, C., Bazan, V., Gebbia, N. and Russo, A. (2002) Ras family genes: an interesting link between cell cycle and cancer. *J. Cell Physiol.* **192**: 125–130.

28. Evanko, S.P. and Wright, T.N. (1999) Intracellular localization of hyaluronan in proliferating cells. *J. Histochem. Cytochem.* **47**: 1331–1341.

Determination of Cell Number

2

Ian Zachary

Contents

2.1 Introduction

Cell division and cell proliferation are fundamental processes in all living organisms. In mammals, cell proliferation and the growth factors, growth inhibitors and cytokines which regulate it, are essential during embryonic development, tissue and organ growth, and for several physiological processes in the adult state such as haematopoiesis, wound healing and pregnancy. For much of their lifespan, the differentiated cells of many adult mammalian tissues exist in a viable, non-proliferating state. Many such cells retain the capacity to proliferate and will do so during the repair of damaged tissue, or can be stimulated to reinitiate DNA synthesis and cell division when placed in culture and challenged with suitable growth factors [1]. The abnormal hyperplastic or neoplastic multiplication of normally non-proliferating adult cells *in vivo* also plays a central role in tumourigenesis, atherosclerosis and

Cell Proliferation and Apoptosis, David Hughes and Huseyin Mehmet (Eds)
© 2003 BIOS Scientific Publishers Ltd, Oxford

other disease processes. Consequently, techniques for the accurate, reliable and rapid evaluation of cell proliferation are among the most widespread and important in clinical and basic biological research.

It is increasingly realised that in cell culture and in disease, the balance between cell survival and programmed cell death or apoptosis is as important as cell proliferation in determining cell number. This makes it important to distinguish between changes in cell number due to proliferation and those resulting from increased survival or apoptosis. This problem is exacerbated by the need to perform many studies in defined and serum-free or reduced-serum media. Many established or immortalised cell lines are resistant to the apoptotic effects of serum or growth factor withdrawal for extended periods of time (24 hours or more), making experiments in defined serum-free medium possible without the potential complication of an increased rate of cell death. In contrast, primary cell cultures are often much more sensitive to apoptosis resulting from serum deprivation. In this chapter, approaches and methods for the measurement of cell number and proliferation only will be considered, methods for determination of apoptosis being described elsewhere in this volume. All the methods described here are suitable for measuring mitogenesis or number in virtually any primary cell culture or cell line with suitable modifications. Access to standard facilities and equipment required for cell culture, such as laminar flow cabinets, bench centrifuges and microscopes, is assumed. Measurement of cell proliferation in *in vivo* animal models poses special problems and a general consideration of some of the most commonly used approaches is given in the final section of this chapter. No detailed method is provided for *in vivo* determination of cell proliferation because these methods may vary significantly between animal models and species. Furthermore, such methods involve specialised skills and techniques in animal experimentation which lie beyond the scope of this chapter.

2.2 Measurement of cell number

The most accurate method for determining the number of viable cells in a given cell population is direct cell counting. Because these methods simply determine the numbers of viable cells at any given time point they are suitable for cell propagation, measurement of cell population doubling times, and accurate determination of experimental cell sample sizes, but used alone do not measure rates of cell proliferation or death. For the purpose of determining the proliferative activity of a growth factor, direct cell counting should always be used in conjunction with another method for estimating the rate of cell death or changes in cell viability particularly under serum-free or reduced-serum conditions.

Two methods of cell counting are widely employed. The most accurate and cheapest but more time-consuming technique is the microscopic counting of a cell suspension on a haemocytometer (see *Table 2.1*). Because cells are counted by eye, experienced laboratory workers can distinguish viable from dead cells or cellular debris, making the haemocytometer the most reliable method for determining the viable cell count. Trypan blue exclusion can also be easily incorporated in this method in order to verify cell viability (Protocol 2.1). The second method uses an electronic Coulter counter to count cells in suspension. The advantage of this approach is speed, which may be a considerable benefit when handling large numbers of different samples. The main disadvantage of the Coulter counter is that it may be more difficult to differentiate viable from apoptotic or necrotic cells.

Table 2.1   Summary of methods for measurement of cell proliferation *in vitro*

Method	Advantages	Disadvantages
Direct cell counting by haemocytometer	Simple, inexpensive, accurate measure of cell number	Does not measure DNA synthesis or, by itself, is sensitive to cell death
Direct cell counting by Coulter-type counter	Simple, rapid technique suitable for large numbers of cells	As for haemocytometer. Less easy to distinguish dead or damaged cells
MTT-type assays	Simple, rapid, suitable for high throughput screening	Measures cell viability and cytotoxicity rather than mitogenesis; sensitive to changes in cell size and hypertrophy
[^{3}H]thymidine incorporation into TCA-insoluble material	Simple, sensitive, specific measure of DNA synthesis; suitable for quite large sample numbers	May be sensitive to changes in thymidine transport; does not measure cell proliferation
Autoradiography of [^{3}H]thymidine-labelled nuclei	Accurate, specific measure of cells in S phase; less sensitive to thymidine transport	Lengthy procedure; cells do not progress to M phase; does not measure cell proliferation
BrdU labelling	Simple, specific measure of DNA synthesis; useful for rapid screening of large numbers of samples	Does not measure cell proliferation
Flow cytometry	Measures cells at different cell cycle stages; versatile technique for determining cell cycle-dependent changes in cellular components and apoptosis	Requires expensive equipment and some training; cell death or damage is source of error

2.2.1   Cell counting using a Coulter counter

The Coulter counter and similar instruments such as the CDA-500 particle counter and analyser (Sysmex Corporation, Japan), use the direct current electronic resistance method for particle (cell) counting and sizing (Protocol 2.2). The counter contains a probe comprising a transducer with an aperture of an exact diameter, the standard being 100 μm, and an internal and external electrode. The probe is immersed in the cell suspension diluted into an electrically conductive diluent, which allows a direct current to pass through the aperture between the electrodes. Cells or particles in the suspension are non-conductive and when they pass through the aperture, they decrease the flow of current and increase the resistance. In the Coulter counter cells are drawn up into the probe using a mercury manometer pump, but more recent instruments (e.g. the Sysmex CDA-500 particle counter) use a non-mercury diluent-filled ball-float manometer pump. Increased resistance due to cells in the aperture causes a change in the voltage between the electrodes which is proportional to the volume of the particle or cell. Since a constant current is maintained, the voltage change will be directly proportional to the change in resistance according to Ohm's Law: E (voltage) $= I$ (current) $\times R$ (resistance).

Protocol 2.1: Cell counting using a haemocytometer

EQUIPMENT, MATERIALS AND REAGENTS

Haemocytometer with improved Neubauer ruling.

Coverslip.

Phase-contrast microscope.

Syringe and 19-gauge needle.

20–200 μl Gilson pipette and tip.

0.4% Trypan blue in Hank's buffered saline solution (optional).

METHOD

This method applies to adherent cell cultures which need to be trypsinised. For suspension cultures, this step is omitted.

1. Prepare the haemocytometer by washing with detergent, rinsing thoroughly with distilled water, and wiping off with alcohol. Ensure that alcohol has been completely removed before use.

2. Trypsinise cells as normal for the given cell type and collect by centrifugation at ~400 g for 5 min or as recommended. For detailed description of cell culture techniques, readers are referred to specialist books [2, 3]. It is important that cells are inspected microscopically during trypsinisation to ensure that all cells have been removed and that a single cell suspension has been obtained. Cell clumps or aggregates should be avoided. Often this may result from cells being allowed to grow to confluence or superconfluence before trypsinisation. Primary mesenchymal cell types such as fibroblasts and VSMC will grow over each other and will detach in large cell rafts if allowed to become very confluent.

3. After collection, disperse cells in a suitable volume of culture medium to form a single cell suspension by passing the cells ~5 times through a 19-gauge needle fitted to a syringe. Use a small volume to begin with to avoid diluting too much. Avoid foaming of the cell suspension and damaging the cells by suspending the cells too vigorously, or using too many passes through the needle. Note that dispersal of cells with a needle and syringe may not be necessary for all cell types and may be avoided as judged appropriate.

4. The counting of only viable cells can be ensured by making up a small volume of the cell suspension for counting with a 0.4% solution of the dye trypan blue to give a final dye concentration of 0.2%.

5. Wet the edge of the haemocytometer and position the coverslip with downward pressure until Newton's rings are formed. Using a Gilson pipette and pipette tip, place one drop of cell suspension under the coverslip, by

gently touching the end of the tip against the coverslip and carefully ejecting sufficient of the suspension to fill the space between the coverslip and haemocytometer. Avoid adding excess suspension to the haemocytometer.

6. Count the cells in the central 25 squares of the grid, each of which is made up of 16 small squares and is divided by triple ruling from the others. Include cells which fall on the middle lines on the same two sides of each square and omit those falling on the middle lines on the other two sides. Viable cells will appear bright and light refractive and exclude Trypan blue. At least 100 cells should be counted to obtain an accurate count. Avoid either too many or too few cells, and dilute the cell suspension as required.

7. The cell count obtained is the number of cells in an area of 1 mm^2. Since the depth of the counting chamber (i.e. the space between coverslip and haemocytometer) is 0.1 mm, the total volume over the 25 central squares in which cells are counted is 0.1 mm^3. To obtain the cell number in 1 ml, multiply the cell count obtained from the haemocytometer by 10^4, then multiply as required to take into account any dilution of the original suspension.

Protocol 2.2: Cell counting using a Coulter counter

EQUIPMENT, MATERIALS AND REAGENTS

CDA-500 particle counter and analyser and cuvettes (Sysmex Corporation, Japan).

CELLPACK (containing 6.38 g/litre sodium chloride, 0.2 g/litre sodium tetraborate, 0.2 g/litre ethylenediaminetetraacetic acid (EDTA) dipotassium salt, 1.0 g/litre Boric acid) (Sysmex Corporation, Japan).

Syringe and 19-gauge needle.

METHOD

1. Switch on counter and allow to warm up in line with the manufacturer's instructions (~15 minutes).

2. Collect cells by trypsinisation and/or centrifugation, resuspend and disperse into a single cell suspension as described in steps 2 and 3 of the method for haemocytometer counting (Protocol 2.1).

3. Dilute 50 µl cell suspension into 5 ml CELLPACK in a cuvette and cap. Carefully mix by gentle inversion.

4. Count cell suspension according to manufacturer's instructions and multiply by a suitable factor.

2.3 Indirect measurement of cell number

Several indirect methods are commonly used for the measurement of cell viability, cytotoxicity and cell proliferation. These assays are simple, rapid and well suited for the analysis of large numbers of samples in 96-well microtitre plates. These methods are usually based on the measurement of an enzyme activity which reflects the general metabolic status of a cell, and are therefore suitable for evaluating cell viability or cytotoxicity. One of the best known is the 3-[4,5-dimethylthiazol-2yl]-2,5-diphenyltetrazolium bromide (MTT) assay based on the formation and colorimetric quantification of a blue enzyme reaction product [4, 5]. An alternative indirect method for measurement of cell number is the alkaline phosphatase assay which is based on conversion of the non-fluorogenic substrate 4-methyl umbelliferyl phosphate to the fluorescent product 4-methyl umbelliferylone by the widely distributed enzyme alkaline phosphatase. Similar to the MTT assay, the level of fluorescence is directly proportional to the activity of alkaline phosphatase and cell number. Though such assays are often described as suitable for assessment of cell proliferation, it is important to emphasise that they do not directly measure either DNA synthesis or cell division, and are not recommended for the primary characterisation of the mitogenic or proliferative activity of cytokines or growth factors. An additional problem with assays of metabolic markers, particularly when they are coupled to colorimetric quantification, is that they are sensitive to changes in cell size or cell growth as distinct from cell proliferation. This may make it difficult to distinguish between a hypertrophic effect (increase in cell size without DNA synthesis or cell division) and a mitogenic effect. The same consideration also applies to methods that have been reported for assessing cell number based on the binding of dyes such as methylene blue to cellular components [6]. Furthermore, though the data obtained are directly proportional to cell number, if accurate determination of cell number is the key objective, the direct methods of cell counting described in the previous section are preferable. The method described in this section, the MTT assay, is appropriate for rapid screening of large numbers of test substances for cytotoxicity or viability, or as an adjunct to more direct methods for measuring mitogenic activity.

2.3.1 The MTT assay

The MTT assay is a quantitative colorimetric assay based on the cleavage of the yellow water-soluble tetrazolium salt, MTT, to form water-insoluble, dark-blue formazan crystals [4]. MTT cleavage occurs only in living cells by the mitochondrial enzyme succinate dehydrogenase. The formazan crystals are solubilised using a suitable organic solvent, usually isopropanol, and the optical density of the resulting solution is measured using a spectrophotometer. The absorbance is directly proportional to the concentration of the blue formazan solution, which is in turn proportional to the number of metabolically active cells.

The method described in Protocol 2.3 can be performed using standard laboratory equipment and individual reagents. MTT assay kits are available (Roche Molecular Biochemicals) and should be used according to the manufacturer's instructions. The method is described for cells in a 96-well plate, but can be adapted for 48-well or 24-well plates. For larger wells, the volumes of solutions have to be increased. Prior to performing experiments, it is important to establish a standard curve for a given cell type or line by MTT assay of known numbers of cells over a range of approximately 5000–200 000.

Protocol 2.3: The MTT assay

EQUIPMENT, MATERIALS AND REAGENTS

Spectrophotometer ELISA plate reader.

MTT stock solution 5 mg ml^{-1}: dissolve 50 mg MTT in a total volume of 10 ml phosphate-buffered saline. Store at $-20°C$.

Working concentration MTT solution: 0.5 mg ml^{-1} MTT in culture medium.

Isopropanol.

METHOD

1. Incubate cells seeded in the wells of a 96-well plate with test growth factors for the desired time periods. Aspirate medium and add 200 µl per well of 0.5 mg ml^{-1} MTT solution. Incubate for 3 hours at 37°C in the tissue culture incubator.

2. Remove medium and add 200 µl per well of isopropanol. Measure absorbance at 570 nm with a reference wavelength at 690 nm in the spectrophotometric plate reader. Calculate cell numbers from absorbance using the standard curve.

2.4   Direct measurement of mitogenic activity in cell culture

The most reliable and accurate way to assess the effect of a factor on the mitogenic activity or cell cycle status of a cell is to directly measure either DNA synthesis or mitosis. All direct methods for determining mitogenic activity, cell proliferation, division or replication rely on the measurement of either *de novo* DNA synthesis or DNA content. Newly synthesised DNA is measured by specifically labelling DNA with a precursor molecule, usually thymidine or the thymidine analogue, 5-bromo-2'-deoxyuridine (BrdU), and then monitoring its incorporation over time either isotopically or by immunostaining. Changes in DNA content are determined cytofluorimetrically to give a measure of the percentage of cells at each stage of the cell cycle. Four protocols are described in detail in this chapter: incorporation of [³H]thymidine into acid-insoluble material, autoradiography of labelled nuclei, BrdU labelling and staining of nuclei, and cytofluorimetric cell cycle analysis.

2.4.1   Incorporation of [³H]thymidine into acid-insoluble material

Measurement of the incorporation of [³H]thymidine into acid-insoluble material is a widely used method for measuring the mitogenic activity of a growth factor in any adherent or non-adherent cell type, which can be made to arrest in G_0 or the G_1 phase of the cell cycle after a suitable period of cell culture or serum deprivation [7]. This method is reproducible, reliable and simple to perform and can be a sensitive way of detecting mitogenic or anti-mitogenic activity. [³H]thymidine incorporation into acid-insoluble material is sensitive to changes in thymidine transport, and some agents originally thought to be anti-proliferative on the basis of inhibition of [³H]thymidine incorporation were subsequently shown to inhibit thymidine transport. Any effect contributed by thymidine transport can readily be overcome by increasing the total concentration of thymidine from 1 µM to 10 µM, a concentration at which thymidine will enter the cell by simple diffusion rather than by carrier-mediated transport. Results obtained from other direct methods for measuring DNA synthesis such as autoradiography of labelled nuclei or BrdU labelling of nuclei are less sensitive to changes in thymidine transport.

Additional problems can arise from the time period over which incorporation is measured. Pulsing with radiolabelled thymidine for periods of ~3 h is sometimes used to measure the mitogenic activity of growth factors, but this method has to be treated with caution because most cell populations exhibit considerable heterogeneity with regard to the duration of G_1 and the time of re-entry into S phase. Furthermore, while pulsing can be used to measure the rate of entry into S phase, the effect of a factor on the rate of reinitiation of DNA synthesis may differ from its cumulative effect on DNA synthesis measured over the entire duration of S phase. These problems are avoided by measuring the cumulative incorporation of isotopically labelled thymidine for a period of 30 to 40 hours. This period ensures that the whole cell population has completed one cell cycle, allowing for differences in the duration of G_1, but has not yet had time to progress through a second S phase. Pulsing at different periods during S phase is, however, useful for establishing the ability of a factor to decrease or increase the number of cells entering S phase as distinct from an effect on the rate of entry [8].

The method for measuring [³H]thymidine incorporation into acid-insoluble material is essentially identical for all cell types and lines, the major difference usually being how cultures are made quiescent. Some cell lines, such as Swiss 3T3 cells, are made quiescent by growing cells to confluence and allowing the cells to deplete serum of its growth-promoting activity [9]; no pre-incubation in serum-free medium prior to addition of growth factors is required in this case, though the stimulation with mitogenic factors and measurement of [³H]thymidine incorporation is performed under serum-free conditions. In other cell lines and primary cell cultures, a period of serum deprivation is often recommended for ensuring arrest of cells in G_0/G_1. The method described in Protocol 2.4 is for vascular smooth muscle cells (VSMC) [8].

2.4.2 Autoradiography of labelled nuclei

This method measures the incorporation of [³H]thymidine into nuclei autoradiographically by exposing the labelled cells to film [9]. As the sensitivity of this method is less than that of scintillation counting of [³H]thymidine incorporated into acid-insoluble material, a higher specific activity of the radiolabelled precursor is used, usually $5\,\mu\text{Ci ml}^{-1}$ [³H]thymidine. The exposure of cells to such a high level of radioactivity causes DNA damage resulting in arrest of cells after S phase. The decreased sensitivity of autoradiography of ³H-labelled material also means that results may take up to 3 weeks to develop. Quantification of the results involves microscopic counting of labelled and unlabelled nuclei and is very time-consuming if large numbers of samples are being analysed. The advantage of this method is that it provides an accurate and direct measure of DNA synthesis (Protocol 2.5).

2.4.3 BrdU incorporation and staining

This method measures incorporation of the thymidine analogue BrdU into DNA. BrdU is detected immunocytochemically using anti-BrdU antibody, or less commonly using fluorescent DNA-binding dyes such as Hoechst 33258, whose fluorescence is quenched by BrdU. This method can either be used for direct immunocytochemical detection of BrdU-labelled nuclei using fluorochrome-conjugated secondary antibody, or for rapid quantification of BrdU-labelled DNA in a 96-well format using enzyme-conjugated anti-BrdU antibody coupled to enzyme-catalysed cleavage of a substrate to form a coloured reaction product. The method described below (Protocol 2.6) is for BrdU labelling of nuclei and immunocytochemical staining. Exposure of incorporated BrdU to antibody requires denaturation of DNA, which is accomplished using either acid incubation and/or nuclease treatment. For the more rapid quantification of BrdU incorporation into cellular DNA using enzyme-conjugated antibody and detection of coloured reaction product, BrdU detection kits are available (Roche Molecular Biochemicals).

2.4.4 BrdU labelling of biopsy samples; a clinical application

It is often useful to be able to assess cell cycle status in cell sample biopsies for pathological diagnosis. Contrary to the conditions under which experimental cell populations are created, there is often considerable delay between the procurement of the biopsy and BrdU processing in the laboratory. To address this problem, a special portable BrdU incubator has been described that can be taken into the ward or clinic. The incubator maintains previously made-up BrdU medium at 37°C in tubes that are ready to accept small tissue biopsies or cell suspensions. BrdU incubation can commence immediately after the biopsy has been performed. Such an incubator is easily constructed in most institutional workshops [10] and in the past it has been possible to purchase commercially produced models (*Figure 2.1*). BrdU is incubated with the unfixed cells. After BrdU uptake, cryostat sections are made from biopsy tissues and cell suspensions are cytocentrifuged, fixed in ethanol and the DNA denatured in formamide. Sites of BrdU uptake are then visualised using an anti-BrdU monoclonal

Protocol 2.4: Incorporation of [³H]thymidine into acid-insoluble material

EQUIPMENT, MATERIALS AND REAGENTS

Liquid scintillation analyser: Packard Tri-Carb 2100TR.

Liquid scintillation cocktail: Packard Ultima Gold.

[Methyl-³H]thymidine, aqueous solution, 25 Ci mmol^{-1}, 1 mCi ml^{-1} thymidine: to make a 10 mM stock solution, dissolve solid thymidine in sterile deionised H_2O and then sterilise by syringing through a 0.2 µm filter. Store aliquots at –20°C.

Trichloroacetic acid (TCA): to make a 50% stock solution, dissolve 500 g TCA in 500 ml distilled H_2O, and then add distilled H_2O to a final volume of 1 litre. Store at 4°C. Dilute 1:10 in distilled H_2O to make a working concentration of 5%.

70% ethanol.

Phosphate-buffered saline (PBS): 0.15 M NaCl in 0.1 M potassium phosphate buffer, pH 7.4.

0.1 M NaOH, 2% Na_2CO_3, 1% sodium dodecyl sulphate (SDS): store at room temperature.

METHOD

1. VSMC should be grown to confluent density. Cultures should display characteristic 'hills and valleys' appearance. These cells do not form an obligate monolayer, but will grow over each other to an extent. Render quiescent, by incubating in serum-free Dulbecco's Modified Eagle Medium (DMEM) for 24–40 hours.

2. Make up a 100 µCi ml^{-1} [³H]thymidine solution with a total thymidine concentration of 100 µM by adding 1 ml [Methyl-³H]thymidine (1 mCi ml^{-1}) and 100 µl 10 mM thymidine to 8.9 ml sterile deionised H_2O to give a total volume of 10 ml. Sterilise by syringing (without needle) through a 0.2 µm filter. Use 1 ml of this solution per 100 ml of medium to be added to cells. Pre-warm the medium containing [³H]thymidine to be added to cells to 37°C.

3. All media should be sterile and pre-warmed to 37°C. Wash confluent and quiescent VSMC two times with serum-free DMEM (pre-warmed to 37°C) and for cells cultured in 35 mm dishes or 6-well plates incubate in 2 ml per well of DMEM containing test factors and 1 µCi ml^{-1}, 1 µM [³H]thymidine for 40 hours at 37°C. Cells can also be used in 24-well plates using 1 ml of medium.

4. Wash cells twice with ice-cold PBS and precipitate acid-insoluble material by incubating cells for 20 min with 5% TCA at 4°C. Precipitated protein and nucleic acid should appear as a milky film on the dish.

5. Wash cells twice with ethanol in order to remove TCA without solubilising precipitated material and then solubilise precipitate by incubating in 1 ml 0.1 M NaOH, 2% Na_2CO_3, 1% SDS at room temperature for 30 minutes. If precipitate does not dissolve or dissolves slowly, incubate at 37°C for an additional 10 minutes.

6. Count 0.5 ml of each sample in 10 to 20 ml of scintillation cocktail in a liquid scintillation analyser.

Protocol 2.5: Autoradiography of labelled nuclei

EQUIPMENT, MATERIALS AND REAGENTS

[Methyl-^{3}H]thymidine: 1 mCi ml^{-1} aqueous solution as in Protocol 2.4.

Formal saline: BDH Chemicals Ltd.

Isotonic saline: 0.9% sodium chloride in sterile distilled water.

5% TCA: as in Protocol 2.4.

70% ethanol.

Chrome alum solution: dissolve 5 g gelatin in 400 ml distilled H$_2$O by heating or in a microwave and allow to cool. Separately dissolve 0.5 g chrome alum (chromic potassium sulphate, BDH Chemicals Ltd) in 400 ml distilled H$_2$O, mix with dissolved cooled gelatin solution and make up to 1 litre. Store at 4°C.

D19 developer, Polymax fixer, and AR 10 stripping film
(all from Kodak Ltd).

Giemsa stain: 0.4% w/v in buffered methanol solution, pH 6.8.

METHOD

1. Treat quiescent cell cultures for 30–40 hours with medium containing 5 μCi ml^{-1} of undiluted [Methyl-^{3}H]thymidine and growth factors. There is no need to make up a diluted stock of [^{3}H]thymidine and it can be added direct from the original vial.

2. Remove medium and fix cells with formal saline for 20 min at 4°C.

3. Wash cells twice with isotonic saline.

4. Extract acid-soluble [^{3}H]thymidine (thymidine not incorporated into DNA) by two consecutive incubations with 5% TCA at 4°C for 5 min each.

5. Wash three times with 70% ethanol and allow to dry.

6. Coat dried cells with a thin layer of chrome alum solution just sufficient to cover the cell surface and leave to dry.

7. In a dark room, cut film into suitably sized pieces. Fill cell culture dishes with distilled H$_2$O, and lay pieces of film on to dishes emulsion side down to allow film to form a smooth surface. Then pour H$_2$O away leaving film on cell surface. Leave dishes to develop inside a sealed light-safe container in the dark room for 1–3 weeks.

8. Develop film in dishes in the dark room under a safe light with D19 for 4 min, fix film in Polymax for 5 min, and wash in tap water for 30 min.

9. Counterstain cell nuclei with Giemsa, and allow dishes to dry.

10. Quantify results by counting labelled and unlabelled cells under the microscope. The proportion of labelled nuclei is calculated using the following formula:

$$\% \text{ labelled nuclei} = \frac{\text{number of labelled cells} \times 100}{\text{total number of cells}}$$

Protocol 2.6: BrdU incorporation and staining

EQUIPMENT, MATERIALS AND REAGENTS

BrdU labelling and detection kit (e.g. Roche Diagnostics Ltd). This kit provides BrdU, monoclonal anti-BrdU antibody, anti-mouse immunoglobulin conjugated to fluorescein isothiocyanate (Ig-FITC), nuclease solution for denaturation of DNA, and washing and incubation buffers. Immunostaining of BrdU-labelled DNA can be performed easily using individual reagents. BrdU, monoclonal anti-BrdU antibody, and anti-mouse Ig-FITC are all available from a range of commercial sources. Nucleases are widely available.

Fixative: 70% ethanol made up with HCl to give a final concentration of 0.5 M HCl.

ProLong™ anti-fade kit.

13 mm glass coverslips.

Microscope fitted with epifluorescence.

METHOD

This protocol is for cells grown on glass coverslips placed in 24-well tissue culture plates.

1. Wash confluent and quiescent cells cultured on glass coverslips with serum-free medium and incubate as required with medium containing BrdU (as recommended in the manufacturer's instructions), test growth factors and/or growth inhibitors.

2. After 40 hours, remove medium, wash cells three times with ice-cold washing buffer (PBS as prepared in Protocol 2.4 can be substituted).

3. Fix cells with 70% ethanol/0.5 M HCl pre-cooled for 30 min at –20°C.

4. Wash cells as in step 2. Cells can be left safely for up to overnight in final wash medium if necessary.

5. Incubate cells with nuclease solution for 30 min at 37°C and wash three times. Refer to step 6 for details of a convenient protocol for economising on use of solution when staining cells on coverslips.

6. Remove final wash and incubate cells with anti-BrdU solution. To avoid using large amounts of antibody, a small volume (~50 µl) of antibody solution is placed onto a sheet of parafilm attached to a rigid surface such as a glass plate, and the coverslip is then placed onto the solution with cell surface down (next to antibody), so that the cell surface of the coverslip is covered with the solution. The glass plate with parafilm and coverslips attached is then placed inside a humidified chamber, such as a sandwich box containing layers of moistened chromatography paper, and the chamber incubated at 37°C for 30 minutes.

7. Wash cells three times with wash buffer.

8. Incubate cells with Ig-FITC at 37°C for 30 minutes in the dark using the arrangement described in step 6.

9. Wash cells three times with wash buffer.

10. Mount coverslips on to glass slides using ProLong™ anti-fade kit. No sealing is required, and fluorescence should remain stable for up to a week or longer. For best results, however, examine and document results using fluorescence microscopy as soon after staining as possible.

11. Quantify results by counting numbers of stained nuclei in 10 random fields in each of three duplicate coverslips, and plotting means ± S.E.M.

The BrdU incubator from Novocastra Laboratories provides five temperature-controlled chambers. The specimen immediately after harvest is placed in a plastic vial of BrdU medium and inserted in a heated chamber for transport to the laboratory. Reproduced with permission from Novocastra.

antibody incorporated in a standard immunofluorescence procedure (Protocol 2.7) [11]. The monoclonal antibodies against BrdU often only detect antigen in single-stranded DNA so the nuclear DNA has to be denatured before labelling – cell morphology is best preserved using formamide for denaturing rather than strong acids or alkalis. The method gives results comparable to tritiated thymidine but is easier and more rapid to perform. Monoclonal anti-BrdU reagents are now commercially available that contain a deoxyribonuclease (DNAse) secreted by mycoplasma contaminating the hybridoma – consequently no additional DNA denaturing step is needed although with this reagent lower numbers of positive cells may be obtained.

2.4.5 Analysis of the cell cycle using flow cytometry

The passage of cells through S phase, G_2 and through to the end of the cell cycle at the completion of M phase (mitosis), results in changes of DNA content which reflect the position of cells in the cell cycle [12]. Thus cells in G_2 or M phase prior to cytokinesis will contain double the amount of DNA present in G_1 cells, as they have completed S phase. Cells in S phase will have different amounts of DNA between that in G_1 and G_2/M. Cellular DNA content can be determined by measuring the binding of one of several DNA-binding fluorochromes, among the most commonly used being 4',6'-diamidino-2-phenylindole (DAPI), propidium iodide (PI) and acridine orange [13]. Detection of DAPI requires excitation with light in the ultraviolet range, the maximum excitation occurring at 359 nm with emission at 461 nm. PI is maximally excited at 536 nm, though optimal excitation of PI fluorescence is usually obtained using a blue laser light at 488 nm; PI emission occurs at 617 nm. As PI also binds to double-stranded RNA, it is necessary to degrade RNA by treatment of cells with DNase-free ribonuclease (RNAse) A during staining with this fluorochrome. Detection of cell-bound fluorescence and analogue to digital conversion of the electronic signal from a photomultiplier is performed by a flow cytometer, which produces data as a frequency distribution DNA content histogram showing cell number as a function of DNA content. A typical result is shown in *Figure 2.2*. The first peak in a DNA content histogram is usually used to calculate the percentage of cells in G_0/G_1, while the second peak represents cells in G_2/M, and S phase cells are spread over the gap between these peaks. Shifts in the percentage of cells under the two main peaks are induced by factors which either promote or inhibit exit of cells from G_0/G_1, and

Protocol 2.7: Bromodeoxyuridine anti-bromodeoxyuridine demonstration of biopsy samples

EQUIPMENT, MATERIALS AND REAGENTS

Biopsy incubation-labelling medium: RPMI 1640 (Gibco Ltd, UK) supplemented with penicillin/streptomycin solution @ 100 U ml^{-1}/100 µg ml^{-1}, Hepes (N-[2-Hydroxyethyl]piperazine-N′-[2-ethanesulphonic acid]) buffer pH 7.3 @ 20 mM, and L-glutamine @ 2 mM. Adjust osmolality of the medium to 290 mosmol kg^{-1} with sterile distilled water. Store 1.93 ml aliquots in capped plastic tubes for up to 1 week. Immediately before use add to each tube 10 µl each of BrdU and hydrogen peroxide solutions to give final concentrations of 10 µM and 0.015%, respectively.

Absolute ethanol/glacial acetic acid (3:1).

Graded concentrations of ethanol (20%, 40%, 60%, 80%, absolute).

PBS pH 7.6.

Cacodylate buffer: 1 mM sodium cacodylate buffer containing 0.1 mM ethylenediaminetetraacetic acid (EDTA).

0.08 M HCl.

Immunostaining of BrdU-labelled DNA can be performed easily using individual reagents. BrdU, monoclonal anti-BrdU antibody, and anti-mouse Ig-FITC are all available from a range of commercial sources.

ProLong™ anti-fade kit.

METHOD

1. Pre-warm incubator block to 37°C.

2. Incubate for 1 hour with 1 ml culture medium per biopsy. A higher or lower concentration of BrdU may be indicated by experimentation.

3. Wash 5 times with PBS.

4. Cut biopsy sections or make cytocentrifuge slides from cell suspensions.

5. Fix for 1 hour in absolute ethanol/glacial acetic acid.

6. Rehydrate in graded concentrations of ethanol.

7. Bring to cacodylate buffer.

8. DNA denaturation (if needed): Incubate with 0.08 M HCl at 4°C for 30 seconds; wash once with cacodylate buffer; immerse for exactly 5 minutes in cacodylate buffer at 85°C in a water bath; rapidly cool by immersion in cacodylate buffer at 4°C.

9. Rinse thoroughly three times in PBS pH 7.6.

10. Incubate with 1:50 PBS pH 7.6 dilution of mouse anti-BrdU immune serum for 30 minutes – as in Protocol 2.6.

11. Rinse thoroughly three times in PBS pH 7.6.

12. Incubate with 1:50 PBS pH 7.6 dilution of FITC rabbit anti-mouse immune serum for 30 minutes.

13. Rinse thoroughly three times in PBS pH 7.6.

14. Mount coverslips on to glass slides using ProLong™ anti-fade kit. No sealing is required, and fluorescence should remain stable for up to a week or longer. For best results, however, examine and document results using fluorescence microscopy as soon after staining as possible.

15. Finally, place reagents containing BrdU medium in a disposal bottle and dispose of by mixing waste solution with ethanol on a paper towel before destruction in a chemical incinerator, or according to local health and safety regulations.

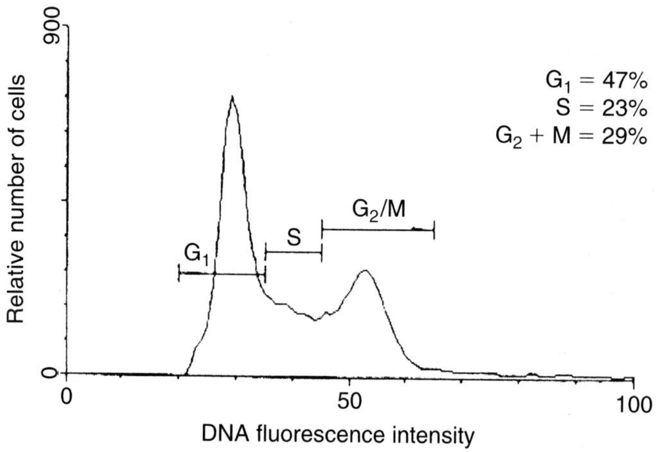

$G_1 = 47\%$
$S = 23\%$
$G_2 + M = 29\%$

Figure 2.2

A frequency distribution DNA content histogram obtained from the defn?HEL cell line stained with PI. The relative number of cells is plotted as a function of the DNA-associated fluorescence intensity. The sub-populations of cells in G_1, S and G_2/M phase and the relative percentages in each are shown. This figure was kindly provided by Dr Ying Hong, Department of Medicine, University College London.

subsequent re-entry into S phase and progression to G_2/M. Theoretically, the fluorescence intensities of all cells in either G_1 or G_2/M should have uniform values, but in practice variation within and between phase-specific populations results in peaks of different widths for G_1 and G_2/M cells. This variation in peak width, or coefficient of variation of the mean DNA-associated fluorescence, will be increased by the presence of apoptotic, necrotic or damaged cells in the cell population.

Flow cytometry is a very versatile methodology which can be used to obtain more refined information about cell cycle status by measuring other markers or characteristics of proliferating cells in addition to DNA content. Such bivariate or multivariate analysis is used to measure DNA content together with either cyclins, proteins which undergo marked changes in expression during the cell cycle, or proliferating cell nuclear antigen (PCNA), a commonly used marker of proliferating cells [13]. For example, cyclin A is present in G_2 cells but absent in M phase, so that measurement of cyclin A and DNA content can be used to determine the percentage of mitotic cells [14]. The extent of chromatin condensation varies in the different stages of the cell cycle and can be used to estimate the distribution of cells in G_0 as opposed to G_1, and G_2 versus M [12]. Since the degree of condensation alters the sensitivity of chromatin to denaturation, it can be measured using fluorochromes such as acridine orange which differentially bind to double-stranded and denatured DNA. The method described in Protocol 2.8 is for single time point univariate analysis of DNA content. Protocols for more sophisticated cytofluorimetric analysis of the cell cycle are described elsewhere [12–15].

Protocol 2.8: Analysis of the cell cycle using flow cytometry

EQUIPMENT, MATERIALS AND REAGENTS

Fluorescence-activated cell sorting (FACS) cytometer.

Lysis buffer: 0.5% Triton X-100, 4 mM $MgCl_2$, 0.6 M sucrose, 10 mM Tris-HCl, pH 7.5.

Nuclei buffer: 5 mM $MgCl_2$, 0.25 M sucrose, 20 mM Tris-HCl, pH 7.4.

PI: stock solution 1 mg ml^{-1} PI in PBS. Use at 20 μg ml^{-1} (10–50 μg ml^{-1} range).

DNase-free RNAse A: stock concentration 500 μg ml^{-1} (Roche Diagnostics Ltd). Use at 100 μg ml^{-1}.

METHOD

1. Incubate quiescent cell cultures with desired combinations of growth-promoting or growth-inhibiting factors and incubate at 37°C for 30–40 hours. If desired, cells can be treated with 1 μM colchicine 20 hours after addition of growth factors in order to arrest cells in M phase. Colchicine is a microtubule-disrupting agent which prevents passage through mitosis by preventing formation of the mitotic spindle. By delaying addition of colchicine for 20 hours after the onset of growth factor treatment, it is ensured that most cells have already entered or completed S phase thereby avoiding complications arising from any effect of colchicine on DNA synthesis. Treatment with colchicine in this way can also be used to estimate the rate of entry of cells into mitosis from the cumulative increase in mitotic cells as a function of the duration of colchicine treatment [15].

2. Trypsinise cells and disperse into a single cell suspension using a syringe fitted with a 19-gauge needle as described in step 3 of Protocol 2.1. After washing and pelting cells, resuspend in lysis buffer and allow to lyse for 5 min at room temperature.

3. Collect pellet, consisting mainly of cell nuclei, by centrifugation at 1000 g at 4°C for 5 min and incubate with 20 μg ml^{-1} PI solution containing 100 μg ml^{-1} DNase-free RNAse A for 20 min at 37°C in the dark.

4. Collect stained nuclei as in step 3, resuspend in cold PBS and analyse DNA content in a FACS cytometer with excitation at 488 nm.

2.5   Measurement of cell proliferation *in vivo*

The measurement of cell proliferation *in vivo* poses special methodological and inter-pretative problems that do not apply, or are less important, in cell cultures. The measurement of cell proliferation in lesions of arteries or other large blood vessels associated with atherosclerosis, vein graft failure and restenosis after angioplasty (*Table 2.2*), illustrates these problems very well [16]. One of the most characteristic features of lesions of blood vessels is the accumulation of cells, usually VSMC and macrophages, in the compartment adjacent to the endothelium, called the intima, to form a neointima. The area of the neointima measured by determining the ratio of neointimal area to that of the media or normal VSMC layer of the artery, usually abbreviated to the intima/media or I/M ratio, is often taken as a surrogate measure of cell proliferation. A major problem with this interpretation is that intimal area or thickness reflects not only proliferation but also cell migration. This problem can be overcome by measuring incorporation of [^{3}H]thymidine or BrdU, but the use of such reagents in animals may be prohibitively expensive, and exposure to radioactivity may be toxic and prevent progression through successive rounds of DNA synthesis thereby leading to an underestimation of proliferation (*Table 2.2*). The use of flow cytometry to determine cell content is possible, but digestion of the whole vessel into a single cell suspension may be problematic and generate artefacts, and will also measure proliferation throughout the vessel wall as opposed to neointimal prolifer-ation. A potentially attractive method is to stain sections of arteries with antibodies to cell proliferation markers such as PCNA or Ki-67. Such markers are not always spe-cific for cells in S or M phase, and consequently can result in considerable overesti-mation of cell proliferation. PCNA, though a component of the DNA polymerase δ complex, is known to stain cells still in G_1 and such cells may not progress to initiate DNA synthesis. The reason for the apparently promiscuous distribution of PCNA, and other 'markers' of cell proliferation, is that the stimulation of many events in mid or late G_1 that are essential for DNA synthesis to occur may not be sufficient to drive cells into S phase in the absence of additional mitogenic stimulation.

Overall, for determining VSMC proliferation in the vessel wall, measurement of BrdU incorporation and staining in conjunction with assessment of neointimal thick-ness is probably the approach which combines the maximum experimental reliabil-ity and methodological simplicity. BrdU staining is also widely used to monitor cell proliferation of tumours and other tissues *in vivo*. If determination of cell prolifer-ation in human tissue samples from biopsy material using BrdU labelling is not an

Table 2.2   Methods for measurement of VSMC proliferation *in vivo*

Method	Advantages	Disadvantages
Intimal area/ medial area	Simple; inexpensive; measure of pathophysiological change	Measures cell migration and matrix as well as proliferation
[^{3}H]thymidine/BrdU incorporation	Specific measure of proliferation	Expensive, may be toxic
Proliferation markers e.g. PCNA, Ki-67	Simple; inexpensive	Not entirely specific, may over-estimate cell proliferation
Flow cytometry	Specific measure of proliferation	Methodologically impractical

option, then staining of proliferation markers such as PCNA or Ki-67 is very useful. Staining for PCNA can also be used in historical tissue samples with the caveats referred to above.

2.6 Tracking cell division *in vitro* and *in vivo* using CFSE dilution and flow cytometry

Fluorescent dyes are being used increasingly for tracking cell migration and proliferation. Of 14 dyes recently reviewed, two stand out as being particularly useful for studying the long-term tracking of cells such as lymphocytes [17]. They are the intracellular covalent coupling dye carboxyfluorescein diacetate succinimidyl ester (CFSE) and the membrane inserting dye PKH26. Both dyes have the advantage that they can be used to track cell division, both *in vivo* and *in vitro*, due to the progressive halving of the fluorescence intensity of the dyes in the cells after each division. However, the CFSE procedure first described in 1994 [18] seems to be superior on the basis of both staining homogeneity and cost. This technique (Protocol 4.2) allows the visualisation of eight to ten discrete cycles of cell division using flow cytometry. Appropriately conjugated antibodies can be used to probe surface marker changes as cells divide, or changes in expression of internal molecules such as cytokines when appropriate fixation and permeabilisation methods are used [19]. The CFSE technique applied in such a way with flow cytometry can be used to determine kinetics of immune responses, track proliferation in minor sub-sets of cells, and follow the acquisition of differentiation markers or internal proteins linked to cell division.

2.7 References

1. Rozengurt, E. (1986) Early signals in the mitogenic response. *Science* **234**: 161–166.
2. Baserga, R. (1990) *Cell Growth and Division: A Practical Approach*. IRL Press, Oxford.
3. Freshney, I.R. (1992) *Animal Cell Culture: A Practical Approach*, 2nd Edn. IRL Press, Oxford.
4. Mosmann, T. (1983) Rapid colorimetric assay for cellular growth and survival: application to proliferation and cytotoxicity assays. *J. Immunol. Methods* **65**: 55–63.
5. Holst-Hansen, C. and Brünner, N. (1998) MTT-Cell proliferation assay. In: *Cell Biology: A Laboratory Handbook*, Vol.1, 2nd Edn (ed. J.E. Celis). Academic Press, San Diego CA.
6. Oliver, M.H., Harrison, N.K., Bishop, J.E., Cole, P.J. and Laurent, G.J. (1989) A rapid and convenient assay for counting cells cultured in microwell plates: application for assessment of growth factors. *J. Cell Sci.* **92**: 513–518.
7. Dicker, P. and Rozengurt, E. (1978) Stimulation of DNA synthesis by tumour promoter and pure mitogenic factors. *Nature* **276**: 723–726.
8. Cospedal, R., Lobo, M. and Zachary, I. (1999) Differential regulation of extracellular signal-regulated kinases (ERK) 1 and 2 by cyclic AMP and dissociation of ERK inhibition from anti-mitogenic effects in rabbit vascular smooth muscle cells. *Biochem. J.* **342**: 407–414.
9. Higgins, T. and Rozengurt, E. (1998) Stimulation of DNA synthesis in quiescent Swiss 3T3 cells: Autoradiography and BrdU staining. In: *Cell Biology: A Laboratory Handbook*, Vol.1, 2nd Edn (ed. J.E. Celis). Academic Press, San Diego CA.
10. Prasad, K.V., Wheeler, J., Robertson, H., MaWhinney, W.H.B., McHugh, M.I. and Morley, A.R. (1994) In vitro bromodeoxyuridine labelling of renal biopsy specimens: correlation between labelling indices and tubular damage. *J. Clin. Pathol.* **47**: 1085–1089.
11. Van Furth, R. and Van Zwet, T.L. (1988) Immunocytochemical detection of 5-bromo-2-deoxyuridine incorporation in individual cells. *J. Immunol. Methods* **108**: 45–51.

12. Darzynkiewicz, Z., Robinson, J.P. and Crissman, H.A. (eds) (1994) *Methods in Cell Biology*, Vol. 41. Academic Press, San Diego CA.

13. Juan, G. and Darzynkiewicz, Z. (1998) Cell cycle analysis by flow and laser scanning cytometry. In: *Cell Biology: A Laboratory Handbook*, Vol.1, 2nd Edn (ed. J.E. Celis). Academic Press, San Diego CA.

14. Gong, J., Traganos, F. and Darzynkiewicz, Z. (1995) Discrimination of G2 and mitotic cells by flow cytometry based on different expression of cyclins A and B1. *Exp. Cell Res.* **220**: 226–231.

15. Darzynkiewicz, Z., Traganos, F. and Kimmel, M. (1987) Assay of cell cycle kinetics by multivariate flow cytometry using the principle of stathmokinesis. In: *Techniques in Cell Cycle Analysis* (eds J.W. Gray and Z. Darzynkiewicz). Humana Press, Clifton NJ, pp. 291–336.

16. Newby, A.C. and George, S.J. (1993) Proposed roles for growth factors in mediating smooth muscle proliferation in vascular pathologies. *Cardiovasc. Res.* **27**: 1173–1183.

17. Parish, C.R. (1999) Fluorescent dyes for lymphocyte migration and proliferation studies. *Immunol. Cell Biol.* **77**: 499–508.

18. Lyons, A.B. and Parish, C.R. (1994) Determination of lymphocyte division by flow cytometry. *J. Immunol. Methods* **171**: 131–137.

19. Lyons, A.B. (2000) Analysing cell division in vivo and in vitro using flow cytometric measurement of CFSE dye dilution. *J. Immunol. Methods* **243**: 147–154.

Nuclear Changes in Dividing Cells

3

Christopher P. Maske and David J. Vaux

Contents

3.1 Introduction

The interphase nucleus is the site of transcription and replication, DNA repair and recombination (*Figure 3.1*). The spectacular changes which occur during mitosis are only part of the dynamic capabilities of the nucleus. As new tools are developed to

Cell Proliferation and Apoptosis, David Hughes and Huseyin Mehmet (Eds)
© 2003 BIOS Scientific Publishers Ltd, Oxford

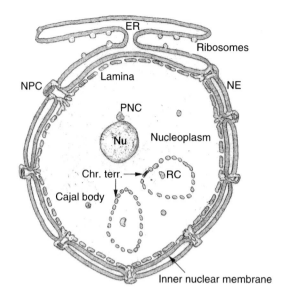

Figure 3.1

A stylised version of interphase nuclear structure is presented. The chromosomes are arranged in interphase territories (Chr. Terr.). Nuclear bodies, including the nucleolus (Nu), peri-nucleolar compartment (PNC), Cajal bodies and replication centres (RC) are arranged throughout the nucleoplasm. The whole is contained within a nuclear lamina made from the filamentous meshwork of the lamin proteins. The nuclear envelope (NE) is specialised on its inner face and contains an enriched population of membrane proteins. The nuclear pores (NPC) allow communication between the nucleoplasm and cytoplasm. The endoplasmic reticulum (ER) extends from the surface of the outer nuclear membrane.

explore with higher resolution the relationships between structure and function, it is becoming clear that, like the cytoplasm, the nucleus contains a host of structures and bodies, functionally defined domains, a predictable organisation, and a dynamic response to cell cycle and metabolic stimuli [1].

Most cells in the mature organism exit the cell cycle and exist in the non-proliferating G_0 phase. Such cells function for years with a stable nucleus and never have to face the challenges of replication or the potential rearrangements of chromatin following mitosis. It would however be incorrect to label these nuclei as static. There are considerable movements of chromatin, nuclear bodies and individual loci in response to metabolic functions like transcription [2, 3]. In contrast, the cycling cell does experience the additional challenges of replication and mitosis, marked by the characteristic biochemical and morphological events which identify these cells as being in the proliferating state.

Mitosis in higher eukaryotes is associated with a complete disassembly of the interphase nuclear structure. The question arises whether any organisation exists when reassembling the nucleus at the conclusion of mitosis, and what the mechanisms may be that are responsible for repositioning chromatin and nuclear bodies in a predictable manner. Just one of the demonstrations of a higher order organisation of chromosomes is the positioning of gene-rich chromosomes in the nuclear interior as opposed to the periphery [4]. Not as obvious as the changes of mitosis are the adjustments in chromatin to accommodate replication. In order for polymerisation to occur, dense heterochromatin must unravel to a certain extent and, in doing so, is repositioned toward the nuclear interior [5].

The changes in chromatin at mitosis are co-ordinated with those in the nuclear envelope and lamina. The role that the components of the nuclear envelope and lamina play in defining the internal structure of the nucleus, and therefore its functioning, has not been determined. This is, however, becoming an ever more urgent field of study as diseases are associated with this compartment. 'Nuclear laminopathies', which include defects in the proteins of the inner nuclear membrane and lamina, have been associated with phenotypes of Emery-Dreyfuss muscular dystrophy, familial cardiomyopathies, lipodystrophy and cardiac conduction defects in various combinations [6]. It is possible that polymorphisms in the genes associated with proteins in this compartment may be implicated in a wider group of individuals with apparently sporadic cardiac and muscular disease.

The changes associated with the cycling nucleus and the functional implications thereof are beginning to be elucidated. The following chapter explores some of the familiar and emerging tools employed in achieving this understanding to equip the researcher with a sound and novel approach to nuclear cell biology.

3.2   Dissecting nuclear dynamics with fluorescent proteins

Spontaneously fluorescent proteins have been used extensively in cell biology to study a variety of pathways, structures and compartments. The great benefit of these genetically encoded fluorophores is the ability to use them in live cells and therefore study dynamics of proteins and even molecular interactions. The green fluorescent protein (GFP) from *Aequoria victoriae* has been modified to include blue, cyan and yellow emission spectra [7]. The red fluorescent protein from coral *Discosoma* species, DsRed, has added an even longer emission wavelength to this repertoire [8].

Fluorescent proteins have been fused to a large number of cellular proteins and signal peptides for targeting to various compartments using molecular biology techniques. The surprising observation is that the added fluorescent proteins rarely alter the behaviour of the parent protein, and localisation and function is often preserved. There remains, of course, the danger of assuming this to be true, and the demonstration that the fusion protein indeed behaves like the endogenous protein remains the key, although often difficult, part of any study involving GFP. Although functioning of a particular protein may not be affected by the fusion to a fluorescent protein, the transfection procedure and ectopic high level expression may produce a stress response in the cell, which makes fluorescent proteins unsuitable for certain applications. The use of inducible expression systems [3] and the introduction of multiphoton excitation microscopy have improved this situation. DsRed, in particular, folds poorly but new mutants have been produced which improve the fluorescence and biological properties [9]. Live cell microscopy has the unwelcome side effect of generating heat and free radicals in the cell. Mitosis is a sensitive indicator of the tolerance of a cell to imaging and fluorescence emission, and the minimisation of laser output and exposure time is essential. Despite this, many studies have focused successfully on mitosis in live cells using fluorescent proteins [10, 11].

The use of fluorescent proteins is not limited to the monitoring of protein and organelle trafficking. Techniques have been developed which expand the use to quantitative measurements and demonstration of molecular interactions at resolutions which exceed those of the light microscope. Fluorescence resonance energy transfer (FRET) describes the transfer of energy from an emitting fluorescent molecule (donor) to another (acceptor) which absorbs at the wavelength being emitted. The probability of the acceptor being excited is inversely proportional to the sixth power of the distance between the molecules. Excitation of the acceptor is an indication that the donor molecule is within nanometers of the acceptor [12].

Fluorescence lifetime imaging (FLIM) measures the decay in time of fluorescence intensity emitting from fluorescent molecules. The fluorescence lifetime of a molecule is dependent on the molecular environment and is altered by the presence of a FRET acceptor molecule. Measuring lifetime can therefore be another powerful technique for measuring molecular interactions and changes in the cellular environment in response to stimuli [13].

FRAP exploits the finite number of photons which can be emitted from any fluorescent molecule. The recovery of fluorescence in a compartment following bleaching is dependent on the diffusional mobility of the molecule of interest, the intensity of the pool from which it is diffusing and the degree to which the bleached compartment is connected with any other pool. Photobleaching is assumed to be non-reversible and recovery of fluorescence represents diffusion into the area. This is a useful technique for measuring turnover and mobility of proteins within compartments [13].

Applications making use of fluorescent proteins rely on standard molecular biology techniques for construction of the fusion protein. The use of FRET, FLIM and FRAP applications is performed with some difficulty by even the most experienced microscopists. The examples described below represent some of the key studies using these techniques for nuclear applications. *Plate 1* demonstrates the differential mobilities of proteins in two different compartments using FRAP in a double-labelled live ECV 304 cell. DsRed is targeted to the nuclear envelope by fusion to the lamin B receptor, and GFP is fused to the polypyrimidine tract binding protein (PTB) accumulating in the nucleoplasm and the peri-nucleolar compartment (PNC).

3.2.1 Nuclear envelope dynamics through the cell cycle

GFP fused to the nucleoplasmic first transmembrane domain of the lamin B receptor (LBR) demonstrated that these domains were sufficient for targeting of LBR to the inner nuclear membrane during interphase [10]. The view that the nuclear envelope vesiculates during mitosis was challenged by the observation that LBR-GFP appeared to be part of the bulk endoplasmic reticulum (ER). Using FRAP, the diffusional mobility of interphase LBR was shown to be slow indicating a low turnover, but the mitotic LBR had a diffusional mobility equal to resident ER membrane proteins. The results from this study have been supported with studies of the lamina associated polypeptides (LAP) proteins, another group of inner nuclear membrane proteins [14].

3.2.2 The role of nuclear lamina proteins in assembly of the nuclear envelope

The lamin proteins are integrated into the stable interphase structure of the lamina. Phosphorylation at the onset of mitosis causes the lamina to disassemble. A-type lamins are soluble during mitosis, whereas B-type lamins remain attached to membrane as the result of a carboxyl terminus farnesyl moiety. These different fates have been shown to influence the recruitment of the lamins to chromatin and the reforming nuclear envelope at the end of mitosis [11]. The B-type lamins assemble first, being recruited to chromatin along with the inner membrane proteins due to their membrane association. A further study of post-mitotic events demonstrated that the components of the nuclear pore complex are recruited to the nuclear membrane from their cytoplasmic location during mitosis even before lamin B1 [15]. The formation of transport competent nuclear pore complexes is necessary for import of the A-type lamins, and indeed for nuclear growth during G_1.

3.2.3   Targeting of fluorescent proteins to sub-nuclear compartments

A distinct set of nuclear bodies which are not membrane bound, but may be just as functionally restrictive as organelles, have been characterised using fluorescent protein fusions to their individual marker proteins. GFP fused to the PTB showed an accumulation in the PNC [16]. The PNC dissociates at mitosis and reforms at late telophase. The size and shape change over time, and FRAP studies reveal a high turnover of PTB in the compartment. Cajal Bodies (formerly coiled bodies) are characterised by the presence of p80 coilin and show some remarkable dynamics in living cells. Small-scale movements similar in scale to chromatin are followed by large-scale movements across micron distances. Fusion and fission events also occur [17]. In contrast to this highly mobile compartment, replication factories labelled with a GFP fusion of PCNA show no movement of individual replication foci during S phase. Rather, foci disappear and appear in different locations in the nucleus co-ordinate with the stage of S phase [18].

The largest of the non-membrane bound compartments of the nucleus is the nucleolus. The reformation of the nucleolus at the conclusion of mitosis has been studied with GFP fusions to fibrillarin, nucleolin, and B23, all RNA processing-related components of the nucleolus [19]. It was shown that processing components of the nucleolus exist during mitosis in transient structures which become incorporated into the reassembling nucleolus during telophase.

3.2.4   Functional microscopy of the genome

The study of individual proteins with GFP has been used extensively to study the trafficking of proteins involved in mitosis, but it is also possible to study the dynamics of chromatin. The use of fluorescent proteins in the visualisation of chromatin and even single locus movement in real time has been a major development in studying higher order nuclear organisation. Belmont and co-workers [20] have developed a method using multiple repeats of the Lac operator sequence (*LacO*) which bind with high affinity the Lac repressor protein. By fusing the Lac repressor to a nuclear localisation signal and GFP, they have been able to monitor the real time dynamics of a single locus. In addition, by affecting transcriptional activation a heterochromatic region has a tendency to move centrally [21].

Fusion of GFP to histone proteins has been a useful way to track chromosomes through mitosis [22, 23]. Mitotic machinery has been studied with GFP fusion proteins in order to understand the mechanisms behind chromosome alignment, the checkpoint associated with anaphase [24] and the subsequent separation of sister chromatids [25].

3.3   Fluorescent *in situ* hybridisation (FISH) of nucleic acid

One of the great scientific achievements of recent years has been the sequencing of the draft human genome. The challenge which has been inherited from this is the functional annotation of this sequence – not only of the proteome, but also to arrange the linear sequence provided by the genome project in the context of the three-dimensional environment in which it exists.

FISH allows denaturation of double-stranded DNA and hybridisation of fluorescent probes to complementary sequence while preserving the three-dimensional structure. FISH can be performed on interphase and metaphase chromosomes and combined with immunofluorescence techniques to produce multiple-labelled samples. An example of a triple label experiment using FISH is given in *Plate 2*. It is clear from this image that interphase chromosomes occupy distinct territories. Studies suggest that the organisation of these territories is not random but part of an ordered arrangement.

Chromosomes have been described as being peripheral or central in the nuclear volume and that this location affects the timing of replication, with late-replicating DNA occurring peripherally [2]. Moreover these early- and late-replicating foci are stable over many cell cycles [26]. Peripheral chromosomes come into contact with the nuclear lamina and envelope, which may have a role in anchoring the chromatin [27]. The lamin B receptor has been shown to interact with heterochromatin-associated proteins [28].

Like the sequencing of all chromosomes in the genome, the three-dimensional map of all human chromosomes has been documented [4]. The findings suggest that gene-rich chromosomes are located centrally while gene-poor chromosomes are located peripherally and associated with heterochromatin. There is no correlation between size of the chromosome and position in the nucleus. This study goes further to analyse the positions of chromosomes in cells lacking the protein emerin, an integral membrane protein of the inner nuclear membrane. No difference in chromosome localisation was found suggesting that the phenotype of emerin-deficient individuals, Emery-Dreyfuss muscular dystrophy, is not due to aberrant gene expression from chromosome mislocalisation. Such disease models will be important in understanding the functional environment of the chromosome.

One of the problems in performing such studies is defining the reference points for chromosome localisation. A central or peripheral localisation may be a crude method of defining position. Nucleoli are arranged around the chromosomes which have NORs containing the genes for ribosomal RNAs [29]. However there are many other nuclear bodies including replication and transcription sites which may define the chromosomal orientation. An example is the association of coiled bodies with the loci for small nuclear RNAs [30].

A method for performing FISH in combination with immunofluorescence labelling is described in Protocol 3.1. *Plate 2* shows a triple label FISH experiment using a whole chromosome paint to chromosome 13. The probes used in hybridisation studies range from oligonucleotides to yeast artificial chromosomes. Single locus probes may be used, for example from a cDNA sequence, or whole chromosome paints which have had repetitive DNA subtracted are available commercially. Labelling of probes in the laboratory may be achieved by using polymerase chain reaction (PCR) methods and fluorescently labelled or biotinylated nucleotides. An example would be a single locus probe which can be amplified by PCR in the presence of biotin-deoxyuridine triphosphate (biotin-dUTP). Counterstaining of DNA is important when measurements of total DNA are to be performed. Cells should be mounted in DAPI mounting medium. DAPI is excited in the UV spectrum making it unsuitable for most confocal laser scanning microscopes. Alternative DNA counterstains excited in the visible spectrum include Topro, which emits in the far red.

Protocol 3.1: FISH of interphase cells

EQUIPMENT, MATERIALS AND REAGENTS

22 mm glass coverslips (No. 1.5). Parafilm. Humidified chamber.

0.1 M HCl.

20% glycerol in PBS.

PFA-Triton: 4% PFA (w/v) in 250 mM Hepes pH 7.4 and 0.1% Triton X-100 (v/v).

8% PFA-Hepes: 8% PFA (w/v) in 250 mM Hepes pH 7.4.

Glycine-PBS: 100 mM glycine in PBS.

Saponin-Triton-PBS: 0.5% saponin (w/v), 0.5% Triton X-100 (v/v) in PBS.

10 mM Tris-HCl buffer.

RNAse-Tris: 100 μg ml^{-1} RNAse A in 10 mM Tris-HCl buffer.

Labelled DNA probes.

1X saline sodium citrate (SSC) buffer: 150 mM sodium chloride, 30 mM sodium citrate, pH 7.0.

70% formamide in 2X SSC, 50 mM sodium phosphate buffer pH 7.0.

50% formamide in 2X SSC, 50 mM sodium phosphate buffer pH 7.0.

DAPI mounting medium: Anti-fade mounting medium containing 0.5 μg ml^{-1} DAPI.

METHOD

1. Cells should be grown on 22 mm glass coverslips in the appropriate medium and to approximately 70% confluence to ensure log phase growth and sufficient cells to withstand the denaturing conditions involved in preparation.

2. Fix the cells and permeabilise simultaneously with PFA-Triton for 30 minutes at 4°C.

3. Aspirate the fixative and replace with 8% PFA-Hepes and incubate at room temperature for 10 minutes.

4. Quench free aldehyde groups with glycine-PBS at room temperature for 20 minutes.

5. Permeabilise the cells with Saponin-Triton-PBS for 60 minutes at room temperature.

6. Wash with 8 changes of PBS over 20 minutes.

7. Spot 0.1 M HCl onto parafilm and incubate the cells in the HCl drop for 20 minutes by inverting the coverslip onto the drop in a humidified chamber.

8. Wash with 6 changes of PBS over 20 minutes.

9. Incubate the cells in 20% glycerol in PBS, changing 3 times over 20 minutes. Immerse the coverslips cell side up in 35 mm Petri dishes containing fresh 20% glycerol in PBS and float the dishes on liquid nitrogen until the glycerol is frozen through. Remove the dishes and thaw.

10. Transfer the coverslips to clean Petri dishes and wash immediately with PBS, changing 6 times over 20 minutes.

11. Incubate in 10 mM Tris-HCl pH 7.4, changing three times over 20 minutes to equilibrate for RNAse treatment and then invert the coverslips onto drops of RNAse-Tris. Incubate at 37°C for 2 hours in a humidified environment. During RNAse treatment the probes should be prepared. This involves denaturation and separation of the strands of DNA before the annealing step can take place. The probe used will depend on the experiment, however probes may include PCR products, cDNAs, yeast artificial chromosomes or repetitive DNA.

12. Prepare the labelled DNA probes by warming to 37°C for 5 minutes. Vortex for 2 seconds and centrifuge briefly for 5 seconds. Aspirate 10 μl of the probe, sufficient for one 22 mm coverslip, and return the rest to storage. The volume of probe to be used should be heated to 72°C for 5 minutes in order to separate the strands and then incubated at 37°C out of direct light for a minimum of 5 minutes and a maximum of 2 hours.

13. Prepare 30 ml of each formamide denaturing solution in a glass beaker. The formamide must be deionised. Heat the solutions in a water bath to above 65°C, measure the pH with indicator strips and adjust if necessary. Finally equilibrate the temperature to 75°C.

14. Following RNAse digestion, wash the coverslips once in 10 mM Tris-HCl pH 7.4, and 4 times in 2X SSC.

15. Place the coverslips in a slot holder and immerse into the 70% formamide solution for 9 minutes, followed immediately by the 50% formamide solution for 1 minute. While the incubation in 70% formamide is being performed the probes can be spotted onto parafilm and kept at 37°C in a humidified chamber.

16. Remove the coverslips from the 50% formamide solution, invert directly onto the probes and incubate at 37°C overnight in the dark to allow annealing to take place. A tissue culture or other equivalent humidified and temperature-controlled incubator is suitable.

17. Following the annealing step, wash the coverslips in 50% formamide in 1X SSC at 37°C for 15 minutes, followed by 2 changes in 0.1X SSC at 65°C for 15 minutes.

18. To continue with immunolabelling wash the cells in PBS and perform immunolabelling according to standard protocols (Protocol 3.2).

3.4   Immunofluorescence microscopy and its applications

The vast libraries of antibodies available against cellular proteins, the improvement in species-specific secondary antibodies and the expansion of the available fluorescent molecules has made routine the use of immunofluorescence techniques (Protocol 3.2). The ability to perform multiple labelling with antibodies raised in different species and to label cultured cells and tissues alike has added to the uses of the technique. Although the technique of immunofluorescence is performed on fixed cells, the technique is not limited to structural and localisation studies of proteins. Immunofluorescence can be used to quantify transcription and replication and has largely replaced the use of radiolabelled nucleotide precursors. Mitotic cells may be labelled in an unsynchronised population where the stage of mitosis can be easily identified by DNA counterstain with DAPI (*Plate 3*). Immunofluorescence may be combined with synchronising of populations in order to stage the cells outside of M phase.

3.4.1   Preparation of samples for immunofluorescence microscopy

The method of preparation of samples for immunofluorescence microscopy will depend on the application, and a careful assessment of factors must be performed. The ultrastructural preservation, the sensitivity of epitopes to fixation and the accessibility of antigens to antibodies will determine the protocol required for fixation. The aim should always be to achieve the highest degree of structural preservation that will allow adequate penetration and binding of the epitopes of interest. *Table 3.1* lists protocols for fixation and their applications. Glutaraldehyde fixation is not recommended for fluorescence microscopy because of autofluorescence of the compound. The best structural preservation for light microscopy is achieved with paraformaldehyde (PFA) buffered with Hepes. Formaldehyde, methanol and acetone are highly extracting and poor at preserving ultrastructure, although they may improve accessibility of certain epitopes. The nucleus is less sensitive to extraction during fixation than the cytosol, however fixation with agents like methanol may precipitate antigens and cause artefacts of localisation. New applications may require substantial optimisation of fixation conditions, blocking and antibody dilutions.

3.4.2   S phase labelling

A particular application of immunofluorescence techniques relevant to proliferating cells is the use of halogenated thymidine analogues in the labelling of replication events. The advantages of fluorescent techniques over radioisotope labelling include the lack of reliance on isotopes, the ability to do quantitative analysis, the three-dimensional resolution possible in the confocal microscope and the ability to do multiple-labelling experiments to correlate replication with other processes or structures. Tritiated thymidine, used because of its specificity for incorporation into DNA rather than RNA and therefore as a quantifiable marker of replication and not transcription, is still used. Autoradiography can generate spatial representation of the incorporated deoxythymidine, however the β-particle emitted can travel several hundred nanometres and therefore produces a low-resolution image [31]. An alternative to this compound as a specific marker of replication is BrdU. BrdU can permeate the cell membrane and incorporates into polymerising DNA during S phase. BrdU-labelled DNA can be detected with specific antibodies and direct or indirect detection methods. One of the problems associated with BrdU is the time taken for labelling to occur. Once the compound has permeated the cell it competes with

Protocol 3.2a: Fixation of cells and tissues for immunofluorescence microscopy

EQUIPMENT, MATERIALS AND REAGENTS

Fresh buffered ammonium chloride: 50 mM NH_4Cl in PBS.

Buffered Triton X: 0.2% Triton X-100 (v/v) in PBS.

Buffered bovine gelatin: 0.2% (w/v) bovine gelatin in PBS.

Buffered bovine serum albumin (BSA): 0.5% (w/v) BSA in PBS.

METHOD

1. Grow adherent cells as monolayers on clean sterile glass coverslips in the appropriate growth medium. Most objectives are corrected for No. 1.5 coverslips. Non-adherent cells should preferably be pelleted and physically sectioned according to the technique described in Protocol 3.4a. Tissue sections cut from fresh frozen tissue should be picked up on glass slides and can be processed for fixation according to the following protocol.

2. Fix the cells with the agent of choice. See *Table 3.1* for fixation protocol and comments.

3. Wash the cells with 3 changes of PBS over 10 minutes.

4. Quench free aldehyde groups with fresh buffered ammonium for 5 minutes.

5. Permeabilise the cells with buffered Triton X for 5 minutes. Permeabilisation may be performed for longer or with higher concentrations of detergent in order to extract more from the cells and expose a difficult epitope.

6. Wash the cells with 3 changes of PBS over 10 minutes.

7. Block the cells with the blocking agent of choice for 30 minutes at room temperature. Blocking should be optimised for the specific application. Generally, buffered bovine gelatin or buffered BSA are used. When using lectins (e.g. concanavalin A), 0.2% fish skin gelatin should be used.

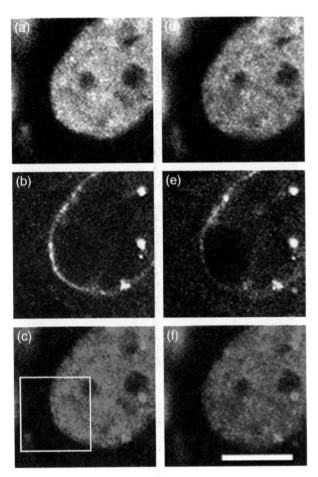

Plate 1

Double-label live-cell immunofluorescence microscopy is used to show the different diffusional mobilities of proteins in the nucleoplasm and inner nuclear membrane. GFP was fused to the N terminus of polypyrimidine tract binding protein (GFP-PTB) and DsRed was fused to the nucleoplastic and first transmembrane domains of the lamin B receptor (LBR-DsRed). ECV 304 cells were transfected with both constructs and imaged with a Bio-Rad MRC-1024 confocal laser scanning microscope. The panels on the left (a–c) show the GFP-PTB, LBR-DsRed and merge respectively, of a single plane from a cell before it is photobleached in the boxed region. The right panels (d–f) show the same cell 1 minute after completion of photobleaching in the boxed region with full laser power for 30 seconds. Recovery of fluorescence is seen in the nucleoplasmic GFP-PTB panel in the area of photobleaching, although there is an overall decrease in fluorescence intensity due to communication with the rest of the pool. In contrast, fluorescence recovery is not seen in the LBR-DsRed indicating a relative immobility of this inner nuclear membrane protein. The signal in the rest of the nuclear envelope is preserved. Scale bar = 10 μm.

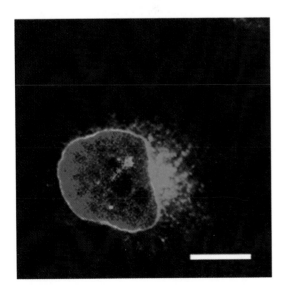

Plate 2

Fluorescent *in situ* hybridisation (FISH) of an ECV 304 cell using whole chromosome paint to chromosome 13 (red) and counterstain of DNA with To-Pro (blue) and ER-resident proteins with FITC-concanavalin A (green). Note the central position and proximity of chromosome 13, which contains nucleolar organiser regions, to the nucleoli. In addition this nucleus contains membrane invaginations of the nuclear envelope which have an apparent association with nucleoli [33]. Scale bar = 10 μm.

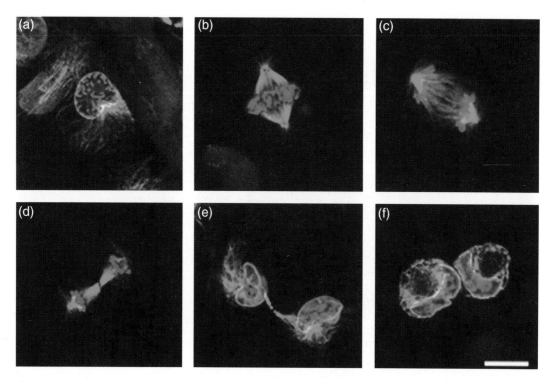

Plate 3

An unsynchronised population of HeLa cells were labelled with antibodies against lamin B1 (red), tubulin (green) and counterstained with To-Pro to label DNA (blue). The different stages of mitosis and the structural changes that accompany them are illustrated with this triple label immunofluorescence experiment.
(a) Prophase is marked by chromosome condensation but the nuclear envelope remains intact and the spindle is not yet formed. (b) Metaphase is marked by the characteristic arrangement of chromosomes on the equator of the cell and attached to the spindle fibres, which form the characteristic aster radiating from the spindle poles. (c) The sister chromatids are separated during anaphase. (d) Late anaphase sees the formation of the mid body at the point of cytokinesis. The nuclear envelope is not yet formed although recruitment of membrane to the surface of chromosomes will have begun. (e) The nuclear envelope assembly around the decondensing chromatin is complete by telophase. (f) G_1 and the beginning of the next cell cycle has begun with a fully competent nuclear envelope and the beginnings of an interphase arrangement of microtubules. The nucleus will grow in size as chromatin decondenses further. Scale bar $= 10\,\mu$m.

Protocol 3.2b: Antibody labelling of cells and tissues

EQUIPMENT, MATERIALS AND REAGENTS

Humidified chamber.

Appropriate range of antisera and reagents for immunofluorescence.

METHOD

1. Following blocking to reduce non-specific binding, cells should be incubated with primary antibodies diluted in the blocking agent used for a minimum of 45 minutes. Dilutions of antibodies should be determined empirically. Antibodies from different species may be used in multiple labelling experiments provided that highly cross-absorbed secondary antibodies are used for detection. Spotting the antibody onto parafilm and incubating the coverslip cell side down on the drop will reduce the amount of reagent needed. All antibody incubations should be performed in a humidified chamber to avoid desiccation.

2. Wash with 3 changes of PBS over 15 minutes.

3. Incubate with the secondary antibody conjugated to a fluorophore, biotin, or enzyme for a minimum of 45 minutes. The antibody should be made up in the same blocking buffer at an optimal dilution. Ensure proper mixing by brief vortex followed by centrifugation for 1 minute at $15\,000\,g$ on the bench top to pellet aggregated material. Fluorescent probes should be incubated in the dark.

4. Wash with 3 changes of PBS over 15 minutes.

5. Rinse briefly in distilled water to avoid formation of salt crystals, drain excess fluid and mount in mounting medium of choice.

Protocol 3.2c: Mowiol mounting medium

EQUIPMENT, MATERIALS AND REAGENTS

Glycerol.

Mowiol 4-88 (Hoechst AG).

Tris-HCl: 0.2 M Tris-HCl pH 8.5.

DAPI: stock solution 100 μg ml^{-1} in distilled water.

METHOD

Mounting media differ in their properties and are optimised for different fluorescent probes. The use of nail varnish to seal coverslips is not recommended because it may impregnate the sample and is autofluorescent in many channels. For most immunofluorescence applications Mowiol, a polymerising agent, with DAPI added to detect nucleic acid is the medium of choice. Mowiol mounting medium can be made up in the following way:

1. To a 50 ml clean tube, add 6 g glycerol (99.5%), 2.4 g Mowiol 4-88 and 6 ml distilled water.

2. Dissolve at 50°C with agitation for 2 hours.

3. Add 12 ml Tris-HCl and dissolve for a further 3 hours at 50°C with agitation.

4. Centrifuge at 2500 r.p.m. for 5 minutes.

5. Add stock solution DAPI to give a final concentration of 0.1 μg ml^{-1}.

6. Aliquot and store at −20°C.

Table 3.1   Fixation reagents and conditions used in the preparation of samples for immunofluorescence microscopy

Fixative	Buffer	Duration	Applications	Limitations
Glutaraldehyde	200 mM sodium cacodylate, pH 7.4	10 minutes at 4°C 20 minutes at RT	EM resin embedding Not for fluorescence microscopy	Autofluorescence Loss of antigenicity
4% PFA/ 8% PFA	250 mM Hepes, pH 7.4	10 minutes at 4°C (4%), followed by 50 minutes at RT (8%)	Cryo-immunoelectron microscopy Immunofluorescence	Rarely, loss of antigenicity
4% PFA	250 mM Hepes, pH 7.4	10 minutes at 4°C 20 minutes at RT	Cryo-immunoelectron microscopy Immunofluorescence	Poorer ultrastructural preservation
4% PFA	PBS	10 minutes at 4°C 20 minutes at RT	Immunofluorescence Immunohistochemistry	Poor ultrastructural preservation
Methanol (−20°C)	None	5 minutes at −20°C	Immunofluorescence Immunohistochemistry	Extreme extraction with very poor ultrastructural preservation
Ethanol	None	10 minutes at 4°C	Immunohistochemistry	Extreme extraction with very poor ultrastructural preservation

Abbreviations: EM, electron microscope; Hepes, N-[2-Hydroxyethyl]piperazine-N′-[2-ethanesulphonic acid]; PFA, paraformaldehyde; PBS, phosphate-buffered saline; RT, room temperature.

a pool of substrate in the cell and therefore has a low specific activity. An alternative to BrdU is therefore use of biotin-dUTP in combination with cell permeabilisation to wash away the endogenous pool of nucleotide and allow access of biotin-dUTP, which can not permeate the cell membrane. As described in Protocol 2.6, the bromouridine derivatives can be used to analyse S phase in single cells, with the benefit that experiments done with immunofluorescence can be translated into cryo-immunoelectron microscopy [32].

3.5   Electron microscopy techniques

The electron microscope (EM) extends the resolution of microscopic techniques and has the added advantage of being able to combine immunolocalisation and double labelling with detailed structural information at a sub-cellular level. Resin embedding of samples provides the best ultrastructural preservation and, because the sample is kept in a defined orientation during physical sectioning, it is possible to perform serial sections and reconstruct three-dimensional data. This technique has been used in defining the nature of intranuclear invaginations of the nuclear envelope [33]. The resolution and structural preservation of resin embedding is demonstrated in *Figure 3.2*, in which an enriched population of mitotic cells were fixed in glutaraldehyde and processed according to the resin embedding protocol below. The recruitment of cisternae from the ER to the surface of the condensed chromosomes in late anaphase can be seen. The mitotic spindle is also clearly visible.

Cryosectioning techniques sacrifice the superb ultrastructure and serial sectioning abilities of resin embedding for the ability to perform immunolabelling. Probes used

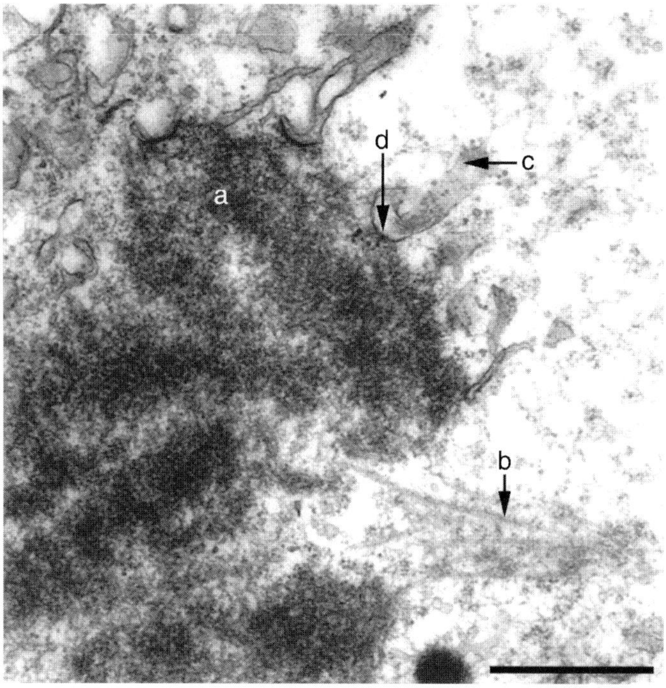

Figure 3.2

Electron micrograph of a HeLa cell in anaphase B or late anaphase. Mitotic cells were enriched by shake-off from an unsynchronised flask of HeLa cells, pelleted and fixed in glutaraldehyde. The pellet was processed for resin embedding in Epon and 100 nm sections cut on an ultramicrotome. The image demonstrates high-resolution structure of mitotic chromosomes (a), the spindle fibres (b), and cisternae of endoplasmic reticulum (c) being recruited to the face of chromosomes (d) by attachment of the inner nuclear membrane proteins and B-type lamins. Scale bar = 1 μm.

include enzymatic methods and colloidal gold labelling. The use of colloidal gold is particularly useful in double-labelling experiments. Although double labels may be performed with enzymatic methods, specialised EM techniques are required to differentiate the signals by energy loss of electrons. Peroxidase and alkaline phosphatase substrates are used to produce electron-dense deposits. The resolution of enzymatic methods is not as good as colloidal gold and obscures the underlying structures which it labels. Cryo-immunoelectron microscopy has many applications in the proliferating cell. It has been used extensively in characterising the replication foci [34] of proliferating cells, demonstrating that replication foci were part of the family of nuclear bodies.

3.5.1 Resin embedding

The combination of glutaraldehyde fixation and resin embedding offers the opportunity to perform serial sections at EM resolution and of the best preservation of ultrastructure. Cell monolayers, cell pellets and tissues can be processed for resin embedding (Protocol 3.3). Tissues from perfusion-fixed animals offer very good ultrastructure and do not require additional fixation. Tissues from biopsy samples require fixation prior to embedding. For optimal fixation, tissue samples should be diced into pieces not larger than 1 mm^3.

Protocol 3.3: Resin embedding

EQUIPMENT, MATERIALS AND REAGENTS

Sodium cacodylate buffer pH 7.2, which may be stored as a 400 mM stock solution at 4°C.

Buffered glutaraldehyde: 0.5% glutaraldehyde in 200 mM sodium cacodylate pH 7.2

Stock 2% (w/v) aqueous solution of osmium tetroxide stored at 4°C.

Potassium ferricyanide 3% (w/v) aqueous stock solution stored at 4°C.

Osmium tetroxide-potassium ferricyanide: 1% (w/v) osmium tetroxide/1.5% (w/v) potassium ferricyanide in distilled water for 1 hour in the dark. The two are mixed 1:1 immediately prior to use.

0.5% magnesium uranyl acetate in distilled water.

70% ethanol, 90% ethanol, 100% ethanol.

Propylene oxide.

Epon is made up by combining 21.2 g of Epon, 9.2 g anhydrous methylnadic acid, and 14.8 g anhydrous dodecenylsuccinic acid. Shake the mixture well and store at −20°C. Prior to use accelerator is added to enhance polymerisation. Add 0.8 g Tris (dimethylaminomethyl) phenol to the Epon stock and shake well.

METHOD

All steps should be performed in a fume hood and protective clothing and eyewear worn. Glutaraldehyde and osmium-containing solutions and labware should be discarded in appropriate containers. Cutting of resin sections by ultramicrotome is a specialised technique and should be performed by or learned from experienced personnel. Briefly, a surface is prepared by grinding the surrounding Epon to a point which can be sectioned. The sample is mounted in an ultramicrotome and sections of the desired thickness can be cut using a diamond knife. The sections are floated onto a water surface and picked up onto a plastic-coated EM grid. The sections can be visualised in the EM immediately.

1. Cells, pellets or tissues should be fixed in buffered glutaraldehyde for 10 minutes at 4°C and 20 minutes at room temperature.

2. Wash three times in sodium cacodylate buffer over 15 minutes. Residual glutaraldehyde may cause precipitation of osmium in step 3.

3. Perform second fixation in osmium tetroxide-potassium ferricyanide.

4. Wash 4 times with distilled water over 15 minutes for monolayers, and over 60 minutes for tissues and cell pellets.

5. Stain protein with magnesium uranyl acetate solution at 4°C overnight. Wash 4 times with distilled water over 15 minutes for monolayers, and over 60 minutes for tissues and pellets.

6. Dehydrate the samples using the following regime: 10 minutes in 70% ethanol; 10 minutes in 90% ethanol; 2 × 30 minutes in 100% ethanol; 2 × 30 minutes in propylene oxide (pellets and tissue samples only).

7. Embedding is performed with graded dilutions of Epon in ethanol (monolayers) or propylene oxide (pellets and tissues). After addition of accelerator the Epon will become viscous with time, and should therefore be used fresh for embedding steps: Ethanol 3:1 Epon for 30 minutes; Ethanol 1:1 Epon for 2 hours; Ethanol 1:3 Epon overnight; 100% Epon for 3 hours with degassing. For pellets and tissue samples, substitute propylene oxide for ethanol in the above regime.

8. Polymerise the Epon at 57°C overnight.

3.5.2   Cryo-immunoelectron microscopy

Immunolabelling of thin sections with gold probes combines the resolution of electron microscopy and physical sectioning with the localisation of antibody labelling used in immunofluorescence techniques. In addition the technique can be quantitative and therefore provides a method for performing measurements at high resolution. The steps in Protocol 3.4a are from the Tokuyasu method [35].

3.5.3   Immunolabelling of cryosections

Once sections have been cut, the following protocol should be used to perform labelling with gold-conjugated probes (Protocol 3.4b). Immunofluorescence labelling should employ the same protocol, the only difference being the secondary antibodies used. Washing steps need not be as stringent for immunofluorescence as for immunogold labelling. EM grids can be incubated on drops on parafilm for all steps. For double labelling experiments with different sizes of colloidal gold probe the antibody incubations are usually done sequentially (Protocol 3.4c).

Protocol 3.4a: Cryo-immunoelectron microscopy sample preparation

EQUIPMENT, MATERIALS AND REAGENTS

Buffered 4% PFA: 4% PFA in 250 mM Hepes pH 7.4.

Buffered 8% PFA: 8% PFA in 250 mM Hepes pH 7.4.

2.1 M sucrose in PBS.

METHOD

1. A 35 mm Petri dish of adherent cells near confluence is sufficient to make a pellet for cryosectioning. Non-adherent cells may be pelleted directly from the medium. Tissue samples which have not been perfusion fixed should be diced into 1mm^3 pieces.

2. Wash cells of medium with cold PBS.

3. Fix the cells for 10 minutes at 4°C in buffered 4% PFA. Monolayer cells should be fixed *in situ*, while pellets of non-adherent cells may be resuspended in 0.5 ml of fixative, transferred to a microfuge tube and pelleted immediately by gentle centrifugation at 1000 g for 1 minute. Tissue samples are immersed in the fixative.

4. Replace the fixative with buffered 8% PFA and fix for 15 minutes at room temperature. The fresh fixative is overlayed on the pellet of non-adherent cells.

5. Scrape the cells from the dish using a rubber policeman and transfer the resulting suspension into a microfuge tube and pellet the cells by centrifuging at 10 000 r.p.m. for 30 seconds. The pellet of non-adherent cells should be centrifuged in the same manner to maintain the integrity of the pellet during subsequent steps.

6. Aspirate the fixative and replace with fresh buffered 8% PFA and incubate at room temperature for 30 minutes.

7. Centrifuge at 15 000 r.p.m. for 30 seconds.

8. Dislodge the pellet from the bottom of the tube with a cocktail stick and aspirate the pellet into a disposable Pasteur pipette, being careful not to disrupt the pellet.

9. Passage the pellet through 3 drops of 2.1 M sucrose in PBS, allowing a 10 minute incubation in each drop. This is most easily done using a disposable Petri dish which can be covered during incubations.

10. Rinse in fresh 2.1 M sucrose in PBS and mount on a cryomicrotome sample holder taking care to remove the sucrose with filter paper. Copper stubs are preferred because they are not expensive and do not heat up quickly.

11. Flash freeze the sample on the holder in liquid nitrogen and store in liquid nitrogen. The sample and stub can be inverted sample-first into a 0.5 ml microfuge tube with the lid removed, which will prevent the sample being damaged and keep it bathed in liquid nitrogen while transferring from storage to cryomicrotome.

Cutting of cryosections on a cryomicrotome is a specialised technique and should be learned from experienced personnel. The exact method depends on the available ultracryomicrotome. The thickness of the sections depends on the application. Immunofluorescence labelling of cryosections generally requires sections from 250 to 500 nm in thickness, although thinner sections may be used for increased resolution. It should be noted that 500 nm sections are equivalent to the resolution capabilities of the confocal laser scanning microscope. The thickness of sections for immunoelectron microscopy should be between 70 and 120 nm. Once sections have been cut they are lifted from the face of the glass or diamond knife with a drop of 2.1 M sucrose in PBS which is suspended in a wire loop. The sections are placed onto a Formvar-coated EM grid for gold labelling, or gelatin-coated glass slide or coverslip for immunofluorescence labelling.

Protocol 3.4b: Immunolabelling of cryosections

EQUIPMENT, MATERIALS AND REAGENTS

Buffered ammonium chloride solution: 50 mM NH_4Cl in PBS.

5% fetal calf serum (FCS) in PBS.

10% FCS in PBS.

Appropriate range of antisera for protein A immunocytochemistry.

Methyl cellulose/uranyl acetate: 2% (w/v) methyl cellulose containing 0.3% magnesium uranyl acetate in triple-distilled water.

METHOD

1. Quench free aldehyde groups with buffered ammonium chloride for 5 minutes.

2. Block with 10% FCS for 30 minutes.

3. Primary antibody should be diluted in 5% FCS. In general the concentration should be 10 times greater than that used for immunofluorescence labelling of whole cells. Incubate the primary antibody for 1 hour at room temperature.

4. Wash with PBS with 5 changes of 5 minutes each.

5. Dilute the secondary reagent in 5% FCS and incubate for 30 minutes. Rabbit antibodies bind protein A directly. All other species of primary antibody require a rabbit secondary antibody for binding to protein A. An alternative is using gold colloids directly conjugated to secondary antibodies, which should be incubated for 1 to 2 hours. See the double labelling protocol in Protocol 3.4c.

6. Wash with PBS with 5 changes of 5 minutes each.

7. If protein A is required, dilute in 5% FCS and incubate for 30 minutes.

8. Wash with PBS with 5 changes of 5 minutes each. This is the final washing step and may be extended to improve signal to noise.

9. Desalt by passage through 5 drops of triple-distilled water.

10. Incubate on a drop of methyl cellulose/uranyl acetate for 10 minutes at 4°C. The magnesium uranyl acetate labels protein and therefore provides contrast to the sample. The methyl cellulose coats the sample.

11. Pick up the grid in a 3.4 mm diameter wire loop, remove excess methyl cellulose by streaking on a piece of hardened filter paper and allow to dry for 10 minutes before electron microscopy is performed.

Protocol 3.4c: Double labelling with colloidal gold, using colloidal gold with a mouse monoclonal and rabbit polyclonal antibody

EQUIPMENT, MATERIALS AND REAGENTS

5% FCS in PBS.

Appropriately diluted immune sera and protein A colloidal gold conjugates.

Methyl cellulose/uranyl acetate: 2% (w/v) methyl cellulose containing 0.3% magnesium uranyl acetate in triple-distilled water.

METHOD

1. Cryosections are prepared, quenched and blocked in the same manner as described in Protocols 3.4a and 3.4b.

2. The mouse monoclonal antibody should be diluted in buffered 5% FCS and incubated for 30 minutes to 1 hour.

3. Wash in PBS for 5 minutes on each of 5 drops.

4. Incubate for 30 minutes with rabbit anti-mouse secondary antibody in buffered 5% FCS.

5. Wash in PBS for 5 minutes on each of 5 drops.

6. Incubate with protein A conjugated to 5 nm colloidal gold for 30 minutes.

7. Wash in PBS for 5 minutes on each of 5 drops.

8. Block free protein A binding sites using $10\,\mu g\,ml^{-1}$ unconjugated protein A for 30 minutes.

9. Wash in PBS for 5 minutes on each of 5 drops.

10. Incubate with the rabbit polyclonal antibody against the second antigen of interest, diluted in buffered 5% FCS and incubated for 30 minutes to 1 hour.

11. Wash in PBS for 5 minutes on each of 5 drops.

12. Incubate with protein A conjugated to 10 nm colloidal gold for 30 minutes.

13. Wash in PBS for 5 minutes on each of 5 drops.

14. Desalt over 5 minutes with 5 changes of deionised water.

15. Incubate for 10 minutes in methyl cellulose/magnesium uranyl acetate at 4°C.

16. Remove excess methyl cellulose by streaking on hardened filter paper and allow to dry for 10 minutes before imaging in the EM.

3.6   Manipulation of the cell cycle

The cell cycle progresses through stages which vary in length for different cell types. The major changes seen in dividing cells occur during S phase where DNA is replicated and represents the commitment of the cell to progress to division, and in mitosis or M phase. *Plate 3* highlights the major structural changes which occur during mitosis. Associated with these morphological changes are a host of biochemical events which co-ordinate the progression of the cell cycle and mediate the structural changes which occur. Immunofluorescence and electron microscopy have been introduced as means for looking at structural changes. Although biochemical events can be visualised in individual cells using microscopic techniques like FRET and FLIM (see Section 3.2), more often traditional approaches are taken requiring large numbers of cells which are more or less biochemically similar. Biochemical similarity can be achieved in a population of cells by manipulation of the cell cycle and synchronisation.

A population of cultured cells is generally asynchronous with regard to the cell cycle. Growing certain cell types to extreme confluency, for example normal rat kidney cells, will arrest cells by contact inhibition and put these cells into G_0 of the cell cycle. Seeding these at lower density will bring about a rapid resumption of the cell cycle, although the progression of cells through the cycle may proceed at different rates. Physical means of purifying cells in a particular phase of the cell cycle is possible by exploiting the differential adherence properties of cells in mitosis, which can be enriched by knocking off from the bottom of the plate. Chemical means for synchronisation may block cells in an identical phase of the cell cycle, but are associated with a loss of viability of a certain proportion of the cells or ineffective blocking of the entire population.

Applications using cell cycle synchronisation include the resolution of replication in time and space. Synchronised cells pulsed with BrdU at time periods after the onset of S phase show a defined positioning of replication depending on the stage of S phase. Early S phase is marked by DNA synthesis in the centre of the nuclear volume, whereas later S phase time points show active replication of the peripheral heterochromatin [26].

The method or combination of methods used for synchronisation depends on the cell type, the percentage of cells required in a particular phase and the numbers of cells needed. The principle of 'block and release' allows enrichment of population in all phases of the cell cycle.

3.6.1   Mitotic shake-off

Mitotic cells tend to round up at prophase and are less adherent than interphase cells. This can be exploited in obtaining highly enriched populations of cells which cover all stages of mitosis and is particularly useful for electron microscopic analysis (Protocol 3.5). Certain cell types do not round up sufficiently to have differential adherence. HeLa cells are amenable to shake-off. The method can be scaled-up to achieve biochemical amounts of cells representing the spectrum of mitotic stages.

3.6.2   Mitotic block and synchronisation

A population of cultured cells can be synchronised by blocking in mitosis using agents which inhibit microtubule polymerisation. Colchicine and nocodazole both bind to tubulin and inhibit its polymerisation and therefore halt mitosis at the point of spindle assembly, namely prometaphase. In a rapidly cycling population a highly

Protocol 3.5: Mitotic shake-off

EQUIPMENT, MATERIALS AND REAGENTS

Ice-cold PBS.

Phase contrast microscopy.

METHOD

1. An unsynchronised population of cells should be grown in large tissue culture flasks (T175) to near confluence in the appropriate medium.

2. Aspirate the medium in order to remove all dead cells. If a washing step with PBS is used, care should be taken not to dislodge mitotic cells and thereby decrease the yield.

3. Add 20 ml of ice-cold PBS to the flask and dislodge the poorly adherent mitotic cells by tapping the sides of the flask. The extent of release can be monitored by phase contrast microscopy. Over-zealous tapping will result in interphase cell contamination.

4. Aspirate the mitotic shake-off in the PBS and pellet the cells at 400 g. The cells can be fixed, lysed for biochemical analysis or seeded into a new flask to achieve a relatively synchronised population.

enriched population can be achieved in a short time. High concentrations of thymidine inhibit the synthesis of deoxyribonucleotide precursors for DNA synthesis and therefore arrest cells at the G_1/S border. The two agents can therefore be used synergistically to achieve a population of cells which are synchronised and highly enriched. Protocol 3.6a is optimised for HeLa cells and has been modified from the protocol of Jackson and colleagues [34]. The use of the so-called double thymidine block (Protocol 3.6b) allows tight synchronisation of a rapidly growing population of cells at the beginning of the S phase.

Protocol 3.6a: Mitotic block and synchronisation

EQUIPMENT, MATERIALS AND REAGENTS

2.5 mM thymidine.

Nocodazole: soluble in dimethylsulphoxide (DMSO) and should be stored at −20°C.

METHOD

1. Culture cells in 2.5 mM thymidine for 22 hours.

2. Release the thymidine block by replacement with fresh growth medium. Culture for 4 more hours.

3. Add nocodazole to a final concentration of 1 μg ml^{-1} and culture for a further 8 hours. Greater than 98% of cells will be arrested in mitosis.

4. Release from the nocodazole block and incubate the cells for desired amount of time. Cells may be assayed at 2 hours (G$_1$), 8.5 hours (early S), 13 hours (mid S), 18 hours (late S), and 20 hours (G$_2$). A small proportion of cells will not be viable after the cell cycle manipulation.

Protocol 3.6b: Double thymidine block

EQUIPMENT, MATERIALS AND REAGENTS

Thymidine.

METHOD

1. Adherent or non-adherent cells (e.g. HeLa spinner S3 cells) should be grown in the appropriate medium until the log phase is achieved.

2. Add thymidine to the medium to a concentration of 2.5 mM. Incubate for a further 12 hours.

3. Remove the thymidine-containing medium, replace with fresh growth medium and incubate for 10 hours.

4. After ten hours add thymidine to a concentration of 2.5 mM and incubate for 12 hours more.

5. Release the cells from the thymidine block by replacement with fresh growth medium and allow to grow under normal conditions. Cells can be analysed at time points after the release.

3.7 Biochemical analysis of nuclear processes

Biochemical analysis of the nucleus is complicated by a number of factors which are unique to study of this compartment. The nuclear lamina forms a stable polymer which is resistant to non-ionic detergent. The lamina may be dissolved by ionic detergents, like SDS, and high salt extraction in 1 M sodium chloride. Even under high salt conditions there remains an insoluble protein component which is defined as the nuclear matrix. Although the extreme rigidity of the nuclear lamina can be exploited in purifying nuclei from other cellular compartments, it can be problematic in biochemical preparations. An additional hindrance is the presence of nucleic acid, which may complicate separation of proteins by two-dimensional electrophoresis.

The ease with which nuclei can be isolated from other cellular compartments makes it a first step in many biochemical preparations. The enrichment of a protein of interest by concentration of large numbers of nuclei is useful for separation techniques. In addition, biochemical assays reconstituting the events of mitosis require the isolation of intact nuclei with competent nuclear envelopes. Isolation of nuclei from whole cells can deplete the cytosolic fraction from nuclear antigens almost completely. Unfortunately, any method used for nuclear isolation inevitably contains contamination from both cytoskeletal elements and resident proteins of the ER. The use of detergent and more stringent methods can minimise this contamination but compromises the integrity of the resulting nuclei. Contamination by cytoplasmic proteins should be taken into account when carrying out downstream analysis of isolated nuclei.

3.7.1 Nuclear isolation from cultured cells

This procedure may be used for the isolation of intact nuclei with low levels of contamination of cytoplasmic components for analysis of processes such as nuclear import and export, nuclear disassembly at mitosis and for preparation of nuclear lysates. Suspension and monolayer cells can be employed using this technique as described in Protocol 3.7 [36].

3.7.2 Preparation of mitotic extracts

The biochemical environment of the mitotic cell can be mimicked *in vitro* by preparing extracts from mitotic cells (Protocol 3.8). These have been used to study nuclear envelope changes and lamina depolymerisation *in vitro*, and can be used as a source of mitotic kinases for *in vitro* experiments [37]. Interphase extracts can be prepared in a similar way. The pellet obtained from the ultracentrifugation contains the so-called mitotic vesicles which are derived from breakdown of the nuclear envelope. The cell biological relevance of these in differentiated cells is not clear as they may represent an artefact of biochemical preparation, however they are enriched in proteins of the inner nuclear membrane and B-type lamins and are competent in assays reconstituting nuclear envelope formation *in vitro* [38].

3.7.3 Preparation of a nuclear lysate for immunoprecipitation and ELISA

The stable structure of the nuclear lamina demands that nuclei are disrupted by physical methods or with ionic detergents. Ionic detergents, for example SDS, are not compatible with many downstream applications including immunoprecipitation and ELISA. A protocol is described using sonication to disrupt nuclear structure (Protocol 3.9).

Protocol 3.7: Nuclear isolation from cultured cells

EQUIPMENT, MATERIALS AND REAGENTS

Ice-cold nuclear isolation buffer: 10 mM Hepes pH 7.5, 2 mM $MgCl_2$, 250 mM sucrose, 25 mM KCl, 1 mM dithiothreitol (DTT), protease inhibitors.

Cytochalasin B.

Dounce homogeniser.

70% glycerol.

METHOD

1. Scrape and pellet cells and wash well with cold PBS.

2. Resuspend the washed cell pellet in 20 volumes of ice-cold nuclear isolation buffer.

3. Add 10 μg ml^{-1} cytochalasin B and incubate on ice for 30 minutes. Cytochalasin B aids the depolymerisation of microfilaments and reduces contamination of nuclei.

4. Homogenise cells in a tight-fitting Dounce homogeniser. A cell cracker set with 15 μm clearance is also very effective at shearing and producing intact nuclei. Approximately 50 to 100 strokes with either instrument is necessary to rid nuclei of cytoplasmic tags.

5. Pellet the nuclei at 400 g for 10 minutes and wash twice with nuclear isolation buffer. Nuclei can be sheared again should the result not be adequate. The nuclei obtained in this way can be stored at −80°C in 70% glycerol. Components of the inner nuclear membrane and nuclear pore complex are preserved. Interphase nuclear envelopes can be prepared from these nuclei by DNase digestion at a concentration of 0.1 mg/ml at room temperature for 20 minutes. Nuclear envelopes can be pelleted by centrifugation and washing in nuclear isolation buffer.

Protocol 3.8: Preparation of mitotic extracts

EQUIPMENT, MATERIALS AND REAGENTS

Ice-cold PBS.

Ice-cold lysis buffer: 20 mM Hepes pH 8.2, 5 mM $MgCl_2$, 10 mM EDTA, 1 mM DTT, 20 $\mu g\,ml^{-1}$ cytochalasin B, 1 $\mu g\,ml^{-1}$ aprotinin, 1 $\mu g\,ml^{-1}$ leupeptin, 100 $\mu g\,ml^{-1}$ phenylmethylsulphonylfluoride (PMSF).

Ultracentrifuge with a Beckman SW55 rotor.

Wash buffer: 250 mM sucrose, 50 mM KCl, 2.5 mM $MgCl_2$, 50 mM Hepes pH 7.5, 1 mM DTT, 1 mM ATP, protease inhibitors.

METHOD

1. Synchronise a HeLa cell population as described in Protocol 3.5 and harvest the synchronised cells by mitotic shake-off.

2. Wash the cells once in ice-cold PBS.

3. Wash once in 20 volumes of ice-cold lysis buffer. If preparing interphase extracts the EDTA should be omitted from the lysis buffer.

4. Pellet the cells at 400 g for 10 minutes at 4°C.

5. Resuspend in 1 volume of lysis buffer and incubate on ice for 30 minutes.

6. Homogenise the cells with two rounds of gentle sonication on ice for 2 minutes each.

7. Centrifuge the homogenised sample at 10 000 g for 15 minutes at 4°C to clear of cell debris.

8. Decant the supernatant into an ultracentrifuge tube and centrifuge at 200 000 g for 3 hours at 4°C in a Beckman SW55 rotor.

9. Aliquot the supernatant and store at −80°C. The concentration of protein is approximately 15 $mg\,ml^{-1}$ for HeLa cells.

10. To harvest these membrane vesicles, resuspend the pellet following ultracentrifugation in wash buffer, using a volume suitable for the centrifuge tube.

11. Pellet the membranes at 100 000 g for 30 minutes in a Beckman SW55 ultracentrifuge rotor.

12. Discard the supernatant and freeze the membranes in liquid nitrogen and store at −80°C.

Protocol 3.9: Preparation of a nuclear lysate for immunoprecipitation and ELISA

EQUIPMENT, MATERIALS AND REAGENTS

Ice-cold physiological buffer: 100 mM potassium acetate, 30 mM potassium chloride, 10 mM di-sodium hydrogen orthophosphate, 1 mM magnesium chloride, 1 mM ATP, 1 mM DTT, 0.2 mM PMSF.

0.1% Triton X-100.

METHOD

1. Adherent cells should be scraped with a rubber policeman and the resulting suspension pelleted and washed in cold PBS. Non-adherent cells can be pelleted directly from growth medium and washed in cold PBS.

2. Resuspend the washed pellet to 10^7 cells ml^{-1} in ice-cold physiological buffer with 0.1% Triton X-100.

3. Sonicate the suspension with two 10-second bursts at medium power.

4. Centrifuge the sample at 15 000 g for 10 minutes at 4°C.

5. Decant the supernatant into a clean tube for use in immunoprecipitation or ELISA.

The gentle buffer conditions allow co-immunoprecipitations under physiological conditions. The technique can be applied to whole cells or purified nuclei. Generally it is not necessary to purify nuclei before immunoprecipitation or ELISA because of the specificity of immunological techniques. The lysate prepared using this protocol is then ready for conventional immunoprecipitation or ELISA. The choice of agent for immunoprecipitation depends on the primary antibody, the quantity of antigen and degree of concentration sought. Available methods include sepharose beads or magnetic beads conjugated with either protein A or G, or species-specific secondary antibodies.

3.7.4 Electrophoretic analysis of molecular changes

The biochemical changes heralding the progression from part of the cell cycle to the next are co-ordinated by the cyclin proteins and the CDKs. Phosphorylation and dephosphorylation cycles are the major molecular switch employed in the control of the cell cycle. The onset of mitosis initiates a series of events in the nuclear envelope and nucleus allowing for the division of chromosomes. Components of the nuclear envelope, including proteins of the inner nuclear membrane, nuclear lamina and nuclear pore complexes, are phosphorylated in key sites which change the interphase affinities. The nuclear pore complexes are solubilised, inner nuclear membrane proteins lose their affinity for the nuclear lamina and chromatin, and the lamina is depolymerised in prophase. The result is the loss of integrity of the nuclear membrane, which is incorporated into the bulk ER [39]. Spindle poles can assemble and condensed metaphase chromosomes arrange on the equator of the cell to be separated at the onset of anaphase.

It is possible to generate phosphospecific antibodies and therefore study phosphorylated targets in single cells using immunofluorescence techniques. In the absence of an antibody specific for a phosphoprotein it is necessary to use biochemical techniques. Two-dimensional (2D) electrophoresis is a powerful method for studying phosphorylation events because of the charge differences which accumulate on the protein with phosphorylation. A single phosphate group, and therefore the addition of a negative charge, will change the isoelectric focusing (IEF) point of a protein sufficiently to identify this new isoform by 2D electrophoresis. Commercially available immobilised pH gradient strips (Amersham Pharmacia Biotech, UK) for IEF have made the technique of two-dimensional microscopy a less-specialised field and increased the resolution significantly. The strips are available in different lengths and covering different pH ranges. The longer the strip and the smaller the pH range distributed over that length, the higher the resolution. Moreover, longer strips can accommodate a greater amount of protein making 2D electrophoresis a useful tool for protein preparation for proteomic analysis. *Figure 3.3* shows interphase lamin A/C proteins separated by high-resolution 2D electrophoresis and western blotted with an anti-lamin A/C antibody. The multiple species are evident, representing post-translational modifications of the parent proteins.

Immobilised pH gradients have standardised the technique of 2D electrophoresis, leaving the emphasis for the researcher on sample preparation. The method used to obtain a sample of nuclear proteins depends on the desired outcome. Two-dimensional gels can be a good method for purifying single proteins or even single isoforms where other methods are not available and, in combination with other proteomic tools like microsequencing and mass spectrometry, can be an excellent tool for studying molecular changes which can not be achieved by gene expression profiling. Isolating enough of a single isoform of a protein from a complex mixture for downstream analysis requires some form of initial enrichment [40].

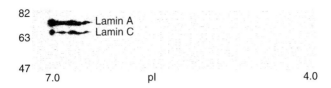

Figure 3.3

Nuclei were prepared from monolayer HeLa cells and lysed in urea lysis buffer. The sample was cleared by centrifugation and dialysed against 9 M urea. Proteins were separated with two-dimensional electrophoresis. The first dimension was run on a 13 cm immobilised pH gradient strip for 40 000 volt hours and the second dimension on a 12% SDS polyacrylamide reducing gel. Proteins were transferred to nitrocellulose and western blot was performed using an anti-human lamin A/C monoclonal antibody and chemiluminescence detection. Lamin A and the smaller splice variant, lamin C, have near-identical IEF patterns. Phosphorylation produces multiple different isoforms, some of which extend beyond pH 7.0 but are not resolved here due to the pH gradient used.

The following guidelines should be followed when preparing samples for IEF. The simpler the protein sample, the better the results will be and the more easily inter-preted. There is a limit to the amount of protein which can be loaded onto the IEF strips, which necessitates an initial enrichment of a species of interest if the 2D gel is being used for preparative purposes. Nucleic acid interferes with IEF and should be removed as much as possible from the sample. This poses a particular problem for studies on nuclear proteins. The solubility of the protein preparation greatly affects the resolution. Precipitation of proteins as a method of purification or enrichment should be avoided. Proteins greater than 100 kDa and membrane proteins are resolved with greater difficulty by IEF. This should be considered when comparing expression patterns using 2D gels, which may contain artefact from large and hydrophobic pro-teins. Steps in the preparation, particularly dialysis, may cause artefactual changes in the levels of certain protein species.

Protocol 3.10 is optimised for tissue culture nuclei and tissue samples, immobilised pH gradient strips and flatbed IEF apparatus. The hardware necessary for performing IEF is available from Amersham Pharmacia Biotech. Following the procedure, western blotting and protein staining are performed according to standard protocols. Proteomic analysis, that is identification of proteins by sequencing or mass spectrometry, may be performed on the individual spots cut from the second dimension gel. Mass spec-trometry has been used with success in identifying proteins in large-scale protein identification projects. This has most recently been done for the nuclear envelope compartment, where 19 uncharacterised proteins were identified [41].

3.7.5 In gel digestion of proteins for proteomic analysis

Protocol 3.11 is optimised for MALDI-MS (matrix-assisted laser desorption ionisa-tion mass spectrometry). Details of mass spectrometry methods have been reviewed elsewhere [42]. A useful search engine for mapping peptide mass data from MALDI-MS can be found at http://prospector.ucsf.edu/.

Protocol 3.10: Electrophoretic analysis of molecular changes

EQUIPMENT, MATERIALS AND REAGENTS

Urea lysis buffer: 8 M urea, 50 mM Tris-HCl pH 7.4, 0.5% 3-[(3-cholamidopropyl)dimethylammonio]-1-propanesulphonate (CHAPS) (w/v), 1 mM DTT.

Beckman JA-21 centrifuge.

Sample loading buffer: 8 M urea, 0.5% CHAPS (w/v), 20 mM DTT, 0.5% ampholyte (v/v), trace of bromophenol blue.

Equilibration buffer: 50 mM Tris-HCl pH 8.8, 6 M urea, 30% (v/v) glycerol, 2% SDS, trace bromophenol blue, with 10 mg ml^{-1} DTT.

Iodoacetamide.

METHOD

1. Interphase nuclei may be fractionated from whole cells prior to solubilisation using one of the techniques described in Section 3.2. Mitotic cells, which may be prepared by mitotic block or shake-off (see Protocols 3.5 and 3.6), may be lysed directly in urea lysis buffer if dialysis is to be performed in step 3. The length of the IEF strip will determine the volume and amount of protein that can be loaded. A rough guideline is 0.5 ml lysis buffer per 10^7 HeLa nuclei. The nuclear lamina and chromatin is completely denatured in 8 M urea. Keep the sample at room temperature to avoid crystallisation of the urea. Tissue samples may be lysed in the same manner, but homogenisation or careful dicing into the smallest possible pieces is essential for solubilisation. In certain cases it may be necessary to use additional detergent. SDS may be added to the sample, but IEF will be compromised. If SDS is added to the sample it is recommended that dialysis be used to remove it following solubilisation of the sample. IEF can take place with trace amounts of SDS, although levels should not be higher than 0.3% and if at all possible should be avoided.

2. Centrifuge the sample at 20 000 g for 30 minutes at room temperature in a Beckman JA-21 centrifuge to pellet the high molecular weight DNA and any insoluble material. Sonication can be employed before this centrifugation step.

3. Dialysis into fresh 9 M urea is recommended for removing ions and improves resolution. The sample is loaded onto the first dimension strip in sample loading buffer. Following dialysis the rest of the sample buffer components may be added directly to the sample. The ampholyte is not critical for IEF but may improve solubility and resolution.

4. The immobilised pH gradient strips are desiccated and are rehydrated in a suitable volume of the sample for a period of 10 hours. The strip and sample are overlayed with mineral oil to prevent evaporation. Rehydration

of the strip at a constant 30 volts may improve the solubility and resolution of certain samples.

5. The total number of volt (V) hours required is approximately $3–4 \times 10^4$. A typical programme for IEF is: 500 V for 500 V hours; 1000 V for 1000 V hours; 8000 V for 40 000 V hours. The IEF time is calculated in terms of volt hours (the product of voltage and time) rather than an absolute time period because the ion content of certain samples reduces the voltage that can be achieved. The IEF apparatus is fitted with a cut-out device to ensure safety during high voltage operation.

6. Equilibrate the strip for 15 minutes in equilibration buffer with 10 mg ml^{-1} DTT.

7. Equilibrate the strip for 15 minutes in fresh equilibration buffer with 25 mg ml^{-1} iodoacetamide.

8. The second dimension is run by overlaying the strip on a standard polyacrylamide gel. If 7 cm strips are being used the second dimension can be run on a minigel apparatus.

Protocol 3.11: In gel digestion of proteins for proteomic analysis

EQUIPMENT, MATERIALS AND REAGENTS

Coomassie blue.

25 mM ammonium bicarbonate pH 8.0 in 50% acetonitrile.

Speedvac centrifuge.

25 mM ammonium bicarbonate pH 8.0.

Sequencing grade methylated trypsin (available from Promega).

50% acetonitrile/5% trifluoroacetic acid.

Mass spectrometry facility.

METHOD

1. Stain proteins in the second dimension gel with Coomassie blue and de-stain thoroughly.

2. Excise the spot of interest with a clean scalpel blade and trim any excess unstained gel from the edges.

3. Dice the gel into the smallest pieces possible and place in a siliconised microfuge tube. Macerating the acrylamide into a fine powder can lead to losses downstream through pipetting.

4. De-stain and dehydrate the gel pieces with 3 to 4 washes of ammonium bicarbonate-acetonitrile. Use gel-loading pipette tips to avoid aspirating the small pieces of gel. The gel pieces will turn opaque white and Coomassie stain will be removed from the sample completely.

5. Aspirate the final wash and dry the gel pieces in a Speedvac centrifuge for approximately 30 minutes.

6. Rehydrate the gel slices in 25 mM ammonium bicarbonate pH 8.0 containing sequencing grade methylated trypsin by addition of 5 μl aliquots until the gel slices are fully hydrated but no excess liquid is present. The amount of trypsin used is one of the most crucial aspects to the digestion. The substrate protein should be in 10-fold excess to the trypsin. The amount of substrate should be judged from the Coomassie stain. Trypsin is diluted fresh into the ammonium bicarbonate buffer. A usual amount of trypsin used is 0.5 to 1.0 μg μl^{-1}.

7. Overlay the gel pieces with the minimum volume of ammonium bicarbonate buffer pH 8.0 that is needed to cover them.

8. Incubate at 37°C for 12 to 20 hours.

9. Harvest the peptides from the sample by aspirating the liquid in the digestion mix. Further extract the peptides by incubating the gel slices with

20 µl acetonitrile/trifluoroacetic acid for 30 minutes with agitation. Repeat the extraction step 3 times and pool all the volumes with the initial harvest of peptides.

10. Lyophilise the peptide sample in a Speedvac. Resuspend in 20 µl of acetonitrile/trifluoroacetic acid and store at −20°C until analysis.

11. Concentrate and desalt the peptide mixture immediately before analysis by mass spectrometry in a C_{18} ZipTip (available from Millipore).

3.8 References

1. Swedlow, J.R. and Lamond, A.I. (2001) Nuclear dynamics: where genes are and how they got there. *Genome Biol.* **2**: reviews 0002.1–0002.7.
2. Croft, J.A., Bridger, J.M., Boyle, S., Perry, P., Teague, P. and Bickmore, W.A. (1999) Differences in the localization and morphology of chromosomes in the human nucleus. *J. Cell Biol.* **145**: 1119–1131.
3. Tsukamoto, T., Hashiguchi, N., Janicki, S.M., Tumbar, T., Belmont, A.S. and Spector, D.L. (2000) Visualization of gene activity in living cells. *Nat. Cell Biol.* **2**: 871–878.
4. Boyle, S., Gilchrist, S., Bridger, J.M., Mahy, N.L., Ellis, J.A. and Bickmore, W.A. (2001) The spatial organization of human chromosomes within the nuclei of normal and emerin-mutant cells. *Hum. Mol. Genet.* **10**: 211–219.
5. Li, G., Sudlow, G. and Belmont, A.S. (1998) Interphase cell cycle dynamics of a late-replicating, heterochromatic homogeneously staining region: precise choreography of condensation/decondensation and nuclear positioning. *J. Cell Biol.* **140**: 975–989.
6. Wilson, K.L. (2000) The nuclear envelope, muscular dystrophy and gene expression. *Trends Cell Biol.* **10**: 125–129.
7. Heim, R. and Tsien, R.Y. (1996) Engineering green fluorescent protein for improved brightness, longer wavelengths and fluorescence resonance energy transfer. *Curr. Biol.* **6**: 178–182.
8. Matz, M.V., Fradkov, A.F., Labas, Y.A., Savitsky, A.P., Zaraisky, A.G., Markelov, M.L. and Lukyanov, S.A. (1999) Fluorescent proteins from non-bioluminescent Anthozoa species. *Nat. Biotechnol.* **17**: 969–973.
9. Verkhusha, V.V., Otsuna, H., Awasaki, T., Oda, H., Tsukita, S. and Ito, K. (2001) An enhanced mutant of red fluorescent protein DsRed for double labeling and developmental timer of neural fibre bundle formation. *J. Biol. Chem.* **276**: 29621–29624.
10. Ellenberg, J., Siggia, E.D., Moreira, J.E., Smith, C.L., Presley, J.F., Worman, H.J. and Lippincott-Schwartz, J. (1997) Nuclear membrane dynamics and reassembly in living cells: targeting of an inner nuclear membrane protein in interphase and mitosis. *J. Cell Biol.* **138**: 1193–1206.
11. Moir, R.D., Yoon, M., Khuon, S. and Goldman, R.D. (2000) Nuclear lamins A and B1: different pathways of assembly during nuclear envelope formation in living cells. *J. Cell Biol.* **151**: 1155–1168.
12. Miyawaki, A. and Tsien, R.Y. (2000) Monitoring protein conformations and interactions by fluorescence resonance energy transfer between mutants of green fluorescent protein. *Methods Enzymol.* **327**: 472–500.
13. Bastiaens, P.I. and Pepperkok, R. (2000) Observing proteins in their natural habitat: the living cell. *Trends Biochem. Sci.* **25**: 631–637.
14. Yang, L., Guan, T. and Gerace, L. (1997) Integral membrane proteins of the nuclear envelope are dispersed throughout the endoplasmic reticulum during mitosis. *J. Cell. Biol.* **137**: 1199–1210.
15. Daigle, N., Beaudouin, J., Hartnell, L., Imreh, G., Hallberg, E., Lippincott-Schwartz, J. and Ellenberg, J. (2001) Nuclear pore complexes form immobile networks and have a very low turnover in live mammalian cells. *J. Cell Biol.* **154**: 71–84.
16. Huang, S., Deerinck, T.J., Ellisman, M.H. and Spector, D.L. (1997) The dynamic organization of the perinucleolar compartment in the cell nucleus. *J. Cell Biol.* **137**: 965–974.
17. Platani, M., Goldberg, I., Swedlow, J.R. and Lamond, A.I. (2000) In vivo analysis of Cajal body movement, separation, and joining in live human cells. *J. Cell Biol.* **151**: 1561–1574.
18. Leonhardt, H., Rahn, H.P., Weinzierl, P., Sporbert, A., Cremer, T., Zink, D. and Cardoso, M.C. (2000) Dynamics of DNA replication factories in living cells. *J. Cell Biol.* **149**: 271–280.
19. Dundr, M., Misteli, T. and Olson, M.O. (2000) The dynamics of post-mitotic reassembly of the nucleolus. *J. Cell Biol.* **150**: 433–446.
20. Robinett, C.C., Straight, A., Li, G., Willhelm, C., Sudlow, G., Murray, A. and Belmont, A.S. (1996) In vivo localization of DNA sequences and visualization of

large-scale chromatin organization using lac operator/repressor recognition. *J. Cell Biol.* **135**: 1685–1700.

21. Tumbar, T., Sudlow, G. and Belmont, A.S. (1999) Large-scale chromatin unfolding and remodelling induced by VP16 acidic activation domain. *J. Cell Biol.* **145**: 1341–1354.

22. Kanda, T., Sullivan, K.F. and Wahl, G.M. (1998) Histone-GFP fusion protein enables sensitive analysis of chromosome dynamics in living mammalian cells. *Curr. Biol.* **8**: 377–385.

23. Kimura, H. and Cook, P.R. (2001) Kinetics of core histones in living human cells: little exchange of H3 and H4 and some rapid exchange of H2B. *J. Cell Biol.* **153**: 1341–1353.

24. Howell, B.J., Hoffman, D.B., Fang, G., Murray, A.W. and Salmon, E.D. (2000) Visualization of Mad2 dynamics at kinetochores, along spindle fibres, and at spindle poles in living cells. *J. Cell Biol.* **150**: 1233–1250.

25. Yucel, J.K., Marszalek, J.D., McIntosh, J.R., Goldstein, L.S., Cleveland, D.W. and Philp, A.V. (2000) CENP-meta, an essential kinetochore kinesin required for the maintenance of metaphase chromosome alignment in Drosophila. *J. Cell Biol.* **150**: 1–11.

26. Zink, D., Bornfleth, H., Visser, A., Cremer, C. and Cremer, T. (1999) Organization of early and late replicating DNA in human chromosome territories. *Exp. Cell Res.* **247**: 176–188.

27. Belmont, A.S., Zhai, Y. and Thilenius, A. (1993) Lamin B distribution and association with peripheral chromatin revealed by optical sectioning and electron microscopy tomography. *J. Cell Biol.* **123**: 1671–1685.

28. Ye, Q. and Worman, H.J. (1996) Interaction between an integral protein of the nuclear envelope inner membrane and human chromodomain proteins homologous to Drosophila HP1. *J. Biol. Chem.* **271**: 14653–14656.

29. Shaw, P.J. and Jordan, E.G. (1995) The nucleolus. *Annu. Rev. Cell Dev. Biol.* **11**: 93–121.

30. Frey, M.R. and Matera, A.G. (2001) RNA-mediated interaction of Cajal bodies and U2 snRNA genes. *J. Cell Biol.* **154**: 499–509.

31. Aherne, W.A., Camplejohn, R.S. and Wright, N.A. (1977) *An Introduction to Cell Population Kinetics.* Edward Arnold, London.

32. Cook, P.R. (2001) *Principles of Nuclear Structure and Function.* Wiley-Liss, New York.

33. Fricker, M., Hollinshead, M., White, N. and Vaux, D. (1997) Interphase nuclei of many mammalian cell types contain deep, dynamic, tubular membrane-bound invaginations of the nuclear envelope. *J. Cell Biol.* **136**: 531–544.

34. Hozak, P., Jackson, D.A. and Cook, P.R. (1994) Replication factories and nuclear bodies: the ultrastructural characterization of replication sites during the cell cycle. *J. Cell Sci.* **107**: 2191–2202.

35. Griffiths, G. (1993) *Fine Structure Immunocytochemistry.* Springer-Verlag, Heidelberg.

36. Collas, P. (1998) Nuclear envelope disassembly in mitotic extract requires functional nuclear pores and a nuclear lamina. *J. Cell Sci.* **111**: 1293–1303.

37. Collas, P., Le Guellec, K. and Tasken, K. (1999) The A-kinase-anchoring protein AKAP95 is a multivalent protein with a key role in chromatin condensation at mitosis. *J. Cell Biol.* **147**: 1167–1180.

38. Collas, I. and Courvalin, J.C. (2000) Sorting nuclear membrane proteins at mitosis. *Trends Cell Biol.* **10**: 5–8.

39. Gant, T.M. and Wilson, K.L. (1997) Nuclear assembly. *Annu. Rev. Cell Dev. Biol.* **13**: 669–695.

40. Garin, J., Diez, R., Kieffer, S., Dermine, J.F., Duclos, S., Gagnon, E., Sadoul, R., Rondeau, C. and Desjardins, M. (2001) The phagosome proteome: insight into phagosome functions. *J. Cell Biol.* **152**: 165–180.

41. Dreger, M., Bengtsson, L., Schoneberg, T., Otto, H. and Hucho, F. (2001) Nuclear envelope proteomics: Novel integral membrane proteins of the inner nuclear membrane. *Proc. Natl Acad. Sci. (USA)* **98**: 11943–11948.

42. Andersen, J.S. and Mann, M. (2000) Functional genomics by mass spectrometry. *FEBS Lett.* **480**: 25–31.

Cell Cycle Proteins

Nicholas C. Lea and N. Shaun B. Thomas

4

Contents

4.1 Overview. How cell cycle proteins regulate the $G_0 \to G_1 \to S$ phase transitions

Proteins that control entry and progression through the cell cycle are regulated by a number of post-translational modifications. These include phosphorylation, glycosylation and acetylation. Each of these modifications can cause changes in protein charge, shape, hydrophilicity and mass, all of which can be exploited in distinguishing the modified from the unmodified protein. This chapter will cover how these

Cell Proliferation and Apoptosis, David Hughes and Huseyin Mehmet (Eds)
© 2003 BIOS Scientific Publishers Ltd, Oxford

post-transcriptional modifications can be analysed, the methods most frequently used and the problems and limitations associated with each. The examples in this chapter are based on work in the Cell Cycle Laboratory at King's College, London, into the molecular mechanisms involved in controlling entry into the cell cycle using members of the retinoblastoma (RB) and E2F families [1]. These are important proteins that are required to respond to a mitogenic stimulus, and protein phosphorylation and acetylation are crucial in regulating their activities. A quiescent (G_0) cell is not capable of duplicating its DNA. It lacks key proteins, such as DNA polymerase α, PCNA and thymidine kinase, that are necessary for DNA synthesis. It also does not have proteins, such as cdc2 and p107, which are required to control the transition through mitosis (M). Such proteins are synthesised as the quiescent cell traverses G_1 for the first time. The levels of many of these proteins are maintained at a constant level with respect to total cell protein content in subsequent cell cycles. Therefore, a cell entering G_1 from G_0 differs from a proliferating cell that enters G_1 from mitosis in that the cell entering G_1 from M phase already has cdc2, PCNA and many other proteins necessary for the next cell cycle.

Entry from G_0 into G_1 in response to activation of mitogenic signalling pathways leads to the synthesis of D-type cyclins. These in turn activate CDKs, which then phosphorylate proteins such as the members of the RB protein family. The retinoblastoma protein (pRB) is a nuclear phosphoprotein that is phosphorylated progressively on up to 16 sites as quiescent cells progress through G_1, reaching a hyperphosphorylated form at the G_1/S border. pRB is phosphorylated on different sites [2] by a succession of CDKs which are activated at different times during G_1: for example, cdk6-cyclin D and cdk4-cyclin D in early G_1, followed by cdk2-cyclin E in mid/late G_1. pRB shares regions of homology with two other proteins, p107 and p130, and like pRB they are phosphoproteins which have consensus sites for phosphorylation by CDKs. It was shown recently that p130 is phosphorylated on up to 21 different sites [3].

Hypophosphorylated pRB binds members of the E2F transcription factor family [4] and it inhibits E2F-dependent gene activation by two methods. The first is by inhibiting the gene that is controlled by the E2F site. The second is by suppressing a region of DNA by recruiting a histone deacetylase (HDAC1) [5]. This causes histone deacetylation, which in turn alters chromatin condensation and represses transcription. pRB is also regulated by acetylation which occurs during the transition from G_1 to S phase. Acetylation of pRB hinders its phosphorylation and alters its ability to bind other proteins such as MDM2 [6].

Each member of the E2F transcription factor family is composed of a heterodimer of an E2F protein (E2F-1 to E2F-6 are known) together with a member of the DP family (DP-1 or DP-2) [7–9]. The other two members of the pRB family, p130 and p107, also bind and repress E2F activity. Different members of the pRB family have some specificity for binding different E2Fs: complexes containing E2F-1, -2 and -3 bind pRB rather than p107, whereas E2F-4 or -5 bind p130 and p107 in preference to pRB. The functional consequences are that E2F-1 over-expression preferentially overcomes cell cycle arrest by pRB, and E2F-4-DP-1 overcomes arrest caused by p130. The E2F factors regulate a number of genes that are necessary for cells to proliferate (e.g. DNA Polymerase α, thymidine kinase, PCNA) or are involved in controlling progression through the cell cycle (e.g. cdc2, p107). Thus the balance between particular E2Fs and pRB or p130 is clearly important in determining whether a cell proliferates or remains quiescent. Furthermore, the timely relief of E2F-dependent transcriptional repression orchestrates the transition through G_1 into S phase without inducing apoptosis. The timing is critical as, for example, over-expression of E2F-1 not only induces cell cycle entry but also causes apoptosis by a p73-dependent mechanism [10].

Members of the E2F family of transcription factors are also regulated by phosphorylation. For example, E2F-1 is a phosphoprotein that can be phosphorylated *in vitro* only by certain CDKs and it is likely that E2F-1 phosphorylation during S phase switches off its transcriptional activation, thereby down-regulating certain E2F-dependent genes that were activated during late G_1 (G_{1B}) when the unphosphorylated E2F was released from repression by pRB. Indeed, there is a checkpoint during S phase that requires E2F-1 phosphorylation by cyclin A-cdk [11]. In contrast to E2F-1, E2F-4 is multiply phosphorylated and the hyperphosphorylated form is predominant in G_0 [12]. Multiple hypophosphorylated forms of E2F-4 replace this hyperphosphorylated form as the cells progress through G_1, and this dephosphorylation may be required for its activation. The DP-1 partner bound by these E2Fs is also a phosphoprotein, which is dephosphorylated as cells enter the cell cycle and this correlates with an increase in E2F activation.

An additional level of transcriptional silencing involving pRB was revealed recently. Histone H3, near the promoter of genes such as *cyclin A* and *E* that are regulated during cell cycle entry, is methylated at lysine 9 by the methylase SUV39H1. It is thought that this localised histone modification occurs as SUV39H1 is bound by pRB and that this causes localised rather than global histone H3 methylation. Such localised histone methylation leads to gene silencing [13]. Post-translational modification is, therefore, highly important in regulating the activity of proteins that are involved in regulating entry into the cell cycle. However, although the interactions of pRB with E2F are the best characterised, pRB binds many other proteins [14]. These include PU.1, c-Abl, p53, PML and all three classes of RNA polymerases.

4.2   Methods for analysing the cell cycle

Before starting to analyse changes in cell cycle proteins it is important to know whether the cells are quiescent, arrested in a particular cell cycle phase or dividing exponentially.

4.2.1   Manual cell counting

Determine the number of viable cells that are present on successive days by manual counting with an improved Neubauer chamber. Count at least 100 cells in the presence of Trypan blue to give an estimate of the number of viable and dead cells. By graphing these values it should be clear whether the number of viable cells is increasing exponentially and the approximate population doubling time can be determined. This may seem too time consuming but this discipline will avoid wasting time on experiments with unhealthy cells. Also, routine inspection of cells is important for detecting changes in morphology that may otherwise be missed.

4.2.2   Automated cell counting

There are more sophisticated automated cell-counting methods that may be more appropriate in certain circumstances. For example, an accurate count of live versus dead cells can be obtained very simply by flow cytometry (Protocol 4.1). If you have very small numbers of cells that cannot be counted accurately by manual methods, their number can be determined by MTT/MTS (3-(4,5-dimethylthiazol-2-yl)-5-(3-carboxymethoxyphenyl)-2-(4-sulphophenyl)-2H-tetrazolium, inner salt) assays and reagents are available commercially. Such assays rely on mitochondrial enzyme cleavage of the substrate to produce a product that can be quantified by a 96-well plate reader. However, care should be taken as they rely on the number of mitochondria per

Protocol 4.1: Flow cytometric method for counting live versus dead cells

EQUIPMENT, MATERIALS AND REAGENTS

Flow cytometer such as FACScan (Beckton Dickenson).

Fluorescent microspheres at known concentration (Coulter).

Propidium iodide (PI): 2 mg ml^{-1} in H$_2$O. Store at 4°C.

METHOD

This method is far more accurate than manual cell counting methods as thousands or tens of thousands of events rather than tens or hundreds can be counted. It is based on the fact that dead cells stain with propidium iodide. The fluorescent microspheres are used as an internal control for cell number and they can be distinguished from cells by flow cytometry by virtue of their size.

1. Add a known number of fluorescent microspheres to a known volume of cells (e.g. 0.5 ml) in culture medium or PBS.

2. To this add 1 μl PI (2 mg ml^{-1} stock).

3. Run a small sample through a flow cytometer and analyse two populations: gate on the small, highly fluorescent beads as well as the cells.

4. Run each sample and set the machine to stop counting as appropriate between 1000 and 10 000 microspheres. If you know the number of microspheres per ml, then the machine will give you a cell count relative to this. Furthermore, viable cells will be PI negative, dead cells PI positive and the number of each in your culture can be determined accurately.

cell and their activity remaining constant. Machines, such as Coulter counters, are also available to automatically count particular cells in a mixed population.

4.2.3 Cell cycle analysis – kinetics of cell division

The methods for cell counting described above give an idea of how the *population* of cells is behaving. However, within this population individual cells may be dividing with very different kinetics. Methods are available that can be used to analyse individual cells.

4.2.4 CFSE staining

The compound carboxyfluorescein diacetate, succinimidyl ester, or CFDA-SE (commonly known as CFSE), enters cells freely and is cleaved in the cytosol by intracellular esterases. The cleaved product is fluorescent and is retained in the cell. Each cell division can be monitored as the amount per cell and hence cell fluorescence halves with each cell division. This can be assayed by flow cytometry and is described in Protocol 4.2. Such analysis can yield a wealth of data. For example, if quiescent primary T cells are loaded with CFSE and then stimulated via CD3 and CD28, some cells divide and so halve their fluorescence before others. Furthermore, after 5 days in culture some cells have divided many times with an estimated time between cell divisions of 5–6 h while others have yet to divide or have undergone only one division [15]. Analysis of cell cycle proteins in the whole population at a given time averages what is occurring in many different sub-populations in which cell cycle control is clearly very different. Flow cytometric sorting can be used to isolate cells in each sub-population and cell cycle proteins in each may then be studied.

4.2.5 Time-lapse microscopy

Just as viewing your cells when counting will give you indications of morphological and adhesion changes at a fixed point in time, so continuous time-lapse microscopy gives a wealth of additional information. In terms of cell proliferation, individual cells can be tracked and the time taken to traverse the cell cycle (inter-mitotic times) can be determined. Furthermore, the fate of each daughter cell, be it to divide further or undergo apoptosis, can be determined. Such analyses have been used to study transition through the restriction point [16] and the cellular response to induction of active c-myc (as a tamoxiphen-regulated ER-fusion protein [17]). These experiments require a specialised microscope stage that is heated to 37°C with a suitable environment, such as a regulated CO_2 supply and timed image capture or video microscopy. When the cells being analysed have been transfected or transduced with a cell cycle protein coupled to GFP or other fluorescent protein, data on its sub-cellular localisation under particular conditions can also be obtained. This is important as such a protein can only interact with other proteins in that location [18]. The interaction of two proteins in living cells can be analysed by FRET methods [19].

4.2.6 DNA synthesis

The incorporation of BrdU or Digoxygenin-11-dUTP (Dig-dUTP) into DNA has become a standard method for detecting cells in S phase. Whereas the incorporation of [^{3}H]thymidine is a good indicator that some cells in the population are in S phase, BrdU or Dig-dUTP incorporation shows whether individual cells are in S phase during the labelling period. A method is given in Protocol 4.3 that analyses the

Protocol 4.2: Counting cell divisions using CFSE staining

EQUIPMENT, MATERIALS AND REAGENTS

Flow cytometer such as FACScan (Becton Dickinson).

CFSE (Molecular Probes).

0.1% (w/v) BSA in PBS.

METHOD

1. Pellet 1×10^7 cells (e.g. primary quiescent T lymphocytes) at $200\,g$ for 10 minutes.

2. Resuspend in 1 ml 0.1% (w/v) BSA in PBS.

3. Add CFSE to a final optimum concentration of 0.5–5 µM, determined experimentally. Too high a concentration of CFSE can interfere with analysis of other molecules, such as cell surface receptors. Decrease the CFSE concentration used if necessary.

4. Incubate for 10 minutes at 37°C.

5. Wash cells twice with 10 ml ice-cold PBS.

6. Resuspend in the appropriate culture medium (e.g. RPMI-1640, 10% (v/v) FCS for T cells) with growth factors if required.

7. Culture for the experimental period and analyse by flow cytometry. The mean cell fluorescence of CFSE halves with each cell division and so numbers of cells that have completed 1, 2, 3 or more cell divisions, or that have not divided during the culture period can be accurately determined. Particular cell types in a mixed population can be identified by staining cell surface molecules (e.g. CD8 or CD4 T cells) with phycoerythrin (PE)-conjugated antibodies. In some cases CFSE staining may be too heterogeneous for further analysis. If this is the case, isolate cells with a narrow, precise CFSE staining by flow cytometric sorting prior to culture.

Protocol 4.3: Flow cytometric analysis of cell cycle by BrdU incorporation

EQUIPMENT, MATERIALS AND REAGENTS

Temperature-controlled CO_2 incubator at 37°C.

Class II cabinet.

Low-speed centrifuge.

Falcon 12 × 75 mm tubes.

Flow cytometer.

PBS.

1% (w/v) BSA in PBS.

0.5% (v/v), Tween 20, 1% (w/v) BSA in PBS.

1 mM BrdU (Sigma) in PBS.

FITC-conjugated anti-BrdU (Beckton Dickenson).

PI solution 5 μg ml^{-1} in PBS.

2 M HCl with 0.5% (v/v) Triton X-100.

0.07 M NaOH.

Aqueous ethanol 70% (v/v) at −20°C.

0.1 M sodium tetraborate pH 8.5.

METHOD

1. Add BrdU to cells in culture to a final concentration of 10 μg ml^{-1} and incubate cells for the desired length of time. Two minutes is sufficient to start to see BrdU incorporation but longer labelling periods are recommended. The period of labelling chosen should be dictated by the question being addressed. For example, 20–30 minutes will show which cells are in S phase at that time. Labelling for 24–48 h will show what proportion of the cell population is actively dividing. An approximate cell cycle time for the population can be determined by a 30 min pulse of BrdU, followed by washing the cells, returning to culture and taking samples for analysis 6, 12, 18 and 24 hours later. However, CFSE staining (Protocol 4.2) gives more information on cell cycle times of different sub-populations.

2. Wash the cells twice in 1% BSA/PBS and centrifuge for 10 minutes at 500 g for 10 min.

3. Resuspend the pellet in 70% (v/v) ethanol at −20°C by gently vortexing and adding ethanol simultaneously.

4. Incubate cells at −20°C for at least 30 min.

5. Centrifuge cells at 500 g for 10 min at 4°C, aspirate the supernatant and loosen the pellet by vortexing gently or by flicking the tube.

6. Slowly add 2 M HCl/Triton X-100 to the cells whilst mixing and incubate at room temperature for 30 min.

7. Centrifuge the cells at 500 g for 10 min, aspirate the supernatant and add 1 ml sodium borate buffer to neutralise the acid. Cells can be stored at −20°C in 70% ethanol at this point for processing later.

8. Centrifuge the cells at 500 g for 10 min, resuspend the cell pellet in Tween 20/BSA/PBS. Adjust the cell density to 1×10^6 cells ml^{-1}.

9. Add 20 μl of anti-BrdU-FITC per 1×10^6 cells and incubate at room temperature for 30 min. Wash cells in 1 ml Tween 20/BSA/PBS.

10. Centrifuge the cells at 500 g for 5 min and resuspend in 5 μg ml^{-1} PI.

11. Analyse by flow cytometry (see *Figure 4.1a*). Cells are usually analysed by FL1-height versus FL3-height. Cells that are in the G_1 phase of the cell cycle have low FL3 staining (PI or 2n DNA content). S phase cells appear with high FL1 staining (BrdU incorporation, i.e. have synthesised DNA within the labelling period). Cells in G_2 and M phases have low FL1 (BrdU incorporation) and high FL3 staining (PI or 4N DNA content).

percentage of cells in S phase by flow cytometry and an example is shown in *Figure 4.1a*. By carrying out pulse-chase experiments the time taken for cells to complete a cell division cycle can also be determined. However, care should be taken with very short pulses of BrdU (e.g. 10 minutes) that you are detecting DNA synthesis and not DNA repair. Similar methods can be used to detect S-phase cells on a microscope slide. Furthermore, a method was described recently for detecting cells in frozen tissue sections that are in S phase. This exploits the fact that DNA synthesis still continues for a short time as the sections thaw and so BrdU or Dig-dUTP can be incorporated [20]. This is a powerful method in which tissue morphology is retained thereby allowing localisation of the cells in S phase. These sections can also be stained with antibodies to distinguish between cell types, or with standard histological stains.

4.2.7 Cell cycle analysis

Cells double their DNA content during S phase and this has led to the development of several methods for determining where a cell is in the cell cycle. DNA can be stained quantitatively with one of a number of stains, for example PI, TOTO3, DAPI or Hoechst 33342. Cellular DNA content can then be determined by flow cytometry, fluorescence microscopy or laser scanning microscopy [21]. The latter enables the DNA content of individual cells to be analysed in combination with morphology and immunostaining. A flow-cytometric method that we use routinely is given in Protocol 4.4 in which DNA staining with PI is combined with total cell protein content (stained with FITC). This has enabled the determination of the percentage of cells in early and late G_1 (G_{1A} and G_{1B}) as well as in S and G_2/M phases of the cell cycle. Other, more complex methods have been used to distinguish many more cell cycle phases [22].

Certain flow cytometers can isolate cells with given characteristics (cell sorting). Therefore, cells in different cell cycle phases can be sorted and proteins in those purified cells can then be analysed by western blotting. We have made use of these methods to analyse the pRB and E2F families in proliferating cells and also in Daudi B cells in response to α-interferon (Protocol 4.4) [23, 24]. Similar methods have also been used to sort viable cells stained with Hoechst 33342 (DNA) and pyronin Y (stains RNA) [25]. Such cells can then be cultured further or cell cycle proteins can be analysed directly.

Analysis of cellular protein or RNA content also gives an indirect measure of cell size, which can also be determined by forward scatter (FSC) measurements. This is important as a cell has to attain a critical size before it divides. Recently, the increase in cell size that accompanies the transition through G_1 was shown to be regulated by cdk4-cyclin D in Drosophila [26, 27] and mice lacking *cdk4*, *cyclin D1* or *cyclin D2* genes are small and display hypoplasia of certain tissues [28, 29]. Our data suggest that a similar cdk4/6-dependent mechanism regulates cell size in human T-lymphocytes [30]. Differences in cell size during the cell cycle have also been exploited in another method that allows the isolation of viable cells by centrifugal elutriation [31], but development and use of this method is beyond the scope of this chapter.

4.2.8 G_0, G_1 and the restriction point

To establish that a cell is in G_0 is difficult as there are no specific markers that are uniquely associated with quiescence. Rather, there are a series of characteristics that indicate the quiescent nature of a cell. Quiescent cells are non-dividing cells that are capable of cell division under the right conditions. This rules out terminally differentiated and senescent cells that have lost the ability to divide. Quiescent haemopoietic cells such as non-activated T and B lymphocytes and pluripotent progenitor cells are small with a lower protein and RNA content than their proliferating counterparts.

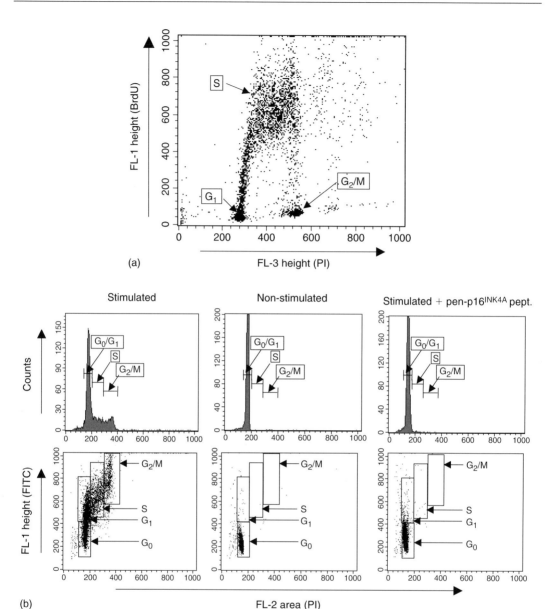

Figure 4.1

(a) BrdU incorporation: A typical cell cycle profile obtained after BrdU pulsing of the exponentially growing cell line U937. Cells were labelled with BrdU for 5 hours prior to fixation and antibody staining (described in Protocol 4.3). Populations of cells in G_1, S and G_2/M phases are indicated. This method has the advantage of allowing the investigator to analyse only the population of cells in S phase during the labelling period.

(b) Two parameter cell cycle analysis: Quiescent (G_0) human T cells isolated from the blood of a normal donor were stimulated for 48 hours with the mitogen phytohaemagglutinin (PHA) either in the presence or absence of a penetratin-p16^{INK4A} peptide that inhibits cdk6/4-cyclin D and prevents entry into the cell cycle. A cell cycle profile was obtained by staining cells with PI (for DNA) and FITC (for protein content) (see Protocol 4.4). This is the method of choice in the cell cycle laboratory since, when the data are plotted as FITC fluorescence height versus PI fluorescence area, cells in G_0 can be distinguished from those in G_1. These data can also be plotted as a histogram. The penetratin-p16^{INK4A} peptidyl mimic effectively blocks the entry of cells into the cell cycle maintaining them in G_0, characterised by their 2n DNA content, small size and low protein content.

Protocol 4.4: Two colour cell cycle analysis and cell sorting

EQUIPMENT, MATERIALS AND REAGENTS

Flow cytometer: FACScan for analysis and FACSvantage for sorting (Becton Dickinson).

Laboratory microscope capable of resolving cells.

Laboratory centrifuges capable of centrifuging tubes up to 50 ml at $>400\,g$.

Improved Neubauer chamber.

PI: $2\,mg\,ml^{-1}$ in H_2O. Store at 4°C.

FITC (unconjugated): $1\,\mu g\,ml^{-1}$ in H_2O. Store at 4°C.

RNAse: $10\,mg\,ml^{-1}$ in 50 mM Tris-HCl, pH 7.5, 15 mM NaCl. Pass through 0.2 μm filter, heat to 100°C for 15 min. Aliquot and store at −20°C.

Staining solution: to 10 ml PBS (containing Ca^{2+}/Mg^{2+}) add 50 μl FITC, 200 μl PI, 500 μl RNAse. Pass through 0.2 μm filter prior to use.

METHOD

Cell preparation and staining procedure

1. Pellet $1 \times 10^5 - 1 \times 10^6$ cells at $400\,g$ for 10 minutes. Aspirate supernatant carefully. Flick the tube containing the cells to disaggregate the pellet.

2. Resuspend cells in 1 ml 70% (v/v) ethanol (at −20°C) per 1×10^6 cells; mixing/vortexing gently works well for primary lymphocytes and most cell lines. If cells are more fragile, flick the tube gently adding ethanol drop by drop. For most applications fixation for about 30 minutes is sufficient. Cells may be stored in 70% (v/v) ethanol at −20°C indefinitely.

3. Pellet cells at $400\,g$ for 10 minutes. Aspirate supernatant and ensure that as much of the ethanol as possible is removed as otherwise we find that the peaks broaden.

4. Resuspend pellet in 1 ml staining solution per 1×10^6 cells. The cell density may affect FITC staining – count cells before fixing and adjust the volume of stain so that the number of cells per ml of stain is the same for each sample and incubate at 37°C for 30 min.

Flow cytometric analysis

Analyse the samples using a flow cytometer (e.g. Becton Dickinson FACScan) excited with a blue laser (488 nm) and with filter sets to detect PI (585 nm) and FITC (530 nm). Doublets of cells in G_1 can appear to be in G_2/M and, as there are usually many more cells in G_1, this can substantially affect the data. We exclude these doublets as far as possible by only including cells that are within linear gates for forward scatter versus forward scatter peak, for FITC staining versus FITC peak, and also for PI staining versus PI peak. The percentage of cells in each cell cycle phase can be determined by setting linear gates. This assumes that the distribution of cells in each phase follows a normal distribution. A more accurate figure can be obtained using proprietary software such as ModFit *LT*. The cells in each cell cycle phase can be isolated by flow cytometric sorting using a machine such as a Becton Dickinson FACSvantage. Proteins in these fixed and sorted cells can then be analysed by western blotting [24].

Sorting cells in different cell cycle phases

1. Sort the cells into a small volume of FCS or medium containing 20% (v/v) FCS. The tube should be coated with FCS in order to prevent cell losses. We have routinely isolated >1×10^6 cells in this way.

2. Count the sorted cells using a haemocytometer (improved Neubauer chamber). This is important as the number of cells you actually get in the tube may not be as many as the machine indicates.

3. Pellet the sorted cells at $400\,g$ for 10 minutes. The cells are fragile and should not be centrifuged at full speed in a microfuge even for a few seconds. Aspirate supernatant carefully.

4. Resuspend the pellet in 1 ml staining solution per 1×10^6 cells.

5. Reanalyse 5×10^4 cells to determine the purity of each sorted population. They should have the same PI and FITC fluorescence as the population from which they were isolated. Never assume that they are pure.

6. Centrifuge the remainder at $400\,g$ for 10 minutes in a screw-top microfuge tube. Add 1 ml SDS-PAGE sample buffer (Protocols 4.5 and 4.6) per 1×10^7 cells and boil immediately for 5–7 min.

7. As controls, fix and FITC/PI stain unsorted cells and proceed through steps 4–6. Fixing in 70% ethanol may not be suitable for all proteins; test other fixation methods before proceeding to sort cells.

8. Proteins in each purified cell sample can then be analysed by western blotting (Protocol 4.6).

This is analysed routinely in our laboratory by flow cytometric analysis as described in the preceding section. During the transition from G_0 into G_1 there is an obligate increase in cell mass and ribosome number [32], which largely accounts for the increase in cellular RNA content. The rate of protein synthesis also increases due, in part, to an increase in initiation factor eIF-4E [33]. In addition to cell cycle profiles we have used analysis of a number of cell cycle regulatory proteins to confirm G_0 status. Quiescent cells contain p130-E2F-4-DP-1 [34] and/or E2F-6-DP-1-c-myc complexes [35] and it was reported recently that they contain dephosphorylated pRB [36]. pRB is phosphorylated by cdk6/4-cyclin D during the transition into early G_1 [30, 36]. Phosphorylation events can be analysed by a number of methods (Section 4.3) such as ^{32}P-labelling, isoelectric focusing or by using phosphorylation site-specific antibodies. Quiescent cells also lack chromatin-associated mini chromosome maintenance proteins 2–7 (MCM 2–7). In fact, cells in most non-proliferating tissues lack MCM proteins altogether, however certain specialised cells such as mammary luminal epithelial acinar cells of the pre-menopausal breast [37] are exceptions that express MCM proteins constitutively. However, these proteins do not form origin replication licensing complexes bound to chromatin. This can be analysed by preparing chromatin-associated protein fractions from cell lysates, followed again by western blot analysis using MCM 2–7 specific antibodies (see Protocol 4.5 and *Figure 4.2*). The absence of MCM proteins in quiescent cells has been used to distinguish between neoplastic and normal cells of the cervix, prostate and bladder [38–41].

As cells enter the cell cycle they become larger (analysed by flow cytometry) and pRB becomes phosphorylated on a number of residues by cdk 4/6-cyclin D complexes. Phosphorylation of pRB on serines 807/811 is an early G_1 event which can be studied by western blotting with a phosphorylation site-specific antibody. Late G_1 is characterised by the loading of MCM 2–7 proteins and Cdc6 onto DNA [42], which can be found in the chromatin-associated fraction of cell lysates (*Figure 4.2*). In addition proteins expressed by E2F-regulated genes, such as p107, cdc2 and E2F1, are induced and can be analysed by western blotting.

The restriction point is defined as the point in G_1 after which cells can enter S phase without growth factors [44]. This point dissects G_1 into two distinct phases G_1-pm (post-mitosis) and G_1-ps (pre-S phase). Passage through the restriction point is associated with an induction of cyclin E, which occurs in G_1-ps and just prior to expression of cyclin A [45]. In this study, time-lapse microscopy data has shown that cyclin E is induced after the restriction point. These data indicate that cdk2-cyclin E activity, and hence the hyperphosphorylation of substrates such as pRB, are not part of the restriction point mechanism as had previously been thought. Cyclins E and A can be assayed either by western blotting or by flow cytometry with appropriate antibodies.

Protocol 4.5: Preparation and analysis of chromatin and non-chromatin associated proteins

EQUIPMENT, MATERIALS AND REAGENTS

Microfuge (cooled or in the cold room).

Heating block (100°C).

Cytoskeletal (CSK) buffer: 10 mM 1,4-piperazinediethanesulphonic acid (PIPES) pH 6.8, 100 mM NaCl, 300 mM sucrose, 1 mM MgCl$_2$, 1 mM EDTA, 1 mM DTT, 0.1% (v/v) Triton X-100, 0.1 mM ATP, 0.2 mM Na$_3$VO$_4$.14H$_2$O, 20 mM NaF, 1 mM PMSF, 1 mM di-isopropyl fluoro phosphate (DIFP), comprehensive protease inhibitor cocktail (Complete, Roche).

PBS, 1 mM DIFP.

1X SDS loading buffer: 125 mM Tris-HCl pH 6.8, 2% SDS, 10% (v/v) glycerol, 0.1% bromophenol blue, 5 mM DTT containing protease and phosphatase inhibitors. In general, add: 50 mM NaF, 2 mM PMSF, 1 mM Na$_3$VO$_4$.14H$_2$O, 5 U ml^{-1} aprotinin, 5 µg ml^{-1} pepstatin A, 5 µg ml^{-1} leupeptin, 1 mM DIFP.

METHOD

This protocol can be used to separate chromatin-bound and free proteins but it does not isolate nuclei and so the proteins that are not chromatin bound are not necessarily nuclear [42].

1. Pellet at least 1 × 10^6 cells in a 1.5 ml Eppendorf microfuge tube (10 minutes, 400 g). Remove all the supernatant. Spin again for a few seconds at 400 g and remove the residual supernatant. Samples can be stored at −85°C as frozen cell pellets and processed at a later date.

2. Resuspend the cell pellet in 20 µl of ice-cold CSK buffer and vortex until the cell pellet disappears and the cells are lysed. If using frozen cells, add ice-cold CSK buffer to cell pellets held on ice until just thawed.

3. Incubate the lysed cell solution on ice for 10 min.

4. Microfuge lysed cell solution for 3 min at 1000 g at 4°C and transfer the supernatant (non-chromatin-bound fraction) to a separate tube. Be careful not to disturb the pellet as this is the non-chromatin-bound fraction.

5. To the pellet add 200 µl of ice-cold PBS containing 1 mM DIFP, vortex to disaggregate the pellet and centrifuge for 3 min at 1000 g.

6. Discard the supernatant ensuring the pellet is dry but not disturbed.

7. Add 50 μl SDS loading buffer to the washed pellet and 30 μl of SDS loading buffer to the non-chromatin-bound fraction and heat to 100°C for 10 min.

8. Centrifuge samples to collect condensation that will be in the lid. Mix and analyse by SDS-PAGE and western blotting (Protocol 4.6).

Following such a separation it is important to check the fidelity of fractionation. Routinely we achieve this by probing western blots of these extracts for a protein known to be predominantly cytoplasmic (e.g. cdk6) or bound to DNA (e.g. histones). No cdk6 should be found in the chromatin fraction and a strong signal is expected for histone (e.g. histone H1). Conversely no histone H1 should be present in the non-chromatin-bound fraction but a strong signal for cdk6 should be observed. A number of cell cycle-associated proteins, such as the origin recognition complex proteins, are active once bound to chromatin. This type of fractionation experiment has been used to analyse the binding of MCM and cdc6 proteins to chromatin during the transition of quiescent T cells into the cell cycle [30] (*Figure 4.2*).

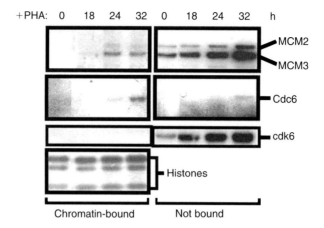

Figure 4.2

Chromatin binding assay: Peripheral blood T lymphocytes were stimulated for 18, 24 or 32 h with the mitogen PHA. Chromatin-bound and non-chromatin-bound proteins were isolated at each time point and separated by SDS-PAGE. The bottom of the gel was cut off and stained with Commassie blue to visualise histones and the remainder was subjected to Western blotting to detect components of the DNA origin replication complex, MCM2, MCM3 and Cdc6. Purity of the chromatin and non-chromatin extracts was determined by staining for histone and cdk6 proteins respectively. All proteins were detected on the same Western blot. (Anti-MCM2 antibody was from Gareth Williams and Kai Stoeber, Cambridge University and the remainder from Santa-Cruz Biotech, CA, USA.)

4.3 Specific methods for analysing cell cycle proteins. Protein modifications: phosphorylation

4.3.1 Gel electrophoresis

Many forms of gel electrophoresis have been used successfully to resolve phosphorylated from unphosphorylated proteins and in some cases proteins phosphorylated on different amino acid sites. These include one-dimensional polyacrylamide gel electrophoresis of proteins that are reduced and denatured by boiling in the presence of SDS (SDS-PAGE) as well as 2D gels. The simplest method for detecting potential phosphorylated forms of the protein is to look for a change in its mobility in SDS-PAGE.

4.3.2 Electrophoresis in one dimension

Western blotting is a standard procedure used in most cell and molecular biology laboratories, however, there are potential pitfalls in sample preparation and gel conditions which can markedly affect the results obtained. For example, selective protein degradation and dephosphorylation of the protein of interest can occur even during lysis in SDS sample buffer. Measures to try and circumvent these problems are detailed in Protocol 4.6. There are many examples where phosphorylation changes alter the migration of a transcription factor, such as E2F-1 [46], E2F-4 [24, 47] and DP-1 [48]. It should be noted that phosphorylation can either cause proteins to be retarded relative to their un- or less-phosphorylated counterparts or to migrate faster [49]. However, if phosphorylation of a given protein does not change its migration by

Protocol 4.6: Analysis by western blotting

EQUIPMENT, MATERIALS AND REAGENTS

Protein gel electrophoresis tank, western blot transfer tank and power packs. Suitable equipment is available from various manufacturers such as the Novex System (Life Sciences).

PBS, 1 mM DIFP.

1X SDS loading buffer: 125 mM Tris-HCl pH 6.8, 2% SDS, 10% (v/v) glycerol, 0.1% bromophenol blue, 5 mM DTT containing protease and phosphatase inhibitors. Add: 50 mM NaF, 2 mM PMSF, 1 mM $Na_3VO_4.14H_2O$, $5\,U\,ml^{-1}$ aprotinin, $5\,\mu g\,ml^{-1}$ pepstatin A, $5\,\mu g\,ml^{-1}$ leupeptin, 1 mM DIFP.

Antibodies.

Enhanced chemiluminescence reagents (e.g. ECL or ECL-Plus, Amersham Pharmacia Biotech Ltd.).

METHOD

The protein of interest should be present in sufficient abundance to be detected in an extract of up to 1×10^6 cells. Otherwise, prior partial purification by immunoprecipitation or other methods should be used.

1. Grow cells in culture or isolate primary cells as appropriate.

2. Pellet 1×10^6 cells at $800\,g$ for 7 min in a 1.5 ml screw-cap microfuge tube (e.g. Sarstedt 72.692.005). Remove the supernatant without disturbing the pellet. Depending on cell type, it may be necessary to put each sample into ice-cold PBS containing 1 mM DIFP to prevent degradation, followed by lysis in boiling 2X SDS sample buffer (containing N-ethyl maleimide (NEM), and protease and phosphatase inhibitors). Note that we have been able to detect and analyse E2F complexes in human primary monocytes but only after pre-incubating cells with DIFP before lysis [50]. Centrifuge again for a few seconds and remove all of the residual supernatant.

3. Flick the tube several times to dislodge the cell pellet and then add 25–50 μl 2X SDS sample buffer (containing protease and phosphatase inhibitors), flick hard or vortex briefly and heat immediately at 100°C for 10 min. See Protocol 4.5 for the list of protease inhibitors. Cells such as neutrophils contain significant protease activity that can degrade many cellular proteins during cell lysis. In this case, pre-incubate the cells on ice in PBS containing 1 mM DIFP before the addition of boiling SDS-sample buffer containing DIFP and NEM. Note that protein degradation during lysis may not be uniform and that the protein of interest may be completely degraded even when others are not. Addition of 2 mM NEM, which covalently modifies SH groups in enzymes including proteases, may prevent such degradation [51].

4. Centrifuge in a microfuge at 10 000 g for 5 seconds, flick to mix the solution and repeat the centrifugation. The sample should not be viscous and is now ready to load. However, if the solution is viscous, either increase the volume of sample buffer or, if this is not possible, shear the DNA through a narrow gauge needle or by sonication.

There are many variations on the original SDS-PAGE method [52, 53] and these are standard in most laboratories.

one-dimensional SDS-PAGE, there are other electrophoretic methods which are more powerful in resolving phosphorylated forms of a protein, such as 2D gels, examples of which are in the next section.

4.3.3 Electrophoresis in two dimensions

The most common technique employs IEF in the first dimension followed by SDS-PAGE in the second [54]. IEF separates proteins on the basis of their isoelectric points either using carrier ampholines to establish a pH gradient or by using an immobilised pH gradient [55]. Therefore, two forms of a protein which differ because they contain different numbers of phosphates can be separated. If IEF is sufficient to separate the different forms of the protein of interest then a second dimension may not be required. 2D gel electrophoresis may be used to analyse the phosphorylation state of the Oct-1 transcription factor during the cell cycle [56]. Two methods of detection may be used. The gels are either subject to western blotting with an anti-Oct-1 antibody, or Oct-1 is immunoprecipitated from ^{32}P-labelled cells with an antibody covalently coupled to protein A-sepharose beads (to avoid overloading the gels with immunoglobulin light and heavy chains) and is detected by autoradiography. Such approaches illustrate the power of 2D gel electrophoresis to resolve different phosphorylated forms of a protein, which are not separated well by SDS-PAGE. Some proteins can be resolved sufficiently well by the IEF first dimension and this has been used for resolving different ^{32}P-labelled forms of pRB [57]. However, it should be noted that some proteins do not focus well and some do not denature completely in the presence of SDS or even remain as dimeric or multimeric complexes. Therefore, alternative gel systems are important in such circumstances and two which could be tried are acid/urea gels [58] or gels which have been used to separate ribosomal proteins [59].

4.3.4 Dephosphorylation *in vitro*

In the examples given above we have given methods whereby different forms of the same protein can be separated by electrophoresis and the implication has been that these are forms of the same protein carrying different post-translational modifications. However, differences in migration could be due to a change in the size of the protein caused by specific cleavage (e.g. caspase-dependent cleavage of pRB when cells are triggered to undergo apoptosis), or to the production of different forms of the protein by alternative splicing or RNA editing. In order to determine that differences in migration in SDS-PAGE are due to changes in phosphorylation, dephosphorylation can be carried out *in vitro*. Phosphatases that are active in a cell lysate can be used to dephosphorylate the protein of interest *in vitro* and we have used such a method to dephosphorylate pRB [60]. However, not all proteins are dephosphorylated. For example, under these conditions the phosphorylation state and migration of another member of the pRB family, p130, does not change. In such circumstances it is necessary, therefore, to carry out dephosphorylation experiments by adding specific phosphatases to the cell extract. A variety of phosphatases may be used to dephosphorylate phosphoproteins *in vitro*, including acid and alkaline phosphatases, λ-protein phosphatase, which is a dual-specificity tyrosine and serine/threonine phosphatase, and other more specific phosphatases. All can be purified from the appropriate source and many are available commercially. λ-protein phosphatase may be used to dephosphorylate E2F-4 and p130 [24], as well as Oct-2 proteins (Protocol 4.7). If changes in migration of the protein caused by λ-phosphatase are prevented by adding NaF and Na_3VO_4, it is probable that these forms are produced by differences in phosphorylation state.

Protocol 4.7: λ-protein phosphatase

EQUIPMENT, MATERIALS AND REAGENTS

Lysis buffer A: 20 mM Hepes pH 7.8, 450 mM NaCl, 25% (v/v) glycerol, 2 mM EDTA, 0.5 mM DTT, 0.5 mM PMSF, 0.5 μg ml^{-1} leupeptin, 0.5 μg ml^{-1} Sigma protease inhibitor, 1 μg ml^{-1} trypsin inhibitor, 0.5 μg ml^{-1} aprotinin, 40 μg ml^{-1} bestatin, 2 mM DIFP.

Tris-MnCl$_2$: 50 mM Tris-HCl pH 7.8 at 25°C, 2 mM MnCl$_2$, 100 μg ml^{-1} BSA.

METHOD

Sample preparation and phosphatase reaction

1. To a pellet of 1×10^7 cells add 100 μl lysis buffer A. It may be necessary to pre-incubate some cells in PBS containing 2 mM DIFP to prevent partial or complete proteolysis (Protocols 4.5 and 4.6).

2. Freeze/thaw rapidly three times in dry-ice/ethanol and at 37°C.

3. Centrifuge at 13 000 g for 2 min and remove the supernatant to a fresh tube.

4. Freeze in aliquots on dry-ice and keep at −80°C or maintain on ice.

5. Measure the protein content of the extract by a suitable method, such as that of Lowry.

6. For the phosphate reaction, aliquot a suitable amount (4 μg protein) into each of eight microfuge tubes on ice containing Tris-MnCl$_2$. Final additions should be made as in *Table 4.1* in order to bring the volumes to 25 μl.

7. The reaction should be stopped by adding an equal volume of 2X SDS sample buffer and heating to 100°C for 5–10 minutes. Particular proteins in the sample can then be analysed by western blotting (Protocol 4.6).

λ-phosphatase is a dual-specificity phosphatase [61] which will dephosphorylate phospho-serine/-threonine as well as phospho-tyrosine. Dephosphorylation of the former is inhibited by NaF and the latter by Na$_3$VO$_4$. It should be noted, however, that λ-phosphatase will also dephosphorylate phospho-histidine [62], which may be important if phospho-amino acid analysis shows that the protein of interest is phosphorylated on amino acids

Table 4.1   Final additions to be made to the λ-protein phosphatase mixture

Sample	1	2	3	4	5	6	7	8
50 mM NaF	−	−	+	−	+	+	−	+
2 mM Na$_3$VO$_4$	−	−	−	+	+	−	+	+
100 U λ-phosphatase	−	+	+	+	+	−	−	−
Incubate at 30°C for up to 60 min								

other than serine, threonine or tyrosine. Cell lysates contain phosphatases and you may not need to add λ-phosphatase to dephosphorylate the protein of interest [60]. To exclude changes in the migration of the protein due to changes in phosphorylation rather than degradation, phosphorylation should be inhibited by NaF, Na_3VO_4 or other phosphatase inhibitors.

4.3.5 Radioactive labelling

The protocols described above are important as they provide simple and robust methods for analysing putative phosphorylation changes. Such methods may be used for characterising changes in the phosphorylation state of E2F-4 and p130 as primary haemopoietic CD34+ progenitor cells enter the cell cycle from G_0 [24]. However, the best test for phosphorylation is labelling the protein with [^{32}P]orthophosphate. Relatively large amounts of radioactivity are usually required for each experiment, typically 5–20 mCi/185–740 MBq, and this has profound effects on the procedures that have to be employed. In order to label the protein of interest, cells should be cultured for a few hours in phosphate-free medium in the presence of [^{32}P]orthophosphate. The duration of labelling will be dictated by the type of experiment you need to perform – as long as possible for mapping, or brief incubations for characterising phosphorylation changes in response to agonists (Protocol 4.8). However, cells require phosphate and so the labelling time has to be a balance between the minimum possible so as not to unduly stress or kill the cells, and a prolonged period to ensure maximum labelling, which has to be determined experimentally. In order to analyse the ^{32}P-labelled protein of interest, this has to be isolated, usually by immunoprecipitation, although one of a number of column chromatography or other physical methods could be used.

4.3.6 Phosphorylation *in vitro*

Labelling a protein in whole cells is important for determining which of perhaps many sites become phosphorylated in response to a physiologically relevant stimulus, but it may be more appropriate for studies on kinase activation to phosphorylate the protein of interest *in vitro*. For example, this has been done for the phosphorylation of Oct-2 by cdc2/cyclin B [65] and c-Jun by p38 [66]. If the substrate is to be the whole protein then it must be cloned, expressed and purified. Expression is usually in bacteria or in insect cells and the expressed protein is frequently engineered with an additional cleavable peptide 'tag' at the N- or C-terminus to aid in purification. There are drawbacks to each system: proteins produced in bacteria may be insoluble except under denaturing conditions and, even if they are soluble, their conformation may not be the same as in eukaryotic cells and this may affect the sites available for phosphorylation. Conversely, proteins produced in insect cells may become partially phosphorylated by endogenous kinases. Alternatively, synthetic peptides corresponding to predicted phosphorylation sites in the protein of interest can be used. This is covered in Protocol 4.9. Active kinases can be isolated from eukaryotic cells by a number of biochemical techniques including immunoprecipitation. However, caution must be exercised as a kinase preparation may be contaminated with other kinases. It may be possible to overcome this by preparing the active kinase by different methods or by using antibodies to different sites in the protein. Alternatively, inhibitors that preferentially inhibit specific kinases can be included in any kinase reaction to try and inhibit contaminating kinases but not the kinase of choice.

4.3.7 Generation and use of phosphorylation-site specific antibodies

Once the sites of phosphorylation have been determined then synthetic phosphopeptides corresponding to these sites can be used to generate antibodies. Specific antibodies, which only recognise the protein when it is phosphorylated at that site, can then be identified [68] (*Figure 4.3*). The antibodies generated are very powerful as they can then be used in western blotting, ELISA or immunocytochemical assays to

Protocol 4.8: Labelling with [^{32}P]orthophosphate

EQUIPMENT (DEDICATED TO RADIOACTIVITY WORK), MATERIALS AND REAGENTS

Laboratory centrifuge capable of centrifuging 50 ml tubes at $>400\,g$.

Microfuge (cooled or in the cold room).

37°C incubator (humidified, 5% CO_2) for culturing eukaryotic cells.

Protein gel tank, blot apparatus and power packs.

50 ml conical screw-capped tubes (e.g. Falcon).

1.5 ml screw-capped tubes with rubber O ring.

Phosphate-free DMEM/10% (v/v) dialysed FCS (Sigma).

Low-salt extraction buffer: 50 mM Tris pH 8.0 at 4°C, 150 mM NaCl, 0.5% (v/v) nonidet P40 (NP40), 5 mM EDTA, 0.1 mM Na_3VO_4, 50 mM NaF, 1 mM Ethylene glycol-bis-(β-amino ethyl ether) N,N,N',N',-tetra acetic acid (EGTA), 5 mM $Na_4P_2O_7 \cdot 10H_2O$, 1 mM β-glycerophosphate, 10 mM K_2HPO_4, 0.5 mM PMSF, 0.5 μg ml^{-1} leupeptin, 0.5 μg ml^{-1} Sigma protease inhibitor, 1 μg ml^{-1} trypsin inhibitor, 0.5 μg ml^{-1} aprotinin, 40 μg ml^{-1} bestatin, 2 mM DIFP.

High-salt extraction buffer: as low-salt buffer, but containing 450 mM NaCl.

[^{32}P]orthophosphate (1–10 mCi/37–370 MBq).

10% (v/v) propan-2-ol, 10% (v/v) acetic acid.

Antibodies for immunoprecipitation and western blotting.

METHOD

1. Pellet 1–10 $\times$ 10^6 cells at 200 g, for 7 min in a disposable 50 ml conical, screw-capped centrifuge tube.

2. Wash three times with 25 ml each phosphate-free DMEM/10% (v/v) dialysed FCS (Sigma). Flick tubes between washes to disaggregate cells. Use phosphate-free media other than DMEM as appropriate to the cells being used. The phosphate-free FCS is produced by dialysing 50 ml heat-inactivated FCS in a 10 000 molecular weight cut-off dialysis membrane against 4 changes of 5 litres 10 mM Hepes pH 7.0, 137 mM NaCl, 2.7 mM KCl to remove small molecules, including orthophosphate. Alternatively, use Tyrode's medium and incubate the cells in a water bath [63, 64].

3. To remove as much unlabelled phosphate from the cells as possible, culture in 20 ml phosphate-free DMEM/10% (v/v) dialysed FCS for 30 minutes at 37°C in a humidified atmosphere of 5% CO_2.

4. Pellet cells as in stage 1 and resuspend in 10 ml of the same media containing 1–10 mCi/37–370 MBq [^{32}P]orthophosphate. The amount of [^{32}P]orthophosphate required for labelling the protein of interest should be determined experimentally and kept to a minimum.

5. Culture for 4 hours at 37°C in a humidified atmosphere of 5% CO_2. Shake to resuspend cells in the medium every 10–30 minutes. An adherent cell line can be left undisturbed during the labelling period. We have used 4 h for labelling pRB in proliferating cells, but a labelling period >5 h is necessary to detect [^{32}P]pRB in quiescent T cells that have been stimulated to enter the cell cycle with phytohaemagglutinin (PHA). Note that phosphate uptake into cells also changes during this period and the way this alters protein labelling can be determined by quantifying total ^{32}P-labelled protein as described in step 8 below. This is crucial if the relative levels of ^{32}P-phosphorylation of a given protein are being compared. The labelling period for other proteins may be minutes rather than hours after stimulation. If possible, leave 30–60 minutes for the [^{32}P]orthophosphate to be incorporated into cellular ATP pools before stimulating the cells.

6. Pellet as in stage 1 and dispose of the supernatant.

7. Transfer the tube to ice and lyse the cells for immunoprecipitation observing local safety regulations. For the analysis of pRB, lyse the cells in 800 μl low-salt buffer, transfer to a screw-capped 1.5 ml microfuge tube with rubber O-ring and centrifuge at 13 000 g for 1 min to pellet nuclei and insoluble matter. Extract proteins in the nuclear pellet with 80 μl high-salt buffer for 20 min with end-over-end mixing, centrifuge at 13 000 g for 3 min and combine the low- and high-salt extracts.

8. Pre-clear the lysates of proteins which stick non-specifically to IgG for 1 h with end-over-end mixing. Use 1–10 μg rabbit IgG, isotype-matched monoclonal antibody, anti-peptide antibody competed with cognate peptide and 20 μl protein A-agarose. Centrifuge at 13 000 g for 5 s, then immunoprecipitate pRB with 1–10 μg of the appropriately titrated antibody and 20 μl protein A-agarose for 1 h (or overnight if necessary). Centrifuge at 13 000 g for 5 s and retain the supernatant then wash the pellets four times with 800 μl each wash of the low-salt buffer before boiling for 10 min in 30 μl 2X SDS sample buffer. The maximum amount of a particular ^{32}P-labelled protein may be obtained by repeated immunoprecipitation from the retained supernatant. Alternatively, additional ^{32}P-labelled proteins can be immunoprecipitated sequentially. If the extent of labelling of a particular protein in cell extracts cultured under different conditions is to be compared, then the incorporation of ^{32}P into total cell protein has to be determined. This is done by precipitating a small volume of the total cell lysate (5–10 μl in triplicate) in 50% (w/v) trichloroacetic acid (with 0.01% (w/v) BSA as carrier), collecting the precipitate on GF/C filters (Whatman) which are washed in 5% (w/v) trichloroacetic acid, then ethanol before quantification by scintillation counting.

9. Separate the proteins by gel electrophoresis. Fix the gel in 25% (v/v) isopropanol, 25% (v/v) acetic acid and wash with the same until the washes are no longer radioactive. Dry the gel and expose to X-ray film or PhosphorImager. Alternatively, proteins in the gel can be western blotted and detected by chemiluminescence before detecting the ^{32}P-labelled protein on the membrane. The gel should have transferred evenly by fixing, drying the gel and then exposing it at the same time as the membrane. With a peroxidase-conjugated second antibody the enhanced chemiluminescence (ECL) signal can be eliminated by adding 0.1% (w/v) sodium azide in PBS to the membrane. The ECL signal will then not interfere with the ^{32}P-detection.

Protocol 4.9: Phosphorylation of synthetic peptides

EQUIPMENT, MATERIALS AND REAGENTS

2X kinase buffer: 50 mM Tris pH 7.4, 10 mM DTT, 2 mM EGTA, 20 mM MgCl$_2$.

ATP and [γ^{32}P]ATP.

p81 paper (Whatman).

75 mM H$_3$PO$_4$.

SDS sample buffer (see Protocol 4.6)

10% (v/v) propan-2-ol, 10% (v/v) acetic acid.

Antibodies for immunoprecipitation.

Protein A or protein G coupled to Sepharose CL4B or agarose beads.

METHOD

1. Immunoprecipitate the kinase of interest (e.g. cdk6-cyclin D) using the appropriate conditions to maintain kinase activity. Start with conditions that have been shown to work for cdk, PKC or PKA, for example, and try immunoprecipitating with a range of different antibodies. Some antibodies will not immunoprecipitate the active form of the kinase, depending on the epitope(s) recognised. This is true of antibodies raised to peptides containing the PSTAIR motif of cdc2 as this site is masked in the active form of the kinase by B-type cyclins.

2. Wash the protein A-agarose-immune complexes twice with 400 μl of kinase buffer.

3. Incubate the immunopurified CDKs at 30°C for 30 min in 1X kinase buffer containing 2.5 μg of peptide substrate, 50 μM ATP and 5 μCi [γ-^{32}P]ATP, 10 mM NaF and 1 mM sodium orthovanadate, in a final volume of 20 μl. The time of incubation should be adjusted to ensure linearity with the peptide and kinase being used. The substrate used for cdk6 kinase assay was RAAPLSPIPHIPR and that used for the cdk2 reaction was AKAKKTPKKAKK [2]. As an alternative, a larger protein such as GST-pRB can be used, though this could contain sites that can be phosphorylated by kinases that may contaminate the immunoprecipitate.

4. The incorporation of ^{32}P onto each peptide can be determined by quantifying the signal obtained from the gel (see step 5) or as follows: remove three 2 μl aliquots of each reaction, spot onto p81 paper, wash six times in 75 mM H$_3$PO$_4$ and quantify by liquid scintillation counting.

5. To the remainder of the reaction, add 10 μl of 2X SDS sample buffer and heat at 100°C for 5 min. Load 15 μl onto each lane of a three-stage tricine 16.5% polyacrylamide gel system, made as described in [67]. Carry out

electrophoresis at 100 V, until the bromophenol blue dye front reaches the bottom edge of the gel. This gel system allows separation of the ^{32}P-labelled peptide in the resolving gel and the cdk in the spacer gel. The separating gel containing the ^{32}P-labelled peptide should be cut away, fixed in four changes of 10% (v/v) propan-2-ol, 10% (v/v) acetic acid, dried in a gel-dryer for 45 min at 80°C and then exposed to pre-flashed X-ray film or to a PhosphorImager plate. Some peptides may elute from the gel under these conditions. If this occurs, alternative fixation conditions should be tried, such as 50% (w/v) trichloroacetic acid. The relative amount of ^{32}P incorporated into the peptide in each sample should be quantified by scanning the film or using image quantification software. In order to ensure that the same amount of kinase was used in each assay, protein in the spacer gel should be transferred to a nitrocellulose membrane for western blotting with the appropriate antibody (to cdk2 or cdk6).

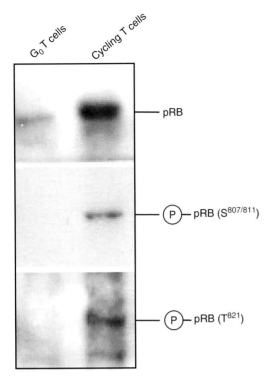

Figure 4.3

Western blot analysis of pRb in quiescent and cycling human T cells. The pRB protein becomes phosphorylated on multiple sites, two of which are shown here (S807/S811 and T821) as cells enter the cell cycle. These modifications can be detected by the use of phosphorylation-specific antibodies. Purified human T cells were either lysed directly in SDS loading buffer or were stimulated to enter the cell cycle with PMA and ionomycin for 30 hours prior to lysis. 2.5×10^5 lysed cells per lane were loaded onto a 4–12% NuPage gel (Novex). After electroblotting, the membrane was blocked with 5% (w/v) non-fat dried milk in PBS/0.05% (v/v) Tween-20 prior to probing with either phosphorylation specific antibodies or an antibody to generic pRB. All antibodies were used at a 1:1000 dilution in 3% (w/v) BSA/PBS/0.05% Tween-20. HRP-conjugated secondary antibodies were used at 1:2000 in 5% milk in PBS/Tween-20. The bound antibodies were detected with ECL-Plus (AP Biotech).

investigate the presence or absence of a particular phosphorylated form of a protein in cell lysates. For example, such a strategy has been used for CREB [69, 70], pRB [2] and for p53 [71].

4.3.8   Other methods for analysing phosphorylated proteins

There are more powerful methods for the analysis of phosphorylated proteins. With the current interest in proteomics their use and applications will increase dramatically in the coming years. However, the techniques require a number of expensive pieces of equipment and specialist knowledge in their use. In brief, peptides are purified by high performance liquid chromatography (HPLC), analysed by multi-dimensional electrospray mass spectrometry (ESMS) and fractions containing phosphorylated peptides are then sequenced [72]. The phosphorylated protein has first to be isolated either from whole cells or phosphorylated *in vitro* with candidate kinases, as described above. The methods for determining the precise sites that are phosphorylated are given in Protocol 4.10.

Protocol 4.10: Alternative methods for determining phosphorylation sites

Starting with 500 pmol of a suitable protein, digest overnight at 37°C with trypsin (such as N-tosyl-L-phenylalanine chloromethyl ketone (TPCK)-Trypsin, Calbiochem or modified Trypsin, Promega) at a ratio of 7:1 (w/w) protein:trypsin. Other suitable enzymes can also be used as appropriate. Methods for analysing the peptides produced are described in detail elsewhere [73]. The peptides produced are then separated by reverse-phase HPLC and detected using an in-line UV detector set at 214 nm. In order to identify which fractions contain phosphopeptides, a small amount (10%) of each fraction is analysed by mass spectroscopy (negative-ion ESMS) optimised to detect PO_3^- ions [74, 75]. The molecular weights of the phosphopeptides in positive fractions can be determined by nanoelectrospray mass spectroscopy. Such accurate molecular weight measurement can yield sufficient data to determine which peptide is in each fraction. The sites phosphorylated can also be determined by sequencing by mass spectroscopy.

4.4 O-linked glycosylation

In the sections above we have described a number of methods which can be used to analyse changes in phosphorylation state. Certain proteins involved in regulating the cell cycle are also modified by O-linked glycosylation on serine and threonine residues. Lysine can also be modified by acetylation. Like phosphorylation, these changes are dynamic.

Over 50 nuclear proteins are known to be modified by O-linked glycosylation. This involves the attachment of the O-linked monosaccharide N-acetylglucosamine (O-GlcNAc) to the hydroxyl groups of either serine or threonine. Many of the proteins which are modified in this way are also phosphorylated and it has been suggested that the sugar may block the sites of phosphorylation and so may have a reciprocal effect to that caused by phosphorylation [76–78]. Attachment of O-GlcNAc is regulated and can turn over rapidly and enzymes capable of adding and removing the sugar from proteins have been purified. Such dynamic changes in glycosylation have been shown to occur in a signal-dependent manner suggesting that they may be crucial for regulating cellular functions [79]. Proteins that control cell cycle progression such as the serum response factor (SRF), p53 and c-Myc are known to be modified with O-GlcNAc [78, 80–82].

For SRF and c-Myc, the modified amino acids are in the transcriptional activation domains and for c-Myc the site modified (Threonine-58) is also known to be phosphorylated *in vivo* [78, 83]. Phosphorylation at this site is important for cell proliferation and c-Myc harbouring a mutation of Threonine-58 has enhanced transformation potential in cell lines and increased tumour-inducing potential in animals. This site is mutated in Burkitt's and other lymphomas, as well as in v-Myc, which provides evidence for the potential importance of reciprocal O-GlcNAcylation/phosphorylation in neoplasia. However, as the same site can be modified by two different mechanisms it is not clear whether phosphorylation and/or O-GlcNAcylation is a critical determinant in cancer.

The protein encoded by the *TP53* tumour suppressor gene, which is mutated or deleted in about 50% of all human cancers, is also O-GlcNAcylated. The DNA affinity of the p53 protein, as well as its activity, is dependent on unmasking a basic region in the C-terminus which normally acts as a repressor domain. Unmasking this region is dependent on O-GlcNAc modification and so dynamic O-GlcNAcylation may well be important in regulating the affinity of p53 for DNA [82]. Although the O-GlcNAc modification of serine and threonine was discovered more than 15 years ago [84], the literature on the role of protein O-GlcNAcylation in the cell cycle remains sparse compared with that for phosphorylation.

4.4.1 Use of lectins

Lectin binding is a common method for isolating proteins modified with specific sugars [85]. Thus different lectins are valuable as probes for different substituents, even though they have overlapping ligand specificity and so cannot be used to determine unequivocally which modification exists on the protein of interest. For example, wheat germ agglutinin (WGA) binds both sialic acid and O-GlcNAc residues. In spite of this, lectin binding has proved very useful. For example, WGA-agarose has been used to purify O-GlcNAc-modified Sp1 [86] and biotinylated-WGA was used as a probe to identify O-GlcNAc-Sp1 on western blots [80] (Protocol 4.11).

4.4.2 Radioactive labelling

Methods using *in vitro* glycosylation of proteins are described elsewhere [88, 89] and are covered briefly in Protocol 4.12. The glycosylated forms of p53 and c-Myc identified to date are O-GlcNAcylated and the methods described above will indicate whether this is also likely to be the case for the protein of interest. However, many different protein glycosylations are possible [85] (Protocol 4.13).

Protocol 4.11: Use of lectins

EQUIPMENT, MATERIALS AND REAGENTS

WGA-agarose column (Vector Laboratories).

Buffer 1:50 mM Tris-HCl pH 7.5, 420 mM KCl, 20% (v/v) glycerol, 10% (w/v) sucrose, 5 mM MgCl$_2$, 0.1 mM EDTA, 1 mM PMSF, 1 mM sodium metabisulphite, 2 mM DTT.

Buffer 2:25 mM Hepes pH 7.6, 100 mM KCl, 12.5 mM MgCl$_2$, 20% (v/v) glycerol, 0.1% (v/v) NP40, 10 µM ZnSO$_4$, 1 mM DTT.

GlcNAc (Sigma).

TBST:10 mM Tris pH 8.0, 150 mM NaCl, 0.05% (v/v) Tween-20.

BSA/TBST:3% (w/v) BSA in TBST.

50 mM sodium citrate pH 5.0.

β-N-acetylglucosaminidase.

β-N-acetylglucosamine or α-methyl-D-mannoside, or

Succinylated WGA or biotinylated concanavalin A (con A).

10 mM sodium phosphate pH 7.4, 0.1% (w/v) SDS, 0.5% (v/v) Triton X-100, 140 mM NaCl.

Avidin-alkaline phosphatase or peroxidase.

METHOD

Isolation using a lectin column and blotting using biotinylated lectin

1. Prepare crude nuclear extract [43].

2. Apply to a WGA-agarose column (~135 mg protein extract per ml column; Vector Laboratories) in Buffer 1.

3. Wash twice with 5 ml Buffer 1 and four times with 5 ml Buffer 2. Elute with free 300 mM GlcNAc (Sigma) in the same buffer.

4. This method is as described [86] for Sp1 purification: ~1% pure Sp1 after WGA-agarose, which can then be purified by binding to a DNA affinity column (yield ~4.5 µg g^{-1} HeLa cells).

5. Separate proteins by SDS-PAGE and blot onto nitrocellulose.

6. Block in 3% (w/v) BSA/TBST.

7. As a control for non-specific binding, remove the O-GlcNAc from proteins on one strip of the membrane as follows: wash in 50 mM sodium citrate pH 5.0, incubate with 2.5 U ml^{-1} β-N-acetylglucosaminidase at 37°C for 24 h; wash in TBST, then BSA/TBST.

8. Incubate the membrane (with or without β-N-acetylglucosaminidase treatment) with 10 µg ml^{-1} biotinylated WGA in TBST. Controls should include blocking biotin-WGA with 200 mM β-N-acetylglucosamine or

α-methyl-D-mannoside, or probing with succinylated WGA or biotinylated con A (add 1 mM CaCl$_2$ to all buffers when using con A).

9. Wash in 10 mM sodium phosphate pH 7.4, 0.1% (w/v) SDS, 0.5% (v/v) Triton X-100, 140 mM NaCl; incubate in BSA/TBST-containing avidin-alkaline phosphatase or peroxidase; wash in TBST, then develop. This method is as described elsewhere [80]. Monoclonal antibodies which recognise O-GlcNAc have been raised [87] and the RL-2 antibody has been used to identify O-GlcNAc-Sp1 by western blotting.

Protocol 4.12: Radioactive galactosylation *in vitro*

EQUIPMENT, MATERIALS AND REAGENTS

Galactosyltransferase (bovine; Sigma).

25 mM Hepes pH 7.3, 5 mM MnCl$_2$, 50% (v/v) glycerol.

UDP-[6-^{3}H]galactose (1–5 μCi (37–185 kBq), 40 Ci mmol^{-1} in 25 mM 5'-AMP).

Galactose buffer: 100 mM galactose, 50 mM Hepes pH 7.3, 150 mM NaCl, 50 mM MnCl$_2$.

METHOD

Glycosylation site(s) can be labelled *in vitro* by treating the glycosylated protein or protein lysates [90] with [^{3}H]galactose and galactosyltransferase:

1. Autoglycosylate galactosyltransferase [84], concentrate by precipitation in 85% ammonium sulphate and store at −20°C in 25 mM Hepes pH 7.3, 5 mM MnCl$_2$, 50% (v/v) glycerol.

2. Prepare the protein of interest or a cell extract.

3. Label the purified transcription factor or protein extract (100 μg) in: 50 mU autoglycosylated galactosyltransferase, galactose buffer, UDP-[6-^{3}H]galactose. Incubate at 37°C for 30–60 min. The GlcNAc residues are modified with [^{3}H]galactose.

4. The [^{3}H]galactose-labelled protein can be separated from unincorporated UDP-[6-^{3}H]galactose by chromatography through Sephadex G50 or immunoprecipitation.

Protocol 4.13: Determination of the type and site of glycosylation

The type of analysis described in brief here is specialised and requires expensive equipment and a detailed knowledge of the procedures being used.

METHOD

1. The type of glycosylated linkage (N- or O-linked) can be determined by enzymatic deglycosylation [91, 92].

2. Carbohydrate analysis may be performed after digesting galactosyltransferase-labelled protein with N-linked-glycopeptide-[N-acetyl β-D-glucosaminyl]-L-asparagine amidohydrolase (PNGase) F (material resistant to digestion is O-linked), separating the released glycans from the protein on a Sephadex G50 column followed by alkali-induced β-elimination [91]. After desalting and size-fractionation by fast performance liquid chromatography (FPLC), the saccharide(s) present may then be identified by anion-exchange HPLC in comparison with known standards [90, 93].

3. The site(s) of glycosylation have been determined as follows [81]: peptides/glycopeptides derived by cyanogen bromide and enzyme proteolysis were separated by reverse-phase HPLC and 10% of each fraction was analysed by mass spectrometry. This allows peptides to be identified by their molecular masses and fractions containing peptides differing by 203 mass units potentially contain a GlcNAc residue. In order to verify the presence of a GlcNAc-modified residue in such fractions they were labelled by galactosyltransferase, as described in Protocol 4.12. Further verification of the modified peptides was by automated sequencing of [3H]Gal-labelled peptides.

4.5 Acetylation

Modification of histones by acetylation has been known for 30 years. This post-translational modification involves the transfer of an acetyl group to the e-amino group of a lysine and the overall level of chromatin acetylation is known to regulate transcription [94]. Histone deacetylation and hence activation of specific genes can be regulated during the cell cycle. For example, pRB associates with a type I HDAC complex via a protein called RBP1 [95]. This is targeted to specific genes by the E2F factor that is also bound to pRB and is thought to repress transcription by causing chromatin condensation. A number of proteins that regulate cell proliferation are also controlled by reversible acetylation [96]. For example, three of the E2F proteins, namely E2F-1, -2 and -3 (but not E2F-4, -5 or -6) as well as pRB, are all acetylated by p300/CBP acetyltransferases [97–99]. Indeed, acetylation of pRB has been reported to decrease its phosphorylation *in vitro* by cdk2-cyclin E [99]. Another example of the importance of acetylation and phosphorylation is p53. This protein is acetylated by p300 once it becomes phosphorylated at Serine-15 [100]. Acetylation is required for p53 activity in promoting growth-arrest or apoptosis and recently deacetylation by hSir2 was shown to reduce its transcriptional activity [101, 102]. Replicative senescence induced by ras-stimulated p53 acetylation is dependent on PML [103]. Thus, a number of growth-inhibitory events are associated with p53 acetylation.

The methods used to investigate the acetylation of proteins such as pRB and p53 are based on the use of an antibody that recognises acetylated lysine and the incorporation of [^{14}C]acetyl-CoA (Protocols 4.14 and 4.15).

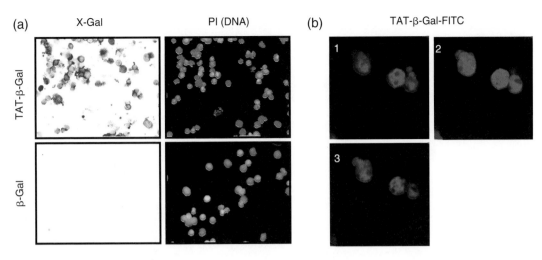

Plate 4

TAT-β-Gal Transduction: (a) CD34+ cells were selected from normal human bone marrow and transduced either with 1 μM TAT-β-Galactosidase or unconjugated β-Galactosidase for 2 hours. Cytospins were prepared, fixed with 0.5% paraformaldehyde and stained with X-Gal (containing 0.2% Triton-X100). Cells on the slide were visualised by fluorescence microscopy after staining DNA with PI (red). Almost all the CD34+ cells were transduced with active TAT-β-Gal (blue) and no staining was observed in the β-Gal control. (b) Although TAT-β-Gal blue staining appeared to be present inside cells, we wished to determine whether TAT-β-Gal staining could be due to cell surface binding. Thus, HL-60 human promyelocytic leukaemia cells were transduced with FITC-conjugated TAT-β-Gal for 2 hours and cytospins were fixed and visualised by fluorescence confocal microscopy. Serial sections taken through a field of cells (labelled 1 to 3) show that the TAT-β-Gal-FITC is inside each cell and is present in the nucleus as well as cytoplasm. In this cell type there was less TAT-β-Gal-FITC in sub-nuclear regions that are probably nucleoli.

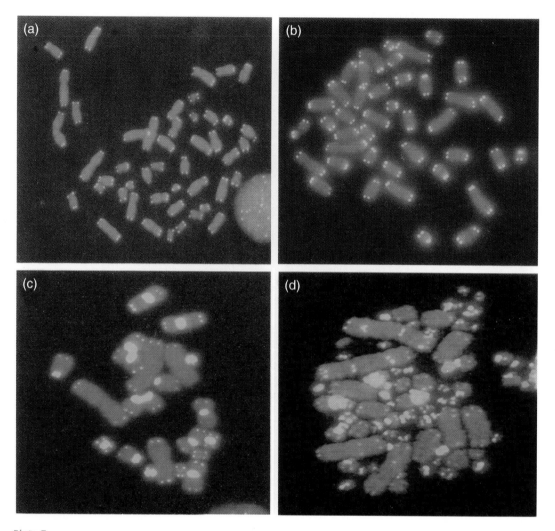

Plate 5

Metaphase spreads from (a) human, (b) mouse, (c) Chinese hamster and (d) chicken cells hybridised with the telomeric PNA oligonucleotides. Chromosomes are stained in blue (DAPI) and telomeres are red, due to Cy3-labelled PNA. Note the presence of interstitial telomeric sequences in Chinese hamster and chicken chromosomes (c, d). Chicken cells contain large numbers of microchromosomes (d).

Protocol 4.14: Detecting acetylated proteins by western blotting

EQUIPMENT, MATERIALS AND REAGENTS

As for western blotting (Protocol 4.5).

METHOD

1. Culture cells under the appropriate conditions and harvest by centrifugation.

2. Immunoprecipitate the protein of interest. A single antibody may not bind to the acetylated form of the protein, so in the first instance it is best to immunoprecipitate with different antibodies or a mixture of antibodies. Also immunoprecipitate a second aliquot with an anti-acetylated lysine antibody (New England BioLabs). This can be coupled directly to protein A-Sepharose [104].

3. Run the samples on two western blots. Probe the first with an antibody to the protein of interest and the second for anti-acetylated lysine.

Protocol 4.15: Acetylation *in vitro*

EQUIPMENT, MATERIALS AND REAGENTS

The method assumes that you can express and purify the proteins of interest in insect cells or in *Escherichia coli* and the appropriate equipment should be available in the laboratory.

Acetylation buffer: 50 mM Tris-HCl pH 8.0, 5–10% glycerol, 50 mM KCl, 10 mM sodium butyrate, 1 mM DTT, 1 mM PMSF, 0.1 mM EDTA, 90 pmol [^{14}C]acetyl-CoA (55 mCi mmol^{-1}).

METHOD

The methods are derived from [105] and have been used in a number of studies to acetylate pRB and E2F proteins [97–99]. Refer to these publications for more specific details for the particular proteins concerned.

1. Express epitope-tagged p300, for example, and your protein of interest in Sf9 cells from a baculovirus vector or in *Escherichia coli*.

2. Isolate the recombinant protein by affinity purification with the appropriate anti-epitope tag antibody coupled to Sepharose or an appropriate affinity resin.

3. Incubate 20–500 ng p300 with 1–2 µg of the protein of interest in acetylation buffer at 30°C for up to 60 min. In a separate reaction, use 9 pmol calf thymus histones (Sigma) as a positive control for your acetylation reaction.

4. Stop the reaction by boiling in SDS sample buffer or as appropriate for the gel system being used.

5. Separate the proteins by gel electrophoresis, stain and dry the gel. Detect ^{14}C-labelled acetylated protein by PhosphorImaging or with X-ray film.

4.6   Functional analysis

4.6.1   Transfection and transduction

To test the functional importance of phosphorylation, glycosylation or acetylation at a specific site, that amino acid has to be mutated in the cDNA of interest. This can be carried out using one of a variety of site-directed mutagenesis strategies and it is usual to mutate the codon(s) encoding serine or threonine to alanine, and tyrosine to phenylalanine. The mutated amino acids have similar structures of the same polarity and it is assumed therefore that the mutated protein will have the same overall conformation as the wild-type protein. However, neither alanine nor phenylalanine can be phosphorylated. These constructs can be used to confirm that phosphorylation does indeed occur at the specific site(s) in whole cells and to determine whether this phosphorylation has a functional consequence. If a cell line is available which lacks the protein of interest or cells can be isolated from the appropriate knockout mouse then cDNAs encoding the mutant or wild-type protein can be stably transfected, otherwise an epitope-tagged construct may have to be used. After treating the transfectants in the appropriate manner, ^{32}P-labelling and mapping can then be carried out. A tryptic peptide map [106] generated from the mutated protein should lack the specific spot that it is believed corresponds to the phosphopeptide of interest. This strategy assumes that mutation of the protein of interest does not cause apoptosis or inhibit cell proliferation, in which case transient transfection (if transfection efficiency is high) or inducible vectors should be used. The transfected cells can also be used to test the effect of the mutated transcription factor on other aspects of cell physiology that will depend on the factor being investigated. Also, the effect on various molecular mechanisms can be tested, such as interaction with other proteins, DNA binding, gene-reporter assays and so on, which are described in detail in other chapters of this book (Chapter 11) and elsewhere [46]. Particular amino acids in proteins such as c-Myc can either be phosphorylated or O-GlcNAcylated under different conditions. Because of this, caution should be exercised in claiming that an effect on cell physiology is due solely to phosphorylation.

4.6.2   Protein transduction

Strategies involving plasmid transfection or viral transduction work particularly well for cell lines in which the efficiency of gene delivery is high. This has led to the adoption of certain cell lines as models in which to test the effects of particular proteins on the cell cycle. Unfortunately, cell lines almost always have an abnormal karyotype and usually contain mutations or deletions that affect key regulators of the cell cycle. Transfecting or transducing primary cells isolated from humans is usually less efficient, and introducing genes into quiescent cells, particularly quiescent haemopoietic cells, is especially difficult. Active proteins can be introduced into a number of different non-dividing haemopoietic cells *in vitro*, including quiescent CD34+ haemopoietic progenitor cells, and most of the cells take up the active protein (*Plate 4* and *Figure 4.4*). This method is based on the fact that the HIV TAT protein can cross cell membranes [107, 108] and that fusing a short sequence from TAT (YGRKKRRQRRR) to another protein acts as a protein transduction domain (PTD) which is sufficient to take it into a cell [109]. This works for a large number of different proteins up to 125 kDa (β-galactosidase), including proteins involved in cell proliferation, such as pRB, p16^{INK4A}, p27^{Kip1} and dn-cdk2 [110]. Further, cells of different lineages can be transduced both *in vitro* and *in vivo* [111]. Thus, this is a method that will allow the investigator to analyse the functions of cell cycle regulators directly by transducing active, TAT fusion proteins into primary cells. This is particularly

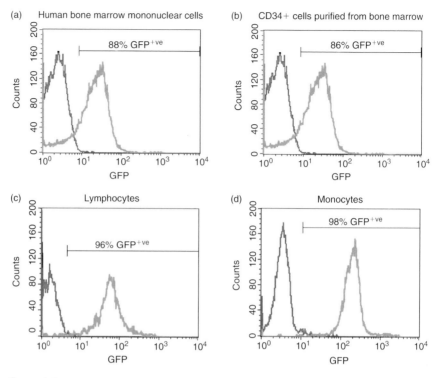

Figure 4.4

Transduction of TAT-GFP into haemopoietic cells: (a) Mononuclear cells were purified from the bone marrow of a normal donor by centrifugation on Ficoll-Paque; (b) CD34+ haemopoietic progenitor cells were purified from these with immuno-magnetic beads (Miltenyi). Mononuclear cells were also purified from human peripheral blood and (c) lymphocytes and (d) monocytes were identified by their forward- and side-scatter characteristics. In each case, cells were transduced with TAT-GFP for 1 h and the percentage of GFP +ve cells (grey) was determined by flow cytometry, using untransduced cells as the negative control (black). Most cells of each cell type are transduced with TAT-GFP.

powerful when applied to non-cycling primary haemopoietic cells that are very difficult if not impossible to transfect using conventional methods. As stated above, we have used these methods successfully to investigate the G_0 to G_{1A} transition and the induction of effector functions in T lymphocytes [30]. We have also used another TAT fusion, TAT-p27^{Kip1}, to investigate the role of p27^{Kip1} in chronic myeloid leukaemia (CML) and B cell apoptosis [112, 113]. TAT-mediated transduction is discussed in Protocol 4.16, but it should be noted that a number of other PTDs have also been identified. One known as Penetratin is comprised of residues 43–58 from the highly conserved 60 amino acid homeodomain of antennapedia homeoprotein transcription factors [114]. In *Figure 4.1b* we show an example of its use. Penetratin coupled to a peptide derived from p16^{INK4A} inhibits cell cycle entry and we have obtained similar data using TAT-p16^{INK4A} [30]. The herpes simplex virus type I (HSV-1) VP22 protein also acts as a PTD [115]. It is a 38 000 molecular weight protein that crosses into the cytoplasm but only translocates to the nucleus when cells are dividing. This should be a consideration when choosing VP22 to deliver cell cycle regulatory proteins, particularly those that act in the nucleus. Other transduction domains have been identified in heat shock protein hsp70 [116], gelanin/melitin fusion proteins [117] and signal peptide-based sequences [118].

Protocol 4.16: TAT-protein transduction

Protocols for the production of TAT-fusion proteins have been published by Dr Stephen Dowdy, Howard Hughes Medical Institute, St Louis, USA [110, 119–123]. Following are methods that we have used to produce active TAT fusion proteins [30, 112, 113].

EQUIPMENT, MATERIALS AND REAGENTS

Class II cabinet and containment facilities for bacteria.

Gel electrophoresis equipment.

Temperature-controlled bacterial shaker.

Class II cabinet and containment facilities for eukaryotic cells and cell lines.

Temperature-controlled CO_2 incubator for culturing eukaryotic cells.

Probe sonicator.

PD10 columns (AP Biotech).

SlideAlyzer (Pierce) or Spectrapore dialysis membrane.

Ultra-filtration apparatus (e.g. Vivaspin 6).

Buffer Z: 8 M urea, 100 mM NaCl, 20 mM Hepes pH 8.0.

1 M imidazole in buffer Z.

Ni^{2+} chromatography media: Qiagen Ni NTA superflow and AP biotech hi trap.

Chelating resins.

Restriction enzymes and general molecular biology reagents.

PBS, 5 mM EDTA.

Bradford assay reagents.

Coomassie blue or equivalent.

METHOD – CLONING FUSION PROTEINS; EXPRESSING TAT FUSION PROTEINS

1. Grow a 200 ml culture of the appropriate expressing clone overnight.

2. Inoculate 1 litre cultures of either Lennox broth (Sigma L3022) or terrific broth (Sigma T9179) supplemented with 100 µg ml^{-1} ampicillin and grow for 6 hours.

3. Harvest the cells by centrifugation and lyse by sonication in a buffer containing either 8 M urea (Buffer Z) or 6 M guanidine HCl. Denaturing conditions are usually required in order to solublise TAT fusion proteins,

which are mostly deposited into inclusion bodies in the bacteria. Denaturing conditions also ensure that the proteins stay soluble throughout subsequent purification steps. However, active TAT-β-galactosidase and TAT-dn-cdk2 are both made by sonication in PBS in the presence of protease inhibitors.

4. Clear the lysate by centrifugation at 12 000 g for 10 min. Dilute with more lysis buffer if viscous.

5. Purify the TAT-fusion protein using immobilised Ni^{2+} chromatography media (Qiagen Ni NTA superflow and AP biotech hi trap chelating). Briefly, apply the cleared lysate to a 5–10 ml column equilibrated in lysis buffer with 5–20 mM imidazole and wash in the same buffer.

6. Elute with a step gradient of imidazole in lysis buffer (100–1000 mM).

7. Identify the peak fractions by using a semi-quantitative Bradford assay and SDS-PAGE followed by staining with Coomassie blue and/or western blotting.

Desalting can be achieved by one of three methods.

8a. Desalting is easily performed by dialysing the pooled fractions against a buffer such as PBS, 5 mM EDTA. This method often results in a large amount of precipitation as the denaturant is removed. However, after centrifugation a portion of the original material stays in solution and is usable.

8b. An alternative method in our hands is the use of PD10 gel filtration columns (AP Biotech). These are capable of desalting up to 2.5 ml of protein solution on one column. The column can be equilibrated with the buffer of choice (either PBS or tissue culture media) and, because small volumes are used, protease inhibitors can also be incorporated. We add 10–20% glycerol to the equilibration buffer, which not only allows proteins to be frozen directly after gel filtration but also helps to stabilise TAT fusions and reduce precipitation. Following gel filtration all fractions must be centrifuged to remove precipitated protein (which is often present) prior to quantitation. Removal of precipitated protein quickly after elution from the PD10 column may also prevent further precipitation by removing aggregates on which precipitation occurs.

8c. The protein can be bound to an ion exchange resin (MonoQ or MonoS) followed by washing in a non-denaturing buffer and elution by increasing the salt concentration to 1M NaCl. The pH should also be optimised. The NaCl in the elution buffer should then be removed by gel filtration.

Since the ultimate aim of these expression and purification protocols is to produce a protein which can be added to cells in culture in order to modulate the cell cycle machinery, then the final preparation of protein should be sterile. This is achieved either by filter sterilisation of the final fusion protein preparation or by using sterile buffers and working in a class II cabinet during the final desalting stage. Quantify the amount of protein present by performing a quantitative Bradford assay using a standard such as BSA. Add 10–20% glycerol to the remainder and store in aliquots at −70°C.

4.7 References

1. Tiwari, S., Jamal, R. and Thomas, N.S.B. (1996) In: *Protein Phosphorylation in Cell Growth Regulation* (ed. M.J. Clemens). Academic Press, London, pp. 255–282.

2. Kitagawa, M., Higashi, H., Jung, H.-K., Suzuki-Takahashi, I., Ikeda, M., Tamai, K., Kato, J., Segawa, K., Yoshida, E., Nishimura, S. and Taya, Y. (1996) The consensus motif for phosphorylation by cyclin D1-Cdk4 is different from that for phosphorylation by cyclin A/E-Cdk2. *EMBO J.* **15**: 7060–7069.

3. Hansen, K., Farkas, T., Lukas, J., Holm, K., Ronnstrand, L. and Bartek, J. (2001) Phosphorylation-dependent and -independent functions of p130 co-operate to evoke a sustained G1 block. *EMBO J.* **20**: 422–432.

4. Lam, E.W.-F. and LaThangue, N.B. (1994) DP and E2F proteins: co-ordinating transcription with cell cycle progression. *Curr. Opin. Cell Biol.* **6**: 859–866.

5. DePinho, R.A. (1998) Transcriptional repression. The cancer-chromatin connection. *Nature* **391**: 533–536.

6. Chan, H.M., Krstic-Demonacos, M., Smith, L., Demonacos, C. and La Thangue, N.B. (2001) Acetylation control of the retinoblastoma tumour-suppressor protein. *Nat. Cell Biol.* **3**: 667–674.

7. LaThangue, N.B. (1994) DP and E2F proteins: components of a heterodimeric transcription factor implicated in cell cycle control. *Curr. Biol.* **6**: 443–450.

8. Peeper, D.S. and Bernards, R. (1997) Communication between the extracellular environment, cytoplasmic signalling cascades and the nuclear cell-cycle machinery. *FEBS Lett.* **410**: 11–16.

9. Nevins, J., Jakoi, L. and Leone, G. (1997) Functional analysis of E2F transcription factor. *Methods Enzymol.* **283**: 205–219.

10. Stiewe, T. and Putzer, B.M. (2000) Role of the p53-homologue p73 in E2F1-induced apoptosis. *Nat. Genet.* **26**: 464–469.

11. Krek, W., Xu, G. and Livingston, D. M. (1995) Cyclin A-kinase regulation of E2F-1 DNA binding function underlies suppression of an S phase checkpoint. *Cell* **83**: 1149–1158.

12. Williams, C.D., Watts, M., Linch, D.C. and Thomas, N.S.B. (1997) Characterization of cell cycle status and E2F complexes in mobilized CD34+ cells before and after cytokine stimulation. *Blood* **90**: 194–203.

13. Ringrose, L. and Paro, R. (2001) Gene regulation. Cycling silence. *Nature* **412**: 493–494.

14. Taya, Y. (1997) RB kinases and RB-binding proteins: new points of view. *Trends Biochem. Sci.* **22**: 14–17.

15. van Stipdonk, M.J.B., Lemmens, E.E. and Schoenberger, S.P. (2001) Naïve CTLs require a single brief period of antigenic stimulation for clonal expansion and differentiation. *Nat. Immunol.* **2**: 423–429.

16. Ekholm, S.V., Zickert, P., Reed, S.I. and Zetterberg, A. (2001) Accumulation of cyclin E is not a prerequisite for passage through the restriction point. *Mol. Cell Biol.* **21**: 3256–3265.

17. Evan, G.I., Wyllie, A.H., Gilbert, C.S., Littlewood, T.D., Land, H., Brooks, M., Waters, M., Penn, L.Z. and Hancock, C. (1992) Induction of apoptosis in fibroblasts by c-myc protein. *Cell* **69**: 119–128.

18. Draviam, V.M., Orrechia, S., Lowe, M., Pardi, R. and Pines, J. (2001) The localization of human cyclins B1 and B2 determines CDK1 substrate specificity and neither enzyme requires MEK to disassemble the Golgi apparatus. *Cell Biol.* **152**: 945–958.

19. Ng, T., Squire, A., Hansra, G., Bornancin, F., Prevostel, C., Hanby, A., Harris, W., Barnes, D., Schmidt, S., Mellor, H., Bastiaens, P.I. and Parker, P.J. (1999) Imaging protein kinase C alpha activation in cells. *Science* **283**: 2085–2089.

20. Mills, A.D., Coleman, N., Morris, L.S. and Laskey, R.A. (2000) Detection of S-phase cells in tissue sections by in situ DNA replication. *Nat. Cell Biol.* **2**: 244–245.

21. Darzynkiewicz, Z., Bedner, E., Li, X., Gorczyca, W. and Melamed, M.R. (1999) Laser-scanning cytometry: A new instrumentation with many applications. *Exp. Cell Res.* **249**: 1–12.

22. Gong, J., Traganos, F. and Darzynkiewicz, Z. (1995) Discrimination of G2 and mitotic cells by flow cytometry based on different expression of cyclins A and B1. *Exp. Cell Res.* **220**: 226–231.

23. Burke, L.C., Bybee, A. and Thomas, N.S. (1992) The retinoblastoma protein is partially phosphorylated during early G1 in cycling cells but not in G1 cells arrested with alpha-interferon. *Oncogene* **7**: 783–788.

24. Thomas, N.S., Pizzey, A.R., Tiwari, S., Williams, C.D. and Yang, J. (1998) p130, p107, and pRB are differentially regulated in proliferating cells and during cell cycle arrest by alpha-interferon. *J. Biol. Chem.* **273**: 23659–23667.

25. Gothot, A., van der Loo, J.C., Clapp, D.W. and Srour, E.F. (1998) Cell cycle-related changes in repopulating capacity of human mobilized peripheral blood CD34(+) cells in non-obese diabetic/severe combined immune-deficient mice. *Blood* **92**: 2641–2649.

26. Datar, S.A., Jacobs, H.W., de La Cruz, A.F., Lehner, C.F. and Edgar, B.A. (2000) The Drosophila cyclin D–cdk4 complex promotes cellular growth. *EMBO J.* **19**: 4543–4554.

27. Meyer, C.A., Jacobs, H.W., Datar, S.A., Du, W., Edgar, B.A. and Lehner, C.F. (2000) Drosophila cdk4 is required for normal growth and is dispensable for cell cycle progression. *EMBO J.* **19**: 4533–4542.

28. Rane, S.G., Dubus, P., Mettus, R.V., Galbreath, E.J., Boden, G., Reddy, E.P. and Barbacid, M. (1999) Loss of cdk4 expression causes insulin-deficient diabetes and cdk4 activation results in β-islet cell hyperplasia. *Nat. Genet.* **22**: 44–52.

29. Tsutsui, T., Hesabi, B., Moons, D.S., Pandolfi, P.P., Hansel, K.S., Koff, A. and Kiyokawa, H. (1999) Targeted disruption of CDK4 delays cell cycle entry with enhanced p27(Kip1) activity. *Mol. Cell Biol.* **19**: 7011–7019.

30. Lea, N.C., Orr, S.J., Stoeber, K., Williams, G.H., Ibrahim, M.A.A., Mufti, G.J. and Thomas, N.S.B. (2003) Commitment point during $G_0 \rightarrow G_1$ that controls entry into the cell cycle. *Mol. Cell Biol.* (in press).

31. Morris, G.F. and Mathews, M.B. (1989) Regulation of proliferating cell nuclear antigen during the cell cycle. *J. Biol. Chem.* **264**: 13856–13864.

32. Zetterberg, A. (1997) Cell growth and cell cycle progression in mammalian cells. In: *Apoptosis and Cell Cycle Control in Cancer* (ed. N.S.B. Thomas). BIOS Scientific Publishers Ltd., Oxford, pp. 17–36.

33. Mao, X., Green, J.M., Safer, B., Lindsten, T., Frederickson, R.M., Miyamoto, S., Sonenberg, N. and Thompson, C.B. (1992) Regulation of translation initiation factor gene expression during human T cell activation. *J. Biol. Chem.* **267**: 20444–20450.

34. Smith, E.J., Leone, G., DeGregori, J., Jakoi, L. and Nevins, J.R. (1996) The accumulation of an E2F-p130 transcriptional repressor distinguishes a G_0 cell state from a G_1 cell state. *Mol. Cell Biol.* **16**: 6965–6976.

35. Ogawa, H., Ishiguro, K., Gaubatz, S., Livingston, D.M. and Nakatani, Y. (2002) A complex with chromatin modifiers that occupies E2F- and Myc-responsive genes in G_0 cells. *Science* **296**: 1132–1136.

36. Ezhevsky, S.A., Ho, A., Becker-Hapak, M., Davis, P.K. and Dowdy, S.F. (2001) Differential regulation of retinoblastoma tumor suppressor protein by G(1) cyclin-dependent kinase complexes in vivo. *Mol. Cell Biol.* **21**: 4773–4784.

37. Stoeber, K., Tlsty, T.D., Happerfield, L., Thomas, G.A., Romanov, S., Bobrow, L., Williams, E.D. and Williams. G.H. (2001) DNA replication licensing and human cell proliferation. *J. Cell Sci.* **114**: 2027–2041.

38. Meng, M.V., Grossfeld, G.D., Williams, G.H., Dilworth, S., Stoeber, K., Mulley, T.W., Weinberg, V., Carroll, P.R. and Tlsty, T.D. (2001) Minichromosome maintenance protein 2 expression in prostate: characterization and association with outcome after therapy for cancer. *Clin. Cancer Res.* **7**: 2712–2718.

39. Stoeber, K., Halsall, I., Freeman, A., Swinn, R., Doble, A., Morris, L., Coleman, N., Bullock, N., Laskey, R.A., Hales, C.N. and Williams, G.H. (1999) Immunoassay for urothelial cancers that detects DNA replication protein Mcm5 in urine. *Lancet* **354**: 1524–1525.

40. Freeman, A., Morris, L.S., Mills, A.D., Stoeber, K., Laskey, R.A., Williams, G.H. and Coleman, N. (1999) Minichromosome maintenance proteins as biological markers of dysplasia and malignancy. *Clin. Cancer Res.* **5**: 2121–2132.

41. Williams, G.H., Romanowski, P., Morris, L., Madine, M., Mills, A.D., Stoeber, K., Marr, J., Laskey, R.A. and Coleman, N. (1998) Improved cervical smear assessment using antibodies against proteins that regulate DNA replication. *Proc. Natl. Acad. Sci. USA* **95**: 14932–14937.

42. Blow, J.J. and Hodgson, B. (2002) Replication licensing – defining the proliferative state? *Trends Cell Biol.* **1**: 72–78.

43. Dignam, J.D., Lebovitz, R.M. and Roeder, R.G. (1983) Accurate transcription initiation by RNA polymerase II in a soluble extract from isolated mammalian nuclei. *Nucleic Acids Res.* **11**: 1475–1489.

44. Zeterberg, A., Larsson, O. and Wilman, K.G. (1995) What is the restriction point? *Curr. Opin. Cell Biol.* **7**: 835–842.

45. Ekholm, S.V., Zickert, P., Reed, S.I. and Zetterberg, A. (2001) Accumulation of cyclin E is not a prerequisite for passage through the restriction point. *Mol. Cell Biol.* **21**: 3256–3265.

46. Fagan, R., Flint, K.J. and Jones, N. (1994) Phosphorylation of E2F-1 modulates its interaction with the retinoblastoma gene product and the adenoviral E4 19 kDa protein. *Cell* **78**: 799–811.

47. Vario, G., Livingston, D.M. and Ginsberg, D. (1995) Functional interaction between E2F-4 and p130: evidence for distinct mechanisms underlying growth suppression by different retinoblastoma protein family members. *Genes Dev.* **9**: 869–881.

48. Bandara, R.L., Lam, E.W., Sorensen, T.S., Zamanian, M., Girling, R. and LaThangue, N.B. (1994) DP-1: a cell cycle-regulated and phosphorylated component of transcription factor DRTF1/E2F which is functionally important for recognition by pRB and the adenovirus E4 orf 6/7 protein. *EMBO J.* **13**: 3104–3114.

49. Gu, Y., Rosenblatt, J. and Moegan, D.O. (1992) Cell cycle regulation of CDK2 activity by phosphorylation of Thr160 and Tyr15. *EMBO J.* **11**: 3995–4005.

50. Williams, C.D., Linch, D.C., Sorensen, T.S., LaThangue, N.B. and Thomas, N.S.B. (1997) The predominant E2F complex in human primary haemopoietic cells and in AML blasts contains E2F-4, DP-1 and p130. *Br. J. Haematol.* **96**: 688–696.

51. Roberts, R.J., Khwaja, A., Lie, A.K.W., Bybee, A., Yong, K., Thomas, N.S.B. and Linch, D.C. (1994) Differentiation-linked changes in tyrosine phosphorylation, functional activity, and gene expression downstream from the granulocyte-macrophage colony-stimulating factor receptor. *Blood* **84**: 1064–1073.

52. Laemmli, U.K. (1970) Cleavage of structural proteins during the assembly of the head of bacteriophage T4. *Nature* **227**: 681–683.

53. Harlow, E. and Lane, D. (1988) *Antibodies: A Laboratory Manual*. CSH Press, New York.

54. O'Farrell, P.H. (1975) High resolution two-dimensional electrophoresis of proteins. *J. Biol. Chem.* **250**: 4007–4021.

55. Gorg, A., Postel, W. and Gunther, S. (1988) The current state of two-dimensional electrophoresis with immobilized pH gradients. *Electrophoresis* **9**: 531–546.

56. Roberts, S.B., Segil, N. and Heintz, N. (1991) Mitotic phosphorylation of the Oct-1 homeodomain and regulation of Oct-1 DNA binding activity. *Science* **253**: 1022–1026.

57. Ezhevsky, S.A., Ho, A., Becker-Hapak, M., Davis, P.K. and Dowdy, S.F. (2001) Differential regulation of retinoblastoma tumor suppressor protein by G(1) cyclin-dependent kinase complexes in vivo. *Mol. Cell Biol.* **21**: 4773–4784.

58. Rovera, G., Magarian, C. and Borun, T.W. (1978) Resolution of hemoglobin subunits by electrophoresis in acid urea polyacrylamide gels containing Triton X-100. *Anal. Biochem.* **85**: 506–518.

59. Cannon, M., Schindler, D. and Davies, J. (1977) Methylation of proteins in 60S ribosomal subunits from *Saccharomyces cerevisiae*. *FEBS Lett.* **75**: 187–191.

60. Burke, L., Bybee, A. and Thomas, N.S.B. (1992) The retinoblastoma protein is partially phosphorylated during early G1 in cycling cells but not in G1 cells arrested with alpha-interferon. *Oncogene* **7**: 783–788.

61. Cohen, P.T.W. and Cohen, P. (1989) Discovery of a protein phosphatase activity encoded in the genome of bacteriophage lambda. Probable identity with open reading frame 221. *Biochem. J.* **260**: 931–934.

62. Zhuo, S., Clemens, J.C., Hakes, D.J., Barford, D., Dixon, J.E. (1993) Expression, purification, crystallization, and biochemical characterization of a recombinant protein phosphatase. *J. Biol. Chem.* **268**: 17754–17761.

63. Cockcroft, S. (1984) Ca^{2+}-dependent conversion of phosphatidylinositol to phosphatidate in neutrophils stimulated with fMet-Leu-Phe or ionophore A23187. *Biochim. Biophys. Acta.* **795**: 37–46.

64. Thomas, N.S.B. (1989) Regulation of the product of a possible human cell cycle control gene CDC2Hs in B-cells by alpha-interferon and phorbol ester. *J. Biol. Chem.* **264**: 13697–13700.

65. Grenfell, S.J, Latchman, D.S. and Thomas, N.S. (1996) Oct-1 and Oct-2 DNA-binding site specificity is regulated in vitro by different kinases. *Biochem. J.* **315**: 889–893.

66. Kyriakis, J.M, Banerjee, P., Nikolakaki, E., Dai, T., Rubie, E.A., Ahmad, M.F., Avruch, J. and Woodgett, J.R. (1994) The stress-activated protein kinase subfamily of c-Jun kinases. *Nature* **369**: 156–160.

67. Schagger, H. and von Jagow, G. (1987) Tricine-sodium dodecyl sulfate-polyacrylamide gel electrophoresis for the separation of proteins in the range from 1 to 100 kDa. *Anal. Biochem.* **166**: 368–379.

68. Alberta, J.A. and Stiles, C.D. (1997) Phosphorylation-directed antibodies in high-flux screens for compounds that modulate signal transduction. *BioTechniques* **23**: 490–493.

69. Ginty, D.D., Kornhauser, J.M., Thompson, M.A., Bading, H., Mayo, K.E., Takahashi, J.S. and Greenberg, M.E. (1993) Regulation of CREB phosphorylation in the suprachiasmatic nucleus by light and a circadian clock. *Science* **260**: 238–241.

70. Hagiwara, M., Brindle, P., Harootunian, A., Armstrong, R., Rivier, J., Vale, W., Tsien, R. and Montminy, M.R. (1993) Coupling of hormonal stimulation and transcription via the cyclic AMP-responsive factor CREB is rate limited by nuclear entry of protein kinase A. *Mol. Cell Biol.* **13**: 4852–4859.

71. Shieh, S.Y., Ikeda, M., Taya, Y. and Prives, C. (1997) DNA damage-induced phosphorylation of p53 alleviates inhibition by MDM2. *Cell* **91**: 325–334.

72. Verma, R., Annan, R.S., Huddleston, M.J., Carr, S.A., Reynard, G., Deshaies, R.J. (1997) Phosphorylation of Sic1p by G1 Cdk required for its degradation and entry into S phase. *Science* **278**: 455–460.

73. Annan, R.S. and Carr, S.A. (1997) The essential role of mass spectrometry in characterizing protein structure: mapping posttranslational modifications. *J. Protein Chem.* **16**: 391–402.

74. Carr, S.A., Huddleston, M.J. and Annan, R.S. (1996) Selective detection and sequencing of phosphopeptides at the femtomole level by mass spectrometry. *Anal. Biochem.* **239**: 180–192.

75. Wilm, M. and Mann, M. (1996) Mass spectrometric sequencing of proteins silverstained polyacrylamide gels. *Anal. Chem.* **68**: 1–8.

76. Holt, G.D., Haltiwanger, R.S., Torres, C.-R. and Hart, G.W. (1987) Erythrocytes contain cytoplasmic glycoproteins. O-linked GlcNAc on Band 4.1 *J. Biol. Chem.* **262**: 14847–14850.

77. Hart, G.W. (1997) Dynamic O-linked glycosylation of nuclear and cytoskeletal proteins. *Annu. Rev. Biochem.* **66**: 315–335.

78. Chou, T.Y., Hart, G.W. and Dang, C.V. (1995) c-Myc is glycosylated at threonine 58, a known phosphorylation site and a mutational hot spot in lymphomas. *J. Biol. Chem.* **270**: 18961–18965.

79. Kearse, K.P. and Hart, G.W. (1991) Lymphocyte activation induces rapid changes in nuclear and cytoplasmic glycoproteins. *Proc. Natl. Acad. Sci. USA* **88**: 1701–1705.

80. Jackson, S.P. and Tjian, R. (1988) O-glycosylation of eukaryotic transcription factors: implications for mechanisms of transcriptional regulation. *Cell* **55**: 125–133.

81. Reason, A.J., Morris, H.R., Pacino, M., Marais, R., Treisman, R.H., Haltiwanger, R.S., Hart, G.W., Kelly, W.G. and Dell, A. (1992) Localization of O-GlcNAc modification on the serum response transcription factor. *J. Biol. Chem.* **267**: 16911–16921.

82. Shaw, P., Freeman, J., Bovey, R. and Iggo, R. (1996) Regulation of specific DNA binding by p53: evidence for a role for O-glycosylation and charged residues at the carboxy-terminus. *Oncogene* **12**: 921–930.

83. Amati, B. and Land, H. (1994) Myc-Max-Mad: a transcription factor network controlling cell cycle progression, differentiation and death. *Curr. Opin. Genet. Dev.* **4**: 102–108.

84. Torres, C.-R. and Hart, G.W. (1984) Topography and polypeptide distribution of terminal N-acetylglucosamine residues on the surfaces of intact lymphocytes. Evidence for O-linked GlcNAc. *J. Biol. Chem.* **259**: 3308–3317.

85. Hart, G.W., Haltiwanger, R.S., Holt, G.D. and Kelly, W.G. (1989) Glycosylation in the nucleus and cytoplasm. *Annu. Rev. Biochem.* **58**: 841–874.

86. Jackson, S.P. and Tjian, R. (1989) Purification and analysis of RNA polymerase II transcription factors by using wheat germ agglutinin affinity chromatography. *Proc. Natl. Acad. Sci. USA* **86**: 1781–1785.

87. Snow, C.M., Senior, A. and Gerace, L. (1987) Monoclonal antibodies identify a group of nuclear pore complex glycoproteins. *J. Cell Biol.* **104**: 1143–1156.

88. Holt, G.D. and Hart, G.W. (1986) The subcellular distribution of terminal N-acetyl-glucosamine moieties. Localization of a novel protein-saccharide linkage, O-linked GlcNAc. *J. Biol. Chem.* **261**: 8049–8057.

89. Roquemore, E.P., Chou, T-Y. and Hart, G.W. (1994) Detection of O-linked N-acetyl-glucosamine (O-GlcNAc) on cytoplasmic and nuclear proteins. *Methods Enzymol.* **230**: 443–460.

90. Haltiwanger, R.S. and Philipsberg, G.A. (1997) Mitotic arrest with nocodazole induces selective changes in the level of O-linked N-acetylglucosamine and accumulation of incompletely processed N-glycans on proteins from HT29 cells. *J. Biol. Chem.* **272**: 8752–8758.

91. Lin, A.I., Philipsberg, G.A. and Haltiwanger, R.S. (1994) Core fucosylation of high-mannose-type oligosaccharides in GlcNAc transferase I-deficient (Lec1) CHO cells. *Glycobiology* **4**: 895–901.

92. Gribben, J.G., Devereux, S., Thomas, N.S.B., Keim, M., Jones, H.M., Goldstone, A.H. and Linch, D.C. (1990) Development of antibodies to unprotected glycosylation sites on recombinant human GM-CSF. *Lancet* **335**: 434–437.

93. Roquemore, E.P., Chevrier, M.R., Cotter, R.J. and Hart, G.W. (1996) Dynamic O-GlcNAcylation of the small heat shock protein alpha B-crystallin. *Biochemistry* **35**: 3578–3586.

94. Kouzarides, T. (1999) Histone acetylases and deacetylases in cell proliferation. *Curr. Opin. Genet. Dev.* **9**: 40–48.

95. Lai, A., Kennedy, B.K., Barbie, D.A., Bertos, N.R., Yang, X.J., Theberge, M.C., Tsai, S.C., Seto, E., Zhang, Y., Kuzmichev, A., Lane, W.S., Reinberg, D., Harlow, E. and Branton, P.E. (2001) RBP1 recruits the mSIN3-histone deacetylase complex to the pocket of retinoblastoma tumor suppressor family proteins found in limited discrete regions of the nucleus at growth arrest. *Mol. Cell Biol.* **21**: 2918–2932.

96. Kouzarides, T. (2000) Acetylation: a regulatory modification to rival phosphorylation? *EMBO J.* **19**: 1176–1179.

97. Martinez-Balbas, M.A., Bauer, U.M., Nielsen, S.J., Brehm, A. and Kouzarides, T. (2000) Regulation of E2F1 activity by acetylation. *EMBO J.* **19**: 662–671.

98. Marzio, G., Wagener, C., Gutierrez, M.I., Cartwright, P., Helin, K. and Giacca, M. (2000) E2F family members are differentially regulated by reversible acetylation. *J. Biol. Chem.* **275**: 10887–10892.

99. Chan, H.M., Krstic-Demonacos, M., Smith, L., Demonacos, C. and La Thangue, N.B. (2001) Acetylation control of the retinoblastoma tumour-suppressor protein. *Cell Biol.* **3**: 667–764.

100. Gu, W. and Roeder, RG. (1997) Activation of p53 sequence-specific DNA binding by acetylation of the p53 C-terminal domain. *Cell* **90**: 595–606.

101. Luo, J., Nikolaev, A.Y., Imai, S., Chen, D., Su, F., Shiloh, A., Guarente, L. and Gu, W. (2001) Negative control of p53 by Sir2alpha promotes cell survival under stress. *Cell* **107**: 137–148.

102. Vaziri, H., Dessain, S.K., Eaton, E.N., Imai, S.I., Frye, R.A., Pandita, T.K., Guarente, L. and Weinberg, R.A. (2001) hSIR2(SIRT1) functions as an NAD-dependent p53 deacetylase. *Cell* **107**: 149–159.

103. Pearson, M., Carbone, R., Sebastiani, C., Cioce, M., Fagioli, M., Saito, S., Higashimoto, Y., Appella, E., Minucci, S., Pandolfi, P.P. and Pelicci, P.G. (2000) PML regulates p53 acetylation and premature senescence induced by oncogenic Ras. *Nature* **406**: 207–210.

104. Harlow, E. and Lane, D. (1988) *Antibodies: A Laboratory Manual.* Cold Spring Harbor Laboratory Press, Cold Spring Harbor, NY.

105. Ogryzko, V.V., Schiltz, R.L., Russanova,V., Howard, B.H. and Nakatani, Y. (1996) The transcriptional coactivators p300 and CBP are histone acetyltransferases. *Cell* **87**: 953–959.

106. Boyle, W.J., van der Geer, P. and Hunter, T. (1991) Phosphopeptide mapping and phosphoamino acid analysis by 2-D separation on thin-layer cellulose plates. *Enzymology* **201**: 110–149.

107. Green, M. and Loewnstein, P.M. (1988) Autonomous functional domains of chemically synthesized human immunodeficiency virus Tat trans-activator protein. *Cell* **55**: 1179–1188.

108. Frankel, A.D. and Pabo, C.O. (1988) Cellular uptake of the Tat protein from human immunodeficiency virus. *Cell* **55**: 1189–1193.

109. Fawell, S., Seery, J., Daikh, Y., Moore, C., Chen, L., Pepinsky, B. and Barsoum, J. (1994) Tat-mediated delivery of heterologous proteins into cells. *Proc. Natl Acad. Sci. USA* **91**: 664–668.

110. Nagahara, H., Vocero-Akbani, A.M., Snyder, E.L., Ho, A., Latham, D.G., Lissy, N.A., Becker-Hapak, M., Ezhevsky, S.A. and Dowdy, S.F. (1998) Transduction of full-length TAT fusion proteins into mammalian cells: TAT-p27Kip1 induces cell migration. *Nat. Med.* **4**: 1449–1452.

111. Schwarze, S., Ho, A., Vocero-Akbani, A. and Dowdy, S.F. (1999) In vivo protein transduction: delivery of a biologically active protein into the mouse. *Science* **285**: 1569–1572.

112. Parada, Y., Banerji, L., Glassford, J., Lea, N.C., Collado, M., Rivas, C., Lewis, J.L., Gordon, M.Y., Thomas, N.S.B. and Lam, E.W.-F. (2001) BCR-ABL and IL3 promotes haematopoietic cell proliferation and survival through modulation of cyclin D2 and p27^{Kip1} expression. *J. Biol. Chem.* **276**: 23572–23580.

113. Banerji, L., Glassford, J., Lea, N.C., Thomas, N.S.B., Klaus, G.B. and Lam, E.W.-F. (2001) BCR signals target p27^{Kip1} and cyclin D2 via the PI3-K signalling pathway to mediate cell cycle arrest and apoptosis of WEHI 231 B cells. *Oncogene* **20**: 7352–7367.

114. Derossi, D., Joliot, A.H., Chassaing, G. and Prochiantz, A. (1994) The third helix of the Antennapedia homeodomain translocates through biological membranes. *J. Biol. Chem.* **269**: 10444–10450.

115. Elliott, G. and O'Hare, P. (2000) Cytoplasm-to-nucleus translocation of a herpes virus tegument protein during cell division. *J. Virol.* **74**: 2131–2141.

116. Fujihara, S.M. and Nadler, S.G. (1999) Intranuclear targeted delivery of functional NF-kappa B by 70 kDa heat shock protein. *EMBO J.* **18**: 411–419.

117. Pooga, M., Hallbrink, M., Zorko, M. and Langel, U. (1998) Cell penetration by transportan. *FASEB J.* **12**: 67–77.

118. Hawiger, J. (1999) Noninvasive intracellular delivery of functional peptides and proteins. *Curr. Opin. Chem. Biol.* **3**: 89–94.

119. Wender, P.A., Mitchell, D.J., Pattabiraman, K., Pelkey, E.T., Steinman, L. and Rothbard, J.B. (2000) The design, synthesis, and evaluation of molecules that enable or enhance cellular uptake: peptoid molecular transporters. *Proc. Natl. Acad. Sci. USA* **97**: 13003–13008.

120. Becker-Hapak, M., McAllister, S.S., Dowdy, S.F. (2001) TAT-mediated protein transduction into mammalian cells. *Methods* **24**: 247–256.

121. Vocero-Akbani, A., Chellaiah, M.A., Hruska, K.A. and Dowdy, S.F. (2001) Protein transduction: delivery of Tat-GTPase fusion proteins into mammalian cells. *Methods Enzymol.* **332**: 36–49.

122. Vocero-Akbani, A., Lissy, N.A. and Dowdy, S.F. (2000) Transduction of full-length Tat fusion proteins directly into mammalian cells: analysis of T cell activation-induced cell death. *Methods Enzymol.* **322**: 508–521.

123. Schwarze, S.R. and Dowdy, S.F. (2000) In vivo protein transduction: intracellular delivery of biologically active proteins, compounds and DNA. *Trends Pharmacol. Sci.* **21**: 2145–2148.

Telomeres and the Control of Cell Division

5

Predrag Slijepcevic

Contents

5.1 Introduction

Mammalian cells have developed complex mechanisms to ensure equal distribution of the genetic material during cell division. For example, the number of chromosomes in dividing somatic cells of a given species is always constant. Key functional elements of chromosomes that must be strictly controlled to maintain species-specific chromosome numbers are centromeres and telomeres [1]. This review focuses on the mechanisms of telomere maintenance in mammalian cells and the methodology used to study telomere maintenance.

As their name implies, telomeres (*telos*, end; *meros*, part) are physical ends of chromosomes. The function of telomeres is to maintain chromosome stability and integrity. Pioneering experiments of H.J. Muller on Drosophila and B. McClintock on maize indicated that chromosome ends behave differently from the rest of the chromosome and that broken chromosomes lose their stability [2, 3]. McClintock realised that chromosome stability may be recovered if broken chromosomes acquire new telomeres [3]. These early observations paved the way for detailed molecular studies of telomeres. In 1978 the first telomeric DNA sequence was identified in *Tetrahymena thermophila* [4]. Several years later the vertebrate telomeric DNA sequence was identified [5]. In addition, it became clear that telomeric DNA sequences are synthesised by a unique enzyme termed telomerase [6, 7]. Telomerase is a reverse transcriptase

Cell Proliferation and Apoptosis, David Hughes and Huseyin Mehmet (Eds)
© 2003 BIOS Scientific Publishers Ltd, Oxford

that uses its own RNA template to synthesise telomeric DNA [8]. Further insights into the structure and function of telomeres revealed links between telomere maintenance and many important cellular processes including the control of cell division [9], DNA repair mechanisms [10], spatial organisation of interphase nuclei [11, 12] and recombination of meiotic chromosomes [13]. Telomeres are also implicated in processes of cellular senescence and carcinogenesis. Numerous experiments carried out on human primary somatic cells revealed telomere shortening that is always proportional to the cell replicative history. This led to the proposal of the telomere hypothesis of cellular ageing [14, 15] which has been revised to take account of recent developments in telomere biology [16]. In addition, about 10 years ago it became clear that telomerase, which is usually inactive in normal human somatic cells, becomes active in cancer cells [17, 18]. This raises the possibility that telomerase activity is a prerequisite for indefinite growth of cancer cells and at the same time opens new avenues for cancer treatment [19]. It has been thought that inactivation of telomerase may permanently arrest growth of cancer cells [20]. However, it is important to emphasise that telomerase is not the only mechanism for telomere maintenance and there is a strong body of evidence suggesting that telomerase-independent mechanisms, collectively termed ALT (alternative lengthening of telomeres), may be operational in human cells that lack telomerase activity [21, 22]. This review examines the importance of telomeres in cellular processes such as regulation of cell division and response to DNA damage.

5.2   Mechanisms of telomere maintenance

Telomeres consist of conserved, species-specific DNA sequences and proteins that bind these sequences known as telomere binding proteins (TBP) [1]. Telomeric DNA sequence in all vertebrates is the same: $(TTAGGG)_n$ [23]. Telomeric DNA together with TBPs forms a unique structure at the chromosome end, the function of which is to preserve chromosomes as single entities. Since each species is characterised by a specific chromosome complement, telomeres are important for preserving karyotypic individuality of eukaryotic species and therefore they are unique structures different from the rest of the chromosome. What makes telomeres unique? In contrast to the rest of the chromosome, telomeres are dynamic structures in which DNA content is not fixed but fluctuates considerably in the cell cycle. This telomere dynamics is mediated by several mechanisms. Firstly, since conventional DNA polymerases cannot replicate the end of the chromosome fully [1], a small amount of telomeric DNA is lost after each cell division. Secondly, telomeric DNA may be lost as a result of the activity of specialised enzymes, exonucleases, again in a cell cycle-dependent manner [24]. The function of exonucleases is to generate 3' overhangs at telomeres, which in turn attract specific enzymes or TBPs [25–29]. The loss of telomeric DNA is compensated by at least two mechanisms: (i) telomerase, a specialised reverse transcriptase [6–8] and (ii) telomerase-independent mechanisms known collectively as ALT [20, 21]. Therefore, in normal cells the loss and gain of telomeric DNA are balanced and the overall length of telomeric DNA remains within a species-specific range (*Figure 5.1*). For example, numerous measurements of telomere length in mouse cells revealed a range between 10 and 60 kilobases [30, 31]. Similarly, telomere length is maintained within a specific range in human [32] and other studied species [33].

However, it is important to emphasise that some human somatic cells lack telomerase activity and their telomeres become progressively shorter [14, 15]. As a result of the loss of telomeric DNA, telomeres in human somatic cells eventually lose their function and can no longer preserve chromosomes as single entities. The first indication of

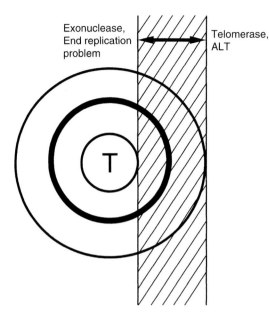

Exonuclease,
End replication
problem

Telomerase,
ALT

T

Figure 5.1

Telomere homeostasis. Concentric circles represent fluctuations in telomere size. The middle circle (thick line) is the median telomere size. Two other circles (thinner lines) represent telomere size fluctuations. The shaded area represents the range of size fluctuation. Each species has a specific range of telomere size fluctuation. Telomere size outside the species-specific range is considered abnormal. Mechanisms mediating telomere size fluctuations are represented by a line with arrows at opposite ends. For example, arrow pointing outside the circle represents growing telomere, whereas arrow pointing inside the circle represents shrinking telomere. (Abbreviation: T, telomere.)

the loss of telomere function is the appearance of dicentric chromosomes due to end-to-end chromosome fusion [17]. Therefore, it has been argued that one of the most important indicators of the loss of telomere function is telomere shortening below the normal range. Experiments in human and mouse cells strongly support this possibility. For example, telomere shortening below the normal range in human somatic cells is associated with the loss of proliferative potential [14, 15]. Mice that lack telomerase, as a result of genetic manipulation, are unable to maintain their telomeres, suffer telomere shortening and eventually their cells lose telomere function [34]. However, telomere shortening is not the only abnormality in the process of telomere maintenance. Sometimes telomeres may become abnormally long. This occurs in mouse DNA repair mutants such as *scid* (severe combined immunodeficiency) mice. In these mice telomeres are on average 1.5 to 2 times longer than normal wild type telomeres [35]. End-to-end chromosome fusions are also present in *scid* cells [36].

Since telomere length is generally maintained within a species-specific range, the mechanisms for telomere length regulation must be extremely precise to maintain this homeostasis. The principal regulators of telomere homeostasis are TBPs. Two key TBPs regulate telomere length in mammalian cells: TRF (telomere repeat factor) 1 and TRF2 [25]. Over-expression of these proteins leads to telomere shortening. This suggests that they act as negative regulators of factors that elongate telomeres, namely telomerase. When the number of TRF1 and TRF2 molecules at telomeres is reduced below the critical level, this signals telomerase to elongate telomeres [37]. Therefore, mammalian cells use a mechanism that involves 'protein counting' to regulate telomere length. A similar mechanism has been described in yeast [38]. Other mammalian telomeric proteins include tankyrase [26], TIN2 [29], hRap1 [27], TANK2 [28] and pot 1 [39]. Some of these proteins work with TRF1 and TRF2 in telomere length regulation.

Function of many cellular components is frequently governed by their molecular structure. Therefore, it is no surprise that telomere function is governed by telomere structure. Both yeast and human telomeres end with long 3′ overhangs [1]. These overhangs, which in principle do not differ from DNA double strand breaks, must be concealed from cellular mechanisms that detect DNA damage. This is achieved in a unique fashion. Examination of mammalian telomeres by electron microscopy revealed the presence of loops at telomeres which are known as T-loops [40]. This model predicts that the 3' overhang folds back and invades telomeric double-stranded DNA forming a T-loop. It has been proposed that the T-loop formation is mediated by TRF2 [40]. Other TBPs have their place within the loop which are likely to interact with TRF2 in a specific fashion.

In summary, telomere length is regulated by a dynamic interplay involving TBPs, telomerase and telomerase-independent mechanisms. The result of this interplay is telomere homeostasis within the range typical for each species (*Figure 5.1*). Telomere function, which is governed by its structure, most likely requires a specific telomere length.

5.3 Abnormalities in telomere maintenance that affect cell division progression

A prerequisite for the proper segregation of chromosomes in the cell division is the presence of functional telomeres. The presence of a chromosome lacking a telomere will cause a cell division block. Experiments in yeast have revealed that when a single telomere was experimentally removed from a chromosome this caused RAD-9-mediated cell division arrest [9]. This eventually caused the loss of the chromosome lacking a telomere [9]. However, in some cases (30%) the chromosome without a telomere was not lost and was propagated in an unstable form for 4–10 cell divisions [9]. Taken together, these results show that telomeres are essential for the stable segregation of chromosomes in the cell division and without telomeres chromosomes are eventually lost. It is very likely that the loss of a telomere will induce a DNA damage response in the cell, the purpose of which is to restore the original telomere. This is probably achieved by first blocking the cell division and then activating an appropriate DNA repair mode that will attempt to restore the original telomere. This process of new telomere formation is usually referred to as chromosome healing [41]. If the cell cannot restore the original telomere, the cell division block cannot be lifted and the alternative is chromosome loss and genomic instability.

The lack of a telomere in a chromosome is not the only abnormality that may cause problems in chromosome segregation during the cell cycle. The same effect may be induced by the presence of non-functional telomeres. As mentioned above, telomere function is mediated by a specific DNA sequence and TBPs binding at that sequence. TBPs and telomeric DNA form a specialised structure that performs telomere function. When the telomeric DNA sequence changes it may cause the loss of telomere function. In their study of telomere structure and function, Kirk *et al.* [42] mutated the telomeric DNA sequence in *Tetrahymena thermophila* by inducing point mutations in the telomerase RNA template. This caused severe cell division difficulties in dividing cells. Sister chromatids in mutant cells failed to separate and this delayed or blocked anaphase. Although sister chromatids were pulled apart as anaphase progressed, they had not been able to separate to daughter poles. These results led Kirk *et al.* [42] to propose that sister chromatid telomeres are normally associated with each other throughout the cell cycle and they are separated only in mitosis. The first cytological evidence for this scenario was provided by Lima-de-Faria and Bose in 1962 in their analysis of plant chromosomes [43]. Therefore, proper telomere function is required to mediate sister chromatid separation during anaphase.

Based on these results it is clear that progression through the cell cycle and telomere maintenance are linked. The most plausible scenario seems to be that telomere maintenance and cell cycle control are integrated into the cellular machinery for DNA

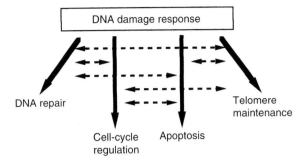

Figure 5.2

Biochemical pathways involved in DNA damage response. Thick lines with arrows at ends represent each pathway. Dashed lines with arrows at both ends represent interactions between individual pathways. In this model each pathway interacts with any other pathway. As a result, changes in one pathway are expected to affect one or more pathways.

damage response. In a recent authoritative review it has been proposed that cellular DNA damage response involves series of interacting biochemical pathways [44]. These pathways may be classified into three larger categories: (i) DNA repair; (ii) cell cycle checkpoints and (iii) apoptosis. Given the likelihood that all these interacting pathways, including the cell cycle, may be affected if telomeres are lost or nonfunctional [9, 42] it seems reasonable to argue that telomere maintenance may be added as a fourth element in DNA damage response (see *Figure 5.2*). This would explain an obvious link between cell cycle progression and telomere maintenance.

5.4 Telomere maintenance in mammalian DNA repair mutants

Telomere maintenance and the cell cycle progression are in some cases linked (see above). The model presented in *Figure 5.2* predicts that biochemical pathways involved in DNA repair, cell cycle regulation, regulation of apoptosis and telomere maintenance interact with each other in a co-ordinated fashion to provide an effective cellular response to DNA damage. If the model is correct then telomere maintenance must also be linked with two remaining pathways, namely that of DNA repair and regulation of apoptosis. The evidence in favour of this model is presented below.

The first indication that DNA repair mechanisms and telomere maintenance may be linked came from yeast studies. Yeast mutants defective in Ku protein showed severe telomere attrition [10]. Ku is a protein involved in the repair of DNA double strand breaks (DSBs). Following this discovery two independent sets of experiments in yeast showed that Ku binds telomeres and that it is distributed from telomeres to the sites of DSBs [45, 46]. It is important to note that eukaryotic cells use two major mechanisms for DSB repair [47–49]. The first is called non-homologous end joining (NHEJ) and the components of this mechanism are DNA dependent-protein kinase (DNA-PK), ligase IV and XRCC4. DNA-PK is composed of Ku and the catalytic subunit DNA-PKcs. NHEJ is used predominantly by mammalian cells. The second mechanism for DSB repair is homologous recombination (HR). HR involves large numbers of proteins including the RAD52 epistasis group and proteins such as Nijmegen breakage syndrome (NBS1), MRE11, BRCA1 and BRCA2. HR is a dominant mechanism for DSB repair in yeast.

Since yeast Ku mutants showed abnormalities in telomere maintenance it was of interest to investigate if mammalian DSB repair mutants show similar phenotypes. The first mammalian DSB repair mutant that has been used to study telomere maintenance is the *scid* mouse. This strain of mice has mutations in the gene encoding DNA-PKcs [50]. As a result of this mutation *scid* mice cannot perform V(D)J recombination and show sensitivity to ionising radiation (IR) [50]. Analysis of telomere maintenance in cell lines derived from *scid* mice [35], as well as directly in *scid* mice [36], showed significant abnormalities in telomere maintenance. The first indication that telomeres may be dysfunctional in *scid* cells was the presence of an unusually high number of end-chromosome fusions or telomeric fusions (TFs) [35]. Interestingly, all TFs identified in *scid* cells had telomeric sequences at fusion points suggesting that they have not been formed as a result of telomere shortening [35]. An independent study confirmed this finding [51]. Telomere length measurements revealed that telomeres in *scid* cells are on average 1.5–2 times longer than telomeres in cells from wild-type mice [36]. Therefore, the *scid* mouse was the first mammalian mutant shown to have abnormally long telomeres. In contrast to *scid* mice, mice that lack DNA-PKcs completely as a result of genetic modification, have normal telomere length but nevertheless show high levels of TFs, further suggesting the role of DNA-PKcs in telomere maintenance [52]. Differences in telomere length between these two strains may be explained, in that *scid* mice have some residual DNA-PKcs activity [50]. Recent studies have revealed that DNA-PKcs is actually present at telomeres [53].

Similarly to *scid* mice, and DNA-PKcs −/− mice, mice deficient in Ku protein also show telomere dysfunction [51, 53, 54]. The first published study revealed that Ku-deficient mice have unusually high levels of TFs [51]. All TFs had telomeric signals at fusion points. Telomere length measurements in several different cell types from Ku-deficient mice revealed either slightly elongated telomeres [54] or severe telomere shortening [53]. In addition, it has been shown that Ku is normally localised at mammalian telomeres [53, 55] suggesting it has a dual function in telomere maintenance as well as DSB repair.

Apart from proteins participating in NHEJ, telomere defects have been associated with proteins involved in HR. For example, cells from NBS1 patients show severe telomere shortening [56] and our results suggest telomere length abnormalities in BRCA2 −/− mice [57]. Also, mice with DNA repair-linked abnormalities that are not directly related to DSB repair show abnormal telomere maintenance. These include mice deficient in ATM [58] (a protein mutated in ataxia telangiectasia (AT) patients [59]) and poly ADP-ribose polymerase (PARP)-deficient mice [60]. However, the role of PARP in telomere maintenance is controversial because a new study revealed apparently normal telomere maintenance in a PARP-deficient strain of mice [61]. In addition, mice that lack functional telomerase as a result of genetic alterations show hypersensitivity to IR [62, 63]. IR triggers a strong DNA damage response in mammalian cells. Furthermore, cells from AT patients show accelerated telomere shortening [64] and altered DNA-nuclear matrix interactions [65]. These cells are hypersensitive to IR. Interestingly, the ATM gene shows homology to the yeast TEL1 gene involved in telomere length regulation [66, 67]. The exact function of ATM is not yet clear, but

Table 5.1 Mammalian models for investigating proteins involved in DNA repair and telomere maintenance

Model	Protein	Mode of DSB repair	Effect on telomere length	Telomeric fusion
Scid mouse	DNA-PKcs	NHEJ	Elongation	+
DNA-PKcs −/− mouse	DNA-PKcs	NHEJ	No effect	+
Ku −/− mouse	Ku	NHEJ	Elongation/shortening	+
NBS patients	NBS1	HR	Shortening	Unknown
BRCA2 −/− mouse	BRCA2	HR	Unknown	+
ATM −/− mouse	ATM	Signalling?	Shortening	+
AT patients	ATM	Signalling?	Shortening	+
PARP −/− mouse	PARP	SSBR	Shortening/no effect	+/−

Abbreviations: SSBR, single strand break repair; +, telomeric fusions present; −, telomeric fusions absent.

it has been thought that this molecule may act as a signalling molecule for the presence of DSBs in the genome [47, 48]. Some proteins involved in the repair of DSBs such as BRCA1 are targets of ATM [68]. A summary of the effects of various proteins involved in DNA repair on telomere length is presented in *Table 5.1*.

Taken together these results argue that DSB repair mechanisms and telomere maintenance are linked. This strongly supports the model presented in *Figure 5.2* predicting that DNA repair mechanisms, cell cycle control, control of apoptosis and telomere maintenance constitute four inter-dependent families of interacting pathways involved in DNA damage response.

5.5 Telomere maintenance and apoptosis

Following the same logic as applied above, if the model presented in *Figure 5.2* is correct, one should expect some connection between telomere maintenance and control of apoptosis. In line with this model, telomerase-deficient mice, which show telomere shortening and TFs, also show an increased level of apoptosis [69]. Apoptosis is observed in highly proliferative tissues including reproductive and haematopoietic systems [70]. In addition, a study employing genetic manipulation of TBPs in human cells revealed the link between telomere maintenance and control of apoptosis. When the gene encoding TRF2 was mutated, the level of apoptosis in human cells became elevated [70]. TRF2 is known to regulate telomere length, prevent TFs and participate in the formation of T-loops [25]. Apoptosis in TRF2-deficient cells was mediated by p53 and ATM [70]. This presents a strong case in support of the above model (*Figure 5.2*). ATM, which most likely signals the presence of DSBs in the genome [47, 48], probably provides a signal for a p53-mediated apoptosis. It is interesting that ATM-deficient mice, as well as AT patients, show severe telomere shortening, increased level of TFs and extra chromosomal telomeric fragments [59]. These observations complete the picture of DNA damage response mechanisms. It is therefore clear that numerous biochemical pathways involved in DNA damage sensing, DNA repair, cell cycle regulation and the regulation of programmed cell death interact with each other. A co-ordinated control of these pathways and their mutual interactions enable mammalian cells to cope with DNA damage. If any of the pathways is compromised this may lead to genomic instability and carcinogenesis. Interestingly, telomere dysfunction *per se* is sufficient to cause genomic instability and promote carcinogenesis [71]. Again, this is in line with the above model (*Figure 5.2*) which predicts that all pathways involved in DNA damage response are highly interactive with each other.

5.6 Methods of telomere length measurements

Detailed understanding of telomere structure and function is dependent upon available molecular methods. Good indicators of functional telomeres include: (a) absence of TFs; (b) telomere length within a species-specific range; (c) proper telomerase activity and (d) integrity of T-loops *et cetera*. Historically, methods that provide accurate estimates of telomere length or detect activity of telomerase, as well as various genetic procedures by which genes involved in telomere maintenance may be manipulated, have proved very useful. This part of the review focuses on the current methods for telomere length measurement and detection of telomerase activity.

The classical methodology for telomere length measurements is based on Southern blot analysis (Protocol 5.1), which measures the length of terminal restriction fragments (TRFs) [14, 15]. In a typical experiment, the genomic DNA is digested using relevant

Protocol 5.1: Telomere length analysis by Southern blot

EQUIPMENT, MATERIALS AND REAGENTS

BioRad CHEF Mapper System.

Mammalian cell lines and tissues.

PBS.

10X SSC (dissolve 175.3 g NaCl and 88.2 g Na citrate in 800 ml double-distilled water, adjust to pH 7.0–7.5 with concentrated HCl and complete the volume to 1 litre).

SDS.

Low melting point (LMP) agarose.

NDS (93 g EDTA, 0.606 g Tris, 50 ml 10% Lauroyl sarcisone; make up to 500 ml with double-distilled water, adjust to pH 9.5 using NaOH, filter sterilise and store at 4°C).

Plug moulds.

Proteinase K.

Restriction enzymes and buffers.

10X Tris-borate EDTA (TBE) buffer.

Tris-EDTA (TE) buffer.

METHODS

Preparation of genomic DNA in agarose plugs

1. In the case of cell lines, trypsinise cells, transfer into 10 ml tubes, centrifuge, remove supernatant and dissolve in 10 ml PBS. In the case of tissues, disaggregate tissue using sterile techniques and make single cell suspension in 10 ml PBS.

2. Allow cell suspension to stand on ice for 5 minutes.

3. Spin the tube at 1200 r.p.m. for 5 minutes.

4. Remove supernatant and resuspend in 10 ml sterile PBS.

5. Spin at 1200 r.p.m. for 5 minutes.

6. Carefully remove supernatant and dilute cells to $6 \times 10^7 \, ml^{-1}$ with sterile PBS.

7. Warm cell suspension to 37°C.

8. Prepare 1% LMP agarose using sterile PBS by melting in a boiling water bath. Once melted transfer to 37°C water bath.

9. Add an equal volume of 1% LMP agarose to the cell suspension, mix and dispense into clean plug mould, ensuring there are no air bubbles trapped in plugs.

10. Place plug moulds on ice to set.

11. Once set, transfer plugs into a sterile 20 ml universal tube containing 10 ml of NDS with 500 μg ml^{-1} proteinase K.

12. Incubate at room temperature for 30 min, then transfer to 50°C overnight.

13. The following day replace the proteinase K and NDS with 10 ml of fresh solution.

14. Incubate at 50°C overnight.

15. Rinse plugs in NDS twice for 2 hours.

16. Store plugs in NDS at 4°C.

Digestion of plugs

1. Carefully take 1 plug for each cell line from NDS solution and place in a sterile 20 ml tube.

2. Wash in 10 ml of TE buffer.

3. Remove TE with sterile pipette and add fresh TE up to 20 ml.

4. Leave overnight at 4°C.

5. Make up 1X buffer for each restriction enzyme to be used (i.e. *Hae* III, *Rsa* I etc.) by mixing 10 μl of 10X enzyme buffer supplied by restriction enzyme manufacturer, 5 μl restriction enzyme and 85 μl H$_2$O.

6. Cut plug into 3 with sterile 22 × 22 mm coverslip and put each into at least 300 μl of appropriate enzyme buffer in an Eppendorf tube.

7. Leave on ice for 2 hours.

8. Remove liquid from plugs and add 100 μl of 1X buffer containing restriction enzyme.

9. Leave 30 min on ice.

10. Incubate overnight at 37°C.

11. Remove liquid and replace with 500 μl of 0.5X TBE buffer. Leave for 30 min at room temperature.

Gel running, blotting and hybridisation

1. Prepare 1% agarose gel, place it in a BioRad CHEF Mapper System and add enough 0.5X TBE buffer to cover the gel. Load plugs into wells.

2. Perform PFGE (pulse-filed gel electrophoresis) according to instructions supplied by manufacturer (BioRad).

3. When PFGE is over, transfer DNA fragments to Hybond N+ membranes (Amersham) using 0.4 M NaOH according to manufacturer instructions.

4. Hybridise overnight with a (TTTGGG)$_4$ telomeric oligonucleotide labelled with [^{32}P]dATP.

5. Wash membranes four times in 4X SSC, 0.1% SDS at 48°C for 15 min and once in 2X SSC, 0.1% SDS at room temperature for 30 minutes.

6. Perform autoradiography overnight using intensifying screens. A typical autoradiogram is shown in *Figure 5.3*.

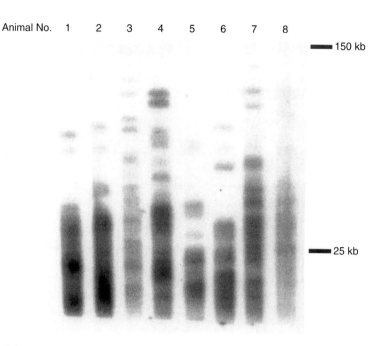

Figure 5.3

Analysis of telomere length by Southern blot. DNA samples are prepared from mouse cells. Each number represents a DNA sample taken from a different animal. Samples 1, 2, 5 and 6 are taken from *scid* mice. Samples 3, 4, 7 and 8 are taken from wild type CB17 animals. Note that some bands in samples from *scid* mice are longer than those of CB17 animals.

combinations of frequently cutting restriction enzymes. After that, genomic fragments are separated by gel electrophoresis and transferred onto nylon membranes. This is followed by hybridisation with radioactive telomeric DNA probes. After autoradiography, length of TRFs is compared against DNA length markers. This method was extremely useful in the past and numerous experiments employing human cells revealed telomere shortening in somatic cells [14, 15]. However, attempts to accurately measure telomere length using this method in mouse cells have failed [34]. This is because mouse telomeres are long (50 kilobases) and it is difficult to distinguish between telomeric DNA sequences and sub-telomeric sequences [30]. Therefore, TRFs measurement by Southern blot may be useful for human cells and cells from the species with relatively short telomeres, which do not exceed 10–20 kilobases. An additional disadvantage of TRFs measurement by Southern blot is that it is relatively crude and it can provide estimates of average telomere length only. In addition, the presence of interstitial telomeric sequences (Its) in any genome would 'contaminate' TRFs and render inaccurate the estimate of telomere length based on Southern blot analysis. In recent years alternative methods that show higher sensitivity than the classical Southern blot analysis have been invented. These include Q-FISH (quantitative fluorescence *in situ* hybridization) and Flow-FISH, as described below.

Q-FISH is based on the use of the peptide nucleic acid (PNA) telomere oligonucleotides [30, 72, 73]. Examples of metaphase chromosomes from various species hybridised with the telomeric PNA oligonucleotides are presented in *Plate 5*. Telomeric PNA oligonucleotides generate stronger and more specific hybridisation signals than the same DNA oligonucleotides. The reason for this is that PNA is electrically neutral whereas DNA is negatively charged. In any hybridisation experiments, two DNA strands will always show a slight repulsion leading to less than 100% hybridisation efficiency. In contrast, repulsion is absent in the case of PNA because of the lack of electrical charge and hybridisation efficiency is close to 100% [30, 72, 73]. The telomeric PNA probe is usually labelled with appropriate fluorescence. The most frequent fluorescence labels are Cy3 and FITC [30, 72, 73]. A prerequisite for Q-FISH is the use of appropriate digital image analysis systems designed to quantify fluorescence signals attached to PNA. This fluorescence is proportional to telomere length [73]. The resolution of Q-FISH is in the region of ~200 base pairs [72]. There are several advantages of Q-FISH in comparison with the classical methodology based on Southern blot analysis. Firstly, all individual telomeres in a given genome can be measured. Secondly, intra-chromosomal distribution of telomere length (i.e. p-arms vs. q-arms) can be determined. Thirdly, the shortest telomere in a given karyotype and its chromosome location can be identified. Finally, accurate average length of all telomeres can be determined. However, there are some disadvantages of Q-FISH in comparison with the classical method. Q-FISH is technically very complex and requires expertise in digital imaging. In addition, it is time consuming.

Any digital imaging requires a fluorescence microscope equipped with a sensitive charge-coupled device (CCD) camera. Also computer programs designed to control acquisition of digital images, as well as to perform fluorescence intensity measurements, are required. Usually, programs for image acquisition and fluorescence intensity measurements are separate and should be compatible with each other in terms of file exchange. In a typical Q-FISH experiment, samples containing metaphase chromosome spreads are hybridised with fluorescently labelled telomeric PNA. Two separate digital images are acquired: a metaphase chromosome spread (usually stained with DAPI) and an image of telomeric signals (usually red, due to Cy3-labelled telomeric oligonucleotide). These two images are combined in a fluorescence measurement program, which assigns telomeres to individual chromosomes and provides values of fluorescence intensity for each individual telomere in arbitrary units. Modern computer programs store original images as a single file. An option is provided to separate DAPI and Cy3 images when required.

One of the most important factors in signal intensity measurement is exposure time during image acquisition. PNA signals are typically very bright and brightness is proportional to exposure time. Any modern image acquisition program will have two options for image acquisition: auto-exposure time and manual exposure time. In the case of auto-exposure, the exposure time is determined automatically by the image acquisition software based on the intensity of the fluorescence signal. In the case of manual exposure, desired exposure time may be selected. If a program for image acquisition does not integrate exposure time and fluorescence intensity, a fixed-time manual exposure should be used in all Q-FISH experiments. The optimal exposure time is determined by trial and error using cell lines with known telomere length. In some cases computer programs for telomere fluorescence intensity measurement and image acquisition may be integrated resulting in correction for exposure time, in which case auto-exposure facilities may be used. This is usually the case with modern commercially available programs.

For the accuracy of the Q-FISH protocol (Protocol 5.2) it is important that proper internal controls are used in all experiments. These controls are required because microscope fluorescence lamp intensity is not constant and varies daily. To avoid inaccuracies due to lamp intensity variations, fluorescence beads of defined size (i.e. 1 μm) should be used at the same time as sample images are acquired. Fluorescence intensity values of beads are then used to correct fluorescence intensities of telomere samples. Alternatively, cell lines with defined telomere length should be used as a calibration standard in all experiments. In the most recent development, which is probably the most precise, another PNA probe is added to the hybridisation mix. This probe usually binds a single internal locus on a chromosome and the sequence on the chromosome to which the probe binds is always constant. Therefore, the signal intensity in all cells reflects exactly the same sequence length providing the best possible calibration standard, and a computer program automatically corrects fluorescence intensity of telomeres. In the original Q-FISH protocol another calibration component was introduced: plasmids containing a defined number of telomeric sequences [30]. This allows fluorescence intensity values to be converted into units of DNA length. However, this step is non-essential as fluorescence measurements in arbitrary units will yield accurate results given that an internal control (i.e. fluorescence beads) is used properly.

Flow-FISH is a new method based on similar principles to Q-FISH [74]. Flow-FISH also uses a telomeric PNA probe but the crucial difference in comparison with Q-FISH is that interphase cells rather than metaphase chromosomes are used for hybridisation, and fluorescence intensity measurements are performed by flow cytometry techniques. Cells are hybridised with the telomeric PNA oligonucleotide in suspension rather than *in situ*. The most important advantage of Flow-FISH compared with Q-FISH is that it is much quicker and completely automated. It takes several hours to analyse 15–20 metaphases by Q-FISH, whereas thousands of cells can be analysed by Flow-FISH in less than an hour. A major disadvantage of Flow-FISH in comparison with Q-FISH is that only average telomere length can be measured and the analysis of telomere length in individual chromosomes is not possible. Ideally, these two techniques should be combined.

Protocol 5.2: Telomere length analysis by Q-FISH

EQUIPMENT, MATERIALS AND REAGENTS

Shaking platform.

Heating block.

Water baths.

A fluorescence microscope equipped with DAPI and Cy3 filters.

A CCD camera.

A computer equipped with: (a) image acquisition software and (b) software for fluorescence intensity measurement.

Any cellular material from which metaphase chromosome spreads may be obtained is suitable.

PBS.

Hypotonic solution (0.075 M KCl).

Fixative (3 parts of methanol and 1 part of acetic acid).

Colcemide.

4% formaldehyde (21.6 ml of 37% formaldehyde, 200 ml PBS, pH 7.0–7.5).

Pepsin 1 mg ml^{-1} (dissolve 200 mg pepsin in acidified (pH 2.0) purified water).

Ethanol.

10X SSC.

Washing solution (70% formamide in 2X SSC (70 ml deionised 100% formamide, 10 ml 20X SSC and 20 ml double-distilled water).

PNA telomeric probe labelled with Cy3-(CCAGGG)$_3$. This oligonucleotide can be obtained from Perceptive Biosystems. In the case of Cy3 labelling ask for 2 linkers (chemical bonds between Cy3 and oligonucleotide) and oligonucleotide purification. Probe should be prepared by dissolving the PNA oligonucleotide in a buffer consisting of 70% formamide, 1 mM Tris and 0.5% blocking reagents (Boehringer Mannheim) to obtain final concentration of 0.6 μg ml^{-1}.

Fluorescence beads (orange beads, size 0.2 μm, Molecular Probes).

DAPI in anti-fade solution.

METHODS

1. Seed the cells into a T75 flask and incubate in an incubator at 37°C until semi-confluent. In the case of lymphocytes follow the standard protocol for chromosome preparation from lymphocytes.

2. Add 10 µl of colcemide into the medium and incubate for 1–2 hours.

3. Trypsinise the cells, transfer them into a 10 ml centrifuge tube. Centrifuge the cells for 5 minutes at 1000 r.p.m. Remove supernatant.

4. Add 10 ml of hypotonic solution into the tube, resuspend the cell pellet and incubate for 20 min at room temperature. Centrifuge for 5 minutes at 1000 r.p.m. and remove supernatant.

5. Add 10 ml of fixative to the cell pellet and resuspend it. Leave for 30 min in the refrigerator. Centrifuge for 5 min at 1000 r.p.m. Repeat fixation two times. In the last fixation add only 0.5–1 ml of fixative and resuspend the cell pellet.

6. Drop 10 µl of cell suspension on to a clean dry microscope slide. Allow to air dry. Incubate the slide on a hot plate (56°C) overnight.

Fluorescence *in situ* hybridisation

1. Place microscope slides (up to five) in a Coplin jar and add 50 ml PBS. Place the Coplin jar on to a shaking platform and incubate for 15 minutes.

2. Remove PBS and replace it with 4% formaldehyde. Incubate for 2 minutes without shaking.

3. Wash in PBS 3 × 5 minutes on a shaking platform.

4. Incubate in a pepsin solution pre-heated at 37°C (water bath) for 10 minutes.

5. Wash 2 × 5 min in PBS on a shaking platform.

6. Fix in formaldehyde for 2 minutes without shaking.

7. Wash 3 × 5 min in PBS on a shaking platform.

8. Dehydrate in ethanol series (70%, 90% and 100%) for 5 minutes each by putting 1 ml of each ethanol concentration directly on the slide.

9. Air dry.

10. Place 20 µl of the probe mix in the centre of the slide and cover with a coverslip (22 × 50 mm).

11. Place the slide on the heating block pre-heated to 80°C. Incubate for 3 minutes. Leave the slide in a dark, damp container for 2 hours.

12. Remove the coverslip and wash the slide in the washing buffer (50 ml of washing buffer in a Coplin jar) 2 × 15 minutes.

13. Wash the slide 3 × 15 minutes in PBS on a shaking platform.

14. Dehydrate in ethanol series.

15. Place 15 µl of DAPI in anti-fade reagent in the centre of the slides and cover the slide with a coverslip. Seal the coverslip with colourless nail varnish.

Image acquisition

1. Place the slide on the microscope platform, select the DAPI filter and observe the slide under 10× magnification. Find a nicely spread metaphase cell. Switch to a 100× objective. Place the metaphase spread in the middle of the field of view. Check the telomeric signal by switching to the Cy3 filter.

2. In the image acquisition program select fixed time exposure. Determine the optimum exposure time by trial and error. Usually 0.4–1 second exposure time should be acceptable. Once a suitable exposure time is determined it should be used permanently. Ideal exposure time should allow acquisition of crisp images without signs of probe saturation. Acquire DAPI and Cy3 images separately (see *Figure 5.4*). Save images as TIFF files. Store images in a suitable storage medium (Zip, Jazz, CD etc.).

Image analysis

1. Analysis of telomere fluorescence intensity depends on the analysis software. There are several commercially available packages for this purpose. Also, there is software called TFL TELO available free of charge from Dr. Peter Lansdorp, Terry Fox Laboratory, Vancouver. This software is highly recommended and instructions for transferring acquired images into this software are described below.

2. Convert TIFF images of chromosomes and telomeres into BMP images using appropriate software. Image size in pixels should be divisible by four.

3. Combine BMP DAPI and Cy3 images in TFL TELO. After 10–15 seconds a combined image will be displayed showing values of telomere fluorescence intensity, in arbitrary units, for each individual telomere. Some editing may be required to distinguish between individual telomeres.

4. Transfer the table containing fluorescence intensities into software for statistical analysis and calculate average telomere fluorescence intensity, as well as telomere fluorescence intensity for each individual telomere. Usually, 10–15 well-spread metaphase spreads should be sufficient for analysis.

Calibration

1. All fluorescence microscopes are known to show strong day-to-day variations in lamp intensity. This means that fluorescence intensity measurements will vary from day to day, leading to unreproducible results. To avoid this, some kind of calibration procedure should be applied. The protocol with fluorescence beads is described below.

2. Dilute orange fluorescence beads in FCS at a ratio 1:25.

3. Place 3 μl of this solution on a coverslip and use a slide to produce a thin homogeneous smear over the glass surface.

4. Air dry the coverslip for several minutes and store in a slide box until use.

5. On the day of sample acquisition prepare a calibration slide by adding 5 μl of slow fade reagent on a glass slide and cover with the bead-coated coverslip. The slide can be used for up to 3 days.

6. Acquire images of beads using fixed time exposure as above.

7. Import bead images into TFL TELO software (see above) and perform calibration according to instructions until an average value of fluorescence intensity of the beads is obtained. Use this number as a correction factor for the fluorescence intensity of telomeres. For example, if metaphase spreads from the same sample or samples that are being compared with each other are acquired on different days, the bead images should be acquired on each day and results corrected accordingly.

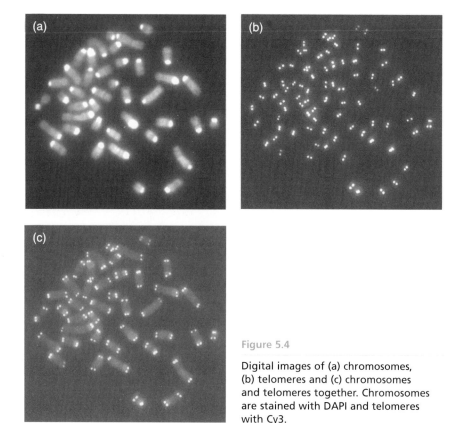

Figure 5.4

Digital images of (a) chromosomes, (b) telomeres and (c) chromosomes and telomeres together. Chromosomes are stained with DAPI and telomeres with Cy3.

5.7 Methods for detection of telomerase activity

Telomeric DNA is synthesised by the enzyme telomerase and this is the most frequently used mechanism for telomere maintenance in mammalian cells (see above). Telomerase is a reverse transcriptase which uses its own RNA template to synthesise telomeric DNA. The most sensitive method for detection of telomerase activity is called TRAP (telomere repeat amplification protocol) and it was developed in 1994 [75]. The method (Protocol 5.3) is based on PCR. In a typical experiment, cell extract is prepared and mixed in a PCR tube with two oligonucleotides, which resemble telomeric DNA. If telomerase is present in the cell extract it will extend one of the oligonucleotides and this telomerase product will then be amplified in a PCR reaction. Radioactively labelled nucleotides are added into PCR tubes so that the reaction product may be detected following autoradiography. The typical product is a ladder containing six base pair repeats. Two samples are always run in parallel, one with the enzyme RNAse and the other one without it. If the reaction product is due to telomerase activity, which uses its own RNA template to synthesise telomeric DNA, the sample with RNAse will be inactivated and will be characterised by the absence of a six base pair repeat ladder. In a more recent improvement, an internal control was included in the TRAP protocol. The purpose of this internal control is to monitor the efficiency of the PCR reaction. TRAP is an extremely sensitive assay and it is relatively straightforward. At the moment there are several commercially available kits for TRAP and each one offers good value for money.

Protocol 5.3: Telomerase detection

EQUIPMENT, MATERIALS AND REAGENTS

Any mammalian cell line or tissue is suitable.

Apparatus for gel electrophoresis.

Centrifuges.

Thermal cycler.

Oligonucleotides: 5' AATCCGTCGAGCAGAGTT 3' (upstream primer, TS); 5' AATCCCATTCCCATTCCCATTCCC 3' (downstream primer, CX).

PBS.

Washing buffer (10 mM Hepes-KOH pH 7.5, 1.5 mM $MgCl_2$, 10 mM KCl, 1 mM DTT).

Lysis buffer (10 mM Tris-HCl pH 7.5, 1 mM $MgCl_2$, 1 mM EGTA, 0.1 mM PMSF, 5 mM β-mercaptoethanol, 0.5% CHAPS, 10% glycerol).

Material for PCR: dNTPs, Taq polymerase, T4g32 protein, [^{32}P]dCTP, TRAP PCR buffer (20 mM Tris-HCl pH 8.3, 1.5 mM $MgCl_2$, 63 mM KCl, 0.005% Tween-20, 1.0 mM EGTA, 0.1 mg ml^{-1} BSA).

RNAse.

Material for polyacrylamide non-denaturing gel.

METHODS

Preparation of the cell extract

1. In the case of established cell lines, sub-culture the cells 2 days before extract preparation. Trypsinise the cells. About 10^6 cells are required per cell line. In the case of tissue preparation, protocols vary. Soft tissues should be minced with sterile scalpels, and homogenised with a motorised pestle. Connective tissues should be frozen by adding liquid nitrogen and then pulverised by grinding with a matching pestle.

2. Wash the cells twice in PBS using a refrigerated centrifuge (4°C).

3. Spin the cells at 1000 r.p.m. for 5 minutes.

4. Resuspend the cells in 20 μl of ice-cold washing buffer. Pellet the cells by centrifuging them at 10 000 g for 20 minutes.

5. Resuspend the cells at a density of 10 000–100 000 cells in 20 μl ice-cold lysis buffer. Incubate the cell suspension on ice for 30 minutes and centrifuge it at 100 000 g for 45 min at 4°C. Store the supernatant at −70°C until required. Measure the protein concentration (usually 2–10 mg ml^{-1} for 100 000 cells).

PCR reaction

1. The reaction is carried out in a volume of 100 μl. Prepare the reaction mix by combining the following ingredients in a PCR tube to make a volume of 99 μl:
10.0 μl TRAP PCR buffer
3.0 μl dATP, dGTP, dTTP (1.0 μl of each, 10 mM each)

1.0 μl dCTP (5 mM)
1.0 μl T4g32 protein
1.0 μl BSA
4.0 μl cell extract
1.0 μl Taq polymerase
1.0 μl upstream primer TS (0.1 mg)
76.0 μl double-distilled water
1.0 μl [^{32}P]dCTP

2. Overlay the reaction mix with mineral oil and incubate the PCR tube with the above mix at 23°C for 20 min to allow primer extension.

3. Add 1.0 μl of downstream primer CX (0.1 μg) into the reaction mix (the reaction volume is now 100 μl), place the tube into a thermal cycler and run 30 cycles (denaturation: 94°C, 30 seconds; annealing: 50°C, 30 seconds; extension: 72°C, 90 seconds). **Important note: the same reaction mix as above should be run in parallel with the same cell extract that has been pre-treated with RNAse for 20–30 minutes.**

Preparation of polyacrylamide gel

1. Clean glass plates using detergent and tap water. Rinse plates with double-distilled water and wipe them with ethanol. Let them air dry.

2. Using clean paper tissue, spread about 5 ml of 2% dimethyldichlorosilane on one side of each glass plate and allow it to dry.

3. Place one glass plate (treated surface facing up) on the bench. Place two spacers on the glass. Place the other glass plate on the top (treated surface facing down). Seal the plates with tape.

4. Prepare 4.5% polyacrylamide solution by combining the following reagents:
9 ml 49% acrylamide solution
10 ml 10X TBE buffer
91 ml double-distilled water
100 μl 20% ammonium persulfate
100 μl TEMED

5. Using a 50 ml syringe inject the above mix between the plates without forming air bubbles. Insert a comb at the desired side and place a clip at the top of the gel.

Gel electrophoresis

1. Wait at least 2 hours until the gel sets. Remove the clip, tape and comb. Position the plates at right side of the electrophoresis apparatus. Add small amount of TBE to the empty chambers and flush small pieces of gel remaining after the comb was removed.

2. Mix desired amount of PCR products with loading buffer.

3. Load desired amount of the above mix.

4. Connect the electrophoresis apparatus to a power supply. Run time should be 1.5 hours at 400 volts.

Autoradiography

1. Disconnect the power supply. Carefully detach plates from the apparatus. Remove glass plates from the gel with great care. Place a piece of Whatman 3 paper on the gel and press gently. Wrap it with Saran wrap.

2. Expose an X-ray film to the gel for several hours at −20°C. Develop the film. If necessary, expose longer.

Interpretation

The TRAP product is a ladder consisting of six base pairs of telomeric DNA (see *Figure 5.5*). Since telomerase contains its own RNA component, treatment of cell extracts with RNAse should inactivate telomerase. As a result, parallel lanes (RNAse treated extracts) should not contain ladders (see *Figure 5.5*). False positive results may include RNAse-resistant products. False negative results may include inhibition of PCR.

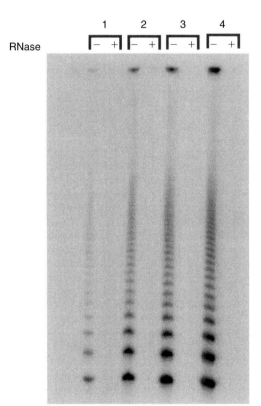

Figure 5.5

TRAP assay. Numbers represent various Chinese hamster cell lines. Each band in the ladder represents a six base pair telomeric repeat. Note that samples pre-treated with RNAse are run in parallel.

5.8 References

1. Zakian, V.A. (1995) Telomeres: beginning to understand the end. *Science* **270**: 1601–1607.
2. Muller, H.J. (1938) The remaking of chromosomes. *Collecting Net* **13**: 181–195.
3. McClintock, B. (1941) The stability of broken ends of chromosomes of *Zea mays*. *Genetics* **23**: 234–282.
4. Blackburn, E. and Gall, J. (1978) A randomly repeated sequence at the termini of the extrachromosomal ribosomal RNA genes in Tetrahymena. *J. Mol. Biol.* **120**: 33–53.
5. Moyzis, R.K., Buckingham, J.M., Cram, L.S., Dani, M., Deaven, L.L., Jones, M.D., Meyne, J., Ratliff, R.L. and Wu, J.R. (1988) A highly conserved repetitive DNA sequence, (TTAGGG)n, present at the telomeres of human chromosomes. *Proc. Natl. Acad. Sci. USA* **85**: 6622–6626.
6. Greider, C.W. and Blackburn, E.H. (1985) Identification of a specific telomere terminal transferase activity in Tetrahymena extracts. *Cell* **43**: 405–413.
7. Greider, C.W. and Blackburn, E.H. (1987) The telomere terminal transferase of Tetrahymena is a ribonucleoprotein enzyme with two kinds of primer specificity. *Cell* **51**: 887–898.
8. Lingner, J., Hughes, T.R., Shehchenko, A., Mann, M., Lundblad, V. and Cech, T.R. (1997) Reverse transcriptase motifs in the catalytic subunit of telomerase. *Science* **276**: 561–567.
9. Sandell, L.L. and Zakian, V.A. (1993) Loss of a yeast telomere: arrest, recovery and chromosome loss. *Cell* **75**: 729–739.
10. Boulton, S.J. and Jackson, S.P. (1996) Identification of a *Saccharomyces cerevisiae* Ku80 homologue: roles in DNA double strand break rejoining and in telomeric maintenance. *Nucleic Acid Res.* **24**: 4639–4648.
11. Bass, H.W., Marshall, W.F., Sedat, J.W., Agard, D.A. and Cande, W.Z. (1997) Telomeres cluster de novo before the initiation of synapsis: a three-dimensional spatial analysis of telomere positions before and during meiotic prophase. *J. Cell Biol.* **137**: 5–18.
12. Luderus, M.E., van Steensel, B., Chong, L., Sibon, O.C., Cremers, F.F. and de Lange, T. (1996) Structure, subnuclear distribution, and nuclear matrix association of the mammalian telomeric complex. *J. Cell Biol.* **135**: 867–881.
13. Cooper, J.P., Watanabe, Y. and Nurse, P. (1998) Fission yeast Taz1 protein is required for meiotic telomere clustering and recombination. *Nature* **392**: 828–831.
14. Harley, C.B., Vaziri, H., Counter, C.M. and Allsopp, R.C. (1992) The telomere hypothesis of cellular aging. *Exp. Gerontol.* **27**: 375–382.
15. Harley, C.B. (1995) Telomeres and aging. In: *Telomeres* (eds E.H. Blackburn and C.W. Greider). Cold Spring Harbor Press, New York, pp. 247–263.
16. Blackburn, E.H. (2000) Telomere states and cell fates. *Nature* **408**: 53–56.
17. Counter, C.M., Avilion, A.A., LeFeuvre, C.E., Stewart, N.G., Greider, C.W., Harley, C.B. and Bacchetti, S. (1992) Telomere shortening associated with chromosome instability is arrested in immortal cells which express telomerase activity. *EMBO J.* **11**: 1921–1929.
18. Counter, C.M., Hirte, H.W., Bacchetti, S. and Harley, C.B. (1994) Telomerase activity in human ovarian carcinoma. *Proc. Natl. Acad. Sci. USA* **91**: 2900–2904.
19. de Lange, T. and Jacks, T. (1999) For better or worse? Telomerase inhibition and cancer. *Cell* **98**: 273–275.
20. Hahn, W.C., Stewart, S.A., Brooks, M.W., York, S.G., Eaton, E., Kurachi, A., Beijersbergen, R.L., Knoll, J.H., Meyerson, M. and Weinberg, R.A. (1999) Inhibition of telomerase limits the growth of human cancer cells. *Nat. Med.* **5**: 1164–1170.
21. Dunham, M.A., Neumann, A.A., Fasching, C.L. and Reddel, R.R. (2000) Telomere maintenance by recombination in human cells. *Nat. Genet.* **26**: 447–450.
22. Bryan, T.M., Englezou, A., Gupta, J., Bacchetti, S. and Reddel, R.R. (1995) Telomere elongation in immortal human cells without detectable telomerase activity. *EMBO J.* **14**: 4240–4248.
23. Meyne, J., Ratliff, R.L. and Moyzis, R.K. (1989) Conservation of the human telomere sequence (TTAGGG)n among vertebrates. *Proc. Natl. Acad. Sci. USA* **86**: 7049–7053.
24. Wellinger, R.J., Ethier, K., Labrecque, P. and Zakian, V.A. (1996) Evidence for a new step in telomere maintenance. *Cell* **85**: 423–433.

25. Smogorzevska, A., van Steensel, B., Bianchi, A., Oelmann, S., Schaefer, M.R., Schnapp, G. and de Lange, T. (2000) Control of telomere length by TRF1 and TRF2. *Mol. Cell Biol.* **20**: 1659–1668.

26. Smith, S., Giriat, I., Schmitt, A. and de Lange, T. (1998) Tankyrase, a poly(ADP-ribose) polymerase at human telomeres. *Science* **282**: 1484–1487.

27. Li, B., Oestreich, S. and de Lange, T. (2000) Identification of human Rap1: implications for telomere evolution. *Cell* **101**: 471–483.

28. Kaminker, P.G., Kim, S.H., Taylor, R.D., Zebarjadian, Y., Funk W.D., Morin, G.B., Yaswen, P. and Campisi, J. (2001) TANK2, a new TRF1-associated PARP, causes rapid induction of cell death upon overexpression. *J. Biol. Chem.* **276**: 35891–35899.

29. Kim, S.H., Kaminker, P. and Campisi, J. (1999) TIN2, a new regulator of telomere length in human cells. *Nat. Genet.* **23**: 405–412.

30. Zijlmans, J.M., Martens, U.M., Poon, S.S.S., Raap, A.K., Tanke, H.J., Ward, R.K. and Lansdorp, P.M. (1997) Telomeres in the mouse have large interchromosomal variations in the number of T_2AG_3 repeats. *Proc. Natl. Acad. Sci. USA* **94**: 7423–7428.

31. Kipling, D. and Cooke, H.J. (1990) Hypervariable ultra-long telomeres in mice. *Nature* **347**: 400–402.

32. Martens, U.M., Zijlmans, J.M., Poon, S.S., Dragowska, W., Yui, J., Chavez, E.A., Ward, R.K. and Lansdorp, P.M. (1998) Short telomeres on human chromosome 17p. *Nat. Genet.* **18**: 76–80.

33. Slijepcevic, P. and Hande, M.P. (1999) Chinese hamster telomeres are comparable in size to mouse telomeres. *Cyto. Genet. Cell Genet.* **85**: 196–199.

34. Blasco, M.A., Lee, H.W., Hande, M.P., Samper, E., Lansdorp, P.M., DePinho, R.A. and Greider, C.W. (1997) Telomere shortening and tumor formation by mouse cells lacking telomerase RNA. *Cell* **91**: 25–34.

35. Slijepcevic, P., Hande, M.P., Bouffler, S.D., Lansdorp, P. and Bryant, P.E. (1997) Telomere length, chromatin structure and chromosome fusigenic potential. *Chromosoma* **106**: 413–421.

36. Hande, P., Slijepcevic, P., Silver, A. Bouffler, S.D., van Buul, P.P., Lansdorp, P. and Bryant, P.E. (1999) Elongated telomeres in *scid* mice. *Genomics* **56**: 221–223.

37. Van Steensel, B. and de Lange, T. (1997) Control of telomere length by the human telomeric protein TRF1. *Nature* **385**: 740–743.

38. Marcand, S., Gilson, E. and Shore, D. (1997) A protein counting mechanism for telomere length regulation in yeast. *Science* **275**: 986–990.

39. Baumann, P. and Cech, T.R. (2001) Pot1, the putative telomere end-binding protein in fission yeast and humans. *Science* **292**: 1171–1175.

40. Griffith, J.D., Comeau, L., Rosenfield, S., Stansel, R.M., Bianchi, A., Moss, H. and de Lange, T. (1999) Mammalian telomeres end in a large duplex loop. *Cell* **97**: 503–514.

41. Slijepcevic, P. and Bryant, P.E. (1998) Chromosome healing, telomere capture and mechanisms of radiation-induced chromosome breakage. *Int. J. Radiat. Biol.* **73**: 1–13.

42. Kirk, K.E., Harmon, B.P., Reichardt, I.K., Sedat, J.W. and Blackburn, E.H. (1997) Block in anaphase chromosome separation caused by a telomerase template mutation. *Science* **275**: 1478–1481.

43. Lima-de-Faria, A. and Bose, S. (1962) The role of telomeres at anaphase. *Chromosoma* **13**: 315–327.

44. Zhou, B.B. and Elledge, S.J. (2000) The DNA damage response: putting checkpoints in perspective. *Nature* **408**: 433–439.

45. Mills, K.D., Sinclair, A. and Guarente, L. (1999) MEC1-dependent redistribution of the Sir3 silencing protein from telomeres to DNA double-strand breaks. *Cell* **97**: 609–629.

46. Martin, S.G., Laroche, T., Suka, N., Grunstein, M. and Gasser, S.M. (1999) Relocalization of telomeric Ku and SIR proteins in response to DNA strand breaks in yeast. *Cell* **97**: 621–633.

47. Hoeijmakers, J.H. (2001) Genome maintenance mechanisms for preventing cancer. *Nature* **411**: 366–374.

48. Karran, P. (2000) DNA double strand break repair in mammalian cells. *Curr. Opin. Genet. Dev.* **10**: 144–150.

49. Scully, R. and Livingston, D.M. (2000) In search of the tumour-suppressor functions of BRCA1 and BRCA2. *Nature* **408**: 429–432.

50. Smith, G.C.M. and Jackson, S.P. (1999) The DNA-dependent protein kinase. *Genes Dev.* **13**: 916–934.

51. Bailey, S.M., Meyne, J., Chen, D.J., Kurimasa, A., Li, G.C., Lehnert, B.E. and Goodwin, E.H. (1999) DNA double-strand break repair proteins are required to cap the ends of mammalian chromosomes. *Proc. Natl. Acad. Sci. USA* **96**: 14899–14904.

52. Goytisolo, F.A., Samper, E., Edmonson, S., Taccioli, G.E. and Blasco, M.A. (2001) The absence of the DNA-dependent protein kinase catalytic subunit in mice results in anaphase bridges and in increased telomeric fusions with normal telomere length and G-strand overhang. *Mol. Cell Biol.* **21**: 3642–3651.

53. d'Adda di Fagagna, F., Hande, M.P., Tong, W., Roth, D., Lansdorp, P.M., Wang, Z. and Jackson, S.P. (2001) Effects of DNA non-homologous end-joining factors on telomere length and chromosomal stability in mammalian cells. *Curr. Biol.* **7**: 1192–1196.

54. Samper, E., Goytisolo, F.A., Slijepcevic, P., van Buul, P. and Blasco, M.A. (2000) Mammalian Ku86 protein prevents telomeric fusions independently of the length of TTAGGG repeats and the G-strand overhang. *EMBO Rep.* **1**: 244–252.

55. Hsu, H, Gilley, D., Blackburn, E.H. and Chen, D.J. (1999) Ku is associated with the telomere in mammals. *Proc. Natl. Acad. Sci. USA* **22**: 12454–12458.

56. Ranganathan, V., Heine, W.F., Ciccone, D.N., Rudolph, K.L., Wu, X., Chang, S., Hai, H., Ahearn, I.M., Livingston, D.M., Resnick. I., Rosen, F., Seemanova, E., Jarolim, P., DePinho, R.A. and Weaver, D.T. (2001) Rescue of a telomere length defect of Nijmegan breakage syndrome cells requires NBS and telomerase catalytic subunit. *Curr Biol.* **11**: 962–966.

57. Finnon, P., Wong, H.P., Silver, A.R., Slijepcevic, P. and Bouffler, S.D. (2001) Long but dysfunctional telomeres correlate with chromosomal radiosensitivity in a mouse AML cell line. *Int. J. Radiat. Biol.* **77**: 1151–1162.

58. Savitsky, K., Bar-Shira, A., Gilad, S., Rotman, G., Ziv, Y., Vanagaite, L., Tagle, D.A., Smith, S., Uziel, T., Sfez, S., et al. (1995) A single ataxia telangiectasia gene with a product similar to PI 3-kinase. *Science* **268**: 1749–1753.

59. Hande, M.P., Balajee, A.S., Tchirkov, A., Wynshaw-Boris, A. and Lansdorp, P.M. (2001) Extra-chromosomal telomeric DNA in cells from Atm($-/-$) mice and patients with ataxia-telangiectasia. *Hum. Mol. Genet.* **10**: 519–528.

60. d'Adda di Fagagna, F., Hande, M.P., Tong, W.M., Lansdorp, P.M., Wang, Z.Q. and Jackson, S.P. (1999) Functions of poly(ADP-ribose) polymerase in controlling telomere length and chromosomal stability. *Nat. Genet.* **23**: 76–80.

61. Samper, E., Goytisolo, F.A., Menissier–de Murcia, J., Gonzalez-Suarez, E., Cigudosa, J.C., de Murcia, G. and Blasco, M.A. (2001) Normal telomere length and chromosomal end capping in poly(ADP-ribose) polymerase-deficient mice and primary cells despite increased chromosomal instability. *J. Cell Biol.* **154**: 49–60.

62. Wong, K.K., Chang, S., Weiler, S.R., Ganesan, S., Chaudhuri, J., Zhu, C., Artandi, S.E., Rudolph, K.L., Gottlieb, G.J., Chin, L., Alt, F.W. and DePinho, R.A. (2000) Telomere dysfunction impairs DNA repair and enhances sensitivity to ionizing radiation. *Nat. Genet.* **26**: 85–88.

63. Goytisolo, F.A., Samper, E., Martin–Caballero, J., Finnon, P., Herrera, E., Flores, J.M., Bouffler, S.D. and Blasco, M.A. (2000) Short telomeres result in organismal hypersensitivity to ionizing radiation in mammals. *J. Exp. Med.* **192**: 1625–1636.

64. Metcalfe, J.A., Parkhill, J., Campbell, L., Stacey, M., Biggs, P., Byrd, P.J. and Taylor, A.M. (1996) Accelerated telomere shortening in ataxia telangiectasia. *Nat. Genet.* **4**: 252–255.

65. Smilenov, L., Dhar, S., Pandita, T.K. (1999) Altered telomere nuclear matrix interactions and nucleosomal periodicity in ataxia telangiectasia cells before and after ionizing radiation treatment. *Mol. Cell Biol.* **19**: 6963–6971.

66. Greenwell, P.W., Kronmal, S.L., Porter, S.E., Gassenhuber, J., Obermaier, B. and Petes, T.D. (1995) TEL1, a gene involved in controlling telomere length in *S. cerevisiae*, is homologous to the human ataxia telangiectasia gene. *Cell* **82**: 823–829.

67. Morrow, D.M., Tagle, D.A., Shiloh, Y., Collins, F.S. and Hieter, P. (1995) TEL1 an *S. cerevisiae* homolog of the human gene mutated in ataxia telangiectasia, is functionally related to the yeast checkpoint gene MEC1. *Cell* **82**: 831–840.

68. Welcsh, P.L., Owens, K.N. and King, M.C. (2000) Insights into the function of BRCA1 and BRCA2. *Trends Genet.* **16**: 69–74.

69. Lee, H.W., Blasco, M.A., Gottlieb, G.J., Horner, J.W., Greider, C.W., DePinho, R.A. (1998) Essential role of mouse telomerase in highly proliferative organs. *Nature* **392**: 569–574.

70. Karlseder, J., Broccoli, D., Dai, Y., Hardy, S. and de Lange, T. (1999) p53- and ATM-dependent apoptosis induced by telomeres lacking TRF2. *Science* **283**: 1321–1325.

71. Artandi, S.E., Chang, S., Lee, S.L., Alson, S., Gottlieb, G.J., Chin, L. and DePinho, R.A. (2000) Telomere dysfunction promotes non-reciprocal translocations and epithelial cancers in mice. *Nature* **406**: 641–645.

72. Lansdorp, P.M., Verwoerd, N.P., van de Rijke, F.M., Dragowska, V., Little, M.T., Dirks, R.W., Raap, A.K. and Tanke, H.J. (1996) Heterogeneity in telomere length of human chromosomes. *Hum. Mol. Genet.* **5**: 685–691.

73. Poon, S.S., Martens, U.M., Ward, R.K. and Lansdorp, P.M. (1999) Telomere length measurements using digital fluorescence microscopy. *Cytometry* **36**: 267–278.

74. Rufer, N., Dragowska, W., Thornbury, G., Roosnek, E. and Lansdorp, P.M. (1998) Telomere length dynamics in human lymphocyte subpopulations measured by flow cytometry. *Nat. Biotechnol.* **16**: 743–747.

75. Kim, N.W., Piatyszek, M.A., Prowse, K.R, Harley, C.B., West, M.D., Ho, P.L., Coviello, G.M., Wright, W.E., Weinrich, S.L. and Shay, J.W. (1994) Specific association of human telomerase activity with immortal cells and cancer. *Science* **266**: 2011–2015

The Nucleolus and Cell Proliferation

6

David Hughes and Antonio Torres-Montaner

Contents

Cell Proliferation and Apoptosis, David Hughes and Huseyin Mehmet (Eds)
© 2003 BIOS Scientific Publishers Ltd, Oxford

6.1　Introduction

The nucleolus was first described by Fontana in 1781, and the term 'nucleolus' or 'small nucleus' was coined by Valentin in 1839. Later in the 1800s it was established that the nucleolus is an unbound area of the nucleus and that it is not a permanent body, disappearing during prophase. During the early 20th century it became apparent that the nucleolus is associated with chromosomes and, in particular, chromosome regions. By 1952, nucleoli had been isolated and shown to contain both DNA and RNA. The second half of the 20th century saw the development of new technologies, including electron microscopy, which led to significant advances in our understanding of nucleolar structure in relation to function. Thiry and Goessens have compiled a comprehensive historical overview of the development of our knowledge of nucleolar structure and function [1].

The nucleolus is a prominent, non-membrane-bound nuclear structure that organises around chromosome segments containing NORs. It is the centre of rDNA transcription and ribosome biogenesis. More recently, additional functions have been attributed to the nucleolus including cell cycle regulation, telomerase activity, p53 metabolism, signal recognition particle biogenesis, small RNA processing, and mRNA transport [2].

Nucleoli are the most obvious, best-characterised, highly compact structural element of eukaryotic interphase cell nuclei. At the light microscope (LM) level, they appear as solitary or multiple spherical bodies which occupy – particularly in actively growing cells – a substantial portion of the total nuclear space [3]. Each nucleolus contains a cluster of transcriptionally active rRNA genes immersed in a 'cloud' of ribosomal precursor particles in various stages of maturation. Thus, a nucleolus may be regarded as a ribosome-producing apparatus where synthesis and processing of the rRNA precursor molecules (pre-RNAs) takes place as well as their co-ordinate assembly with specific ribosomal and non-ribosomal proteins to form pre-ribosomal particles. Nucleoli are highly dynamic structures as is illustrated by their cyclical disappearance and reappearance during mitosis. The nucleolus reforms at the end of mitosis at specific chromosomal sites termed NORs that contain tandemly reiterated rRNA genes. The site-specific assembly of pre-formed nucleolar entities around the NORs is termed 'nucleologenesis'. The nucleolus is sometimes referred to as a nuclear organelle but, as it has no clear membrane-like border, and the protein and RNA transport are controlled at the nuclear envelope rather than the nucleolar periphery, it is probably more accurate to call it a nuclear domain.

Ribosome biogenesis involves rRNA synthesis and maturation, and the assembly of rRNA and ribosomal proteins into the large and small ribosome subunits. The process is regulated throughout the cell cycle, primarily at the level of rRNA synthesis. rDNA synthesis peaks during the S and G_2 phases, stops as the cell enters mitosis and then reactivates as the cell exits from mitosis. The transcription initiation complex consists of upstream binding factor-1 (UBF-1), SL1 factors containing the TATA-binding protein (TBF), and RNA polymerase-1 (RNA pol I). Many factors have been shown to participate in the various steps of rRNA processing, these extensive modifications producing 18S, 5.8S and 28S rRNA which together with a 5S rRNA are processed and assembled into the ribosome subunits [4]. The mature RNAs are subsequently assembled with ribosomal proteins into pre-ribosomal particles in the nucleolus. Comparisons of relative molecular mobilities using GFP fusion proteins with fluorescence recovery after photobleaching (FRAP) and fluorescence loss in photobleaching (FLIP) have revealed that UBF-1, nucleolin, B23, fibrillarin and Rpp29 rapidly exchange between nucleolus and nucleoplasm, while ribosomal proteins move relatively more slowly [5].

The mammalian nucleolus is a large (0.5–1.0 μm) structure that disassembles in late prophase and reforms in telophase. Nucleoli form in response to transcription of ribosomal rDNA repeats that are often randomly arrayed in the NORs. The nucleoli reside next to chromosomes in the interphase nucleus and are of comparable size. Each nucleolus is the cytological manifestation of the activity of the NOR, a genetic locus on the chromosome that consists of repeated genes for rRNA. Nucleoli have very high concentrations of mass per unit volume compared to most other intracellular domains or organelles and were first isolated at high purity in the early 1960s [6].

The nucleolus as the site of rRNA synthesis provides us with one of the best opportunities for unravelling the relationships between transcription and the spatial organisation of chromatin. The nucleolus is essentially a large ribonucleoprotein complex whose structure and morphology is directly related to its synthetic activity. Relationship between nucleolar structure and function is, therefore, intimately linked. Synthesis and processing of precursor pre-ribosomal RNA seems to occur dynamically from the centre of the nucleolus to the periphery. In the human karyotype, NORs are located on each of the short arms of the acrocentric chromosomes 13, 14, 15, 21 and 22. Each locus has varying rRNA outputs under different conditions. Potentially, there could be 20 NORs at a metaphase. However, this is a very transient phenomenon and 10 can generally be regarded as a full complement in diploid cells. Differentiation *per se* affects the number of NORs seen but, in general, the level of proliferation has the major influence. The production of rRNA is complex and requires numerous enzymes, transcription factors, nuclear RNA, and regulatory factors.

Evidence is accumulating that nucleoli functionally interact with coiled bodies and are also involved in the activation of non-ribosomal RNA species. The nucleolus has been implicated in the site of assembly of the HIV-1 Rev protein, with the nuclear export of factor Crm1 (also termed exportin-1) and the partially soluble nucleoporins Nup 98 and 214, raising the possibility that the nucleolus is involved more generally in the export of proteins. The nucleolus has also been implicated in the ageing process and in inherited diseases such as dyskeratosis congenita (DKC) and Treacher Collins syndrome [7].

The methodologies included in this chapter describe histochemical and immunohistochemical procedures that have been central to the study of basic nucleolar function and the involvement of nucleoli in cell proliferation and in a wide range of disease processes. Other procedures that have either been in use for many years (such as nucleolar isolation and relatively simple morphological identification techniques) or have been introduced more recently (such as the modern fluorescence and gene-labelling procedures) are addressed but with suitable reference to the current literature.

6.2 The structure and function of the nucleolus

Ribosome biogenesis is necessary for cellular adaptation, growth, and proliferation, as well as major energetic and biosynthetic demands upon cells; the process requires precise spatial and temporal regulation to balance supply and demand [8]. Advances in microscopical and biochemical technology are now giving us a much greater insight into how the nucleolus functions in relation to its structure.

6.2.1 Nucleolar structure

There are three distinct sub-domains within mammalian nucleoli (*Figure 6.1*), named according to their ultrastructural appearance:

Granular component (GC). This is the main body of the nucleolus and comprises of particles 15–20 nm in diameter. The GC is the area with the greatest mass and volume and is where pre-ribosomes are processed, packaged and stored – they contain rRNA and associated proteins such as B23, nucleolin, ribosomal S6 and No55.

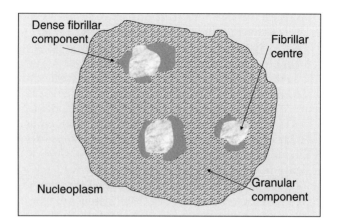

Figure 6.1

Ultrastructural appearance of the nucleolus.

Fibrillar centre (FC). These are found embedded in the GC, as several islets of rounded structures of relatively low contrast. They represent the interphase component or counterpart of chromosomal NORs and are probably the site of primary rRNA transcription. The FC is essentially a storage site for inactive rDNA and proteins such as RNA Pol-I, DNA topoisomerase I and UBF. There are usually several large interconnected FCs in resting phase nucleoli; on stimulation these unravel to become smaller and more numerous, particularly in mitogen-stimulated and malignant cells.

Dense fibrillar component (DFC). These are tightly packed, densely staining fibrils surrounding the FCs. They are the proposed sites of early processing of the rRNA precursor. They stain with monoclonal antibodies to fibrillarin, a protein associated with the small nuclear ribonucleoprotein, SnRNP. The DFC is a site for active transcription and early processing so ribosomal nucleic acids (rDNA and rRNA) and other associated proteins (nucleolin, Nopp 140, NAP57 and DNA topoisomerase I) are found.

Transcription probably occurs near the border of the FC and the DFC. Processing of nascent pre-rRNAs takes place in the DFC. Although in many cases intracellular compartmentalisation relies on boundaries imposed by membranes, nucleoli are among several non-membrane-bound compartments in eukaryotic cells and the emerging view is that molecular confinement in the nucleolus contributes to the control of cellular survival and proliferation [9]. Available data currently confirm the hypothesis that rDNA transcription is connected with the DFC but not necessarily with the presence of an FC. Within the DFC, rDNA transcription is restricted to foci possibly representing single transcribing genes. FCs may serve to store active transcription factors to initiate rDNA transcription and may provide support for transcription. GC can be interpreted as a collection of maturing peri-ribosomes [10]. A novel nucleolar RNA-associated protein (Nrap) is a large molecule (over 1000 amino acids) that appears to be associated with ribosomal biogenesis by interacting with pre-rRNA primary transcript [11].

The image of the stretched rDNA axis with attached polymerases and attached rRNA transcripts with terminal nodes ('Christmas Trees'), and how such structures are arranged in the nucleolus, has been discussed by Mosgoeller *et al.* [12]. Other nucleolus-related bodies have also been described. Nucleolar proteins are found in many locations outside the nucleolus, for example in the pre-nucleolar bodies (PNBs – containing fibrillarin, nucleolin and B23) and coiled bodies (CBs – containing their own specific protein, coilin, as well as fibrillarin). The structure and function of the nucleolus as a site of rRNA synthesis has been well reviewed [13, 14]. Among several described peri-nucleolar structures, the PNC and the Sam68 nuclear body (SNB) are found predominantly in transformed cells. They are dynamic, functional organelles, highly enriched with RNA-binding protein and nucleic acids, whose relationship to each other and the nucleolus is still poorly understood [15]. In eukaryotic cells, rRNA genes are transcribed by RNA pol I in the nucleolus as a 40S pre-rRNA. The mature rRNA species (28S, 18S, 5.8S) are formed after a series of specific endonucleolytic cleavage reactions. Various proteins involved in rRNA gene and rDNA transcription have been reviewed by Jacob [16].

The importance of the nuclear matrix, and the nucleolus in particular, in pathology has been discussed [17]. The existence of anti-nucleolar antibodies has been known for some time in human connective tissue disease – anti-nucleolar auto-antibodies, for example, occur predominantly in scleroderma, with a small overlap in polymyositis.

6.2.2 Nucleolomics – a proteomic inventory of the human nucleolus

The foundation for proteomics was established some 30 years ago with the building of protein databases and the practice of 2D gel electrophoresis [18]. Approaches that exploit protein–protein interactions, such as the yeast two-hybrid system and co-immunoprecipitation analysis, have provided a means for identification of a modest number of nucleolar proteins, but identifying the balance of the nucleolar proteome should provide a basis for future research into these ancillary functions.

Proteomic analyses have already been performed for the nucleolus; for example, a proteomic analysis of nucleolin-binding complexes has been used to clarify the possible role of nucleolin in ribosome biogenesis [19]. Recent advances in mass spectrometry, in combination with database analysis, has allowed the identification of protein components of extremely complex mixtures including spliceosomes, interchromatin granule clusters, centrosomes and spindle pole bodies, nuclear pore complexes and the Golgi apparatus. Such an approach has now been used in a proteomic analysis of the human nucleolus – the largest such analysis to date of a cellular compartment. This first proteomic analysis of nucleoli was performed in the laboratories of Angus Lamond (University of Dundee, UK) and Matthias Mann (University of Southern Denmark, Odense, Denmark) [20]. These researchers isolated nucleoli from HeLa cell nuclei and separated the proteins by one-dimensional and 2D SDS-PAGE. Tryptic peptides were then subjected to mass spectrometry using either matrix-assisted laser desorption-ionisation time-of-flight mass spectrometry or nanolectrospray tandem mass spectrometry. Using a peptide sequence tag search algorithm adapted for scanning the human genome draft sequence, Anderson *et al.* [20] identified 271 gene products in the nucleolar protein preparation. By searching for conserved motifs, the nucleolar proteins were grouped into putative functional categories. Although many of the proteins were expected, some were of unknown significance – an outcome that typifies proteomic studies in the initial, analytical stage [21]. Following treatment with actinomycin D to inhibit transcription, many proteins localise to individual domains as peri-nucleolar caps. These include coilin, fibrillarin, the Cajal body signature protein and PSP2. Such proteins are found simultaneously in both nuclei and at least one other nuclear domain under normal conditions, showing that nucleolar cap formation is a normal process and not just an artefact of the actinomycin D treatment. As a by-product of the proteomic work, a novel nuclear compartment, the paraspeckles, was identified [22, 23]. PSP1 is recruited to the nucleolus after inhibition of transcription. This suggests that, in common with the nucleolus, paraspeckles are dynamic bodies that respond to the changing environment of the cell and may be important in transcription.

Nucleoli are thought to play a role in various other activities, such as the maturation of certain mRNAs, the sequestration of regulatory molecules and the maturation of telomerase and signal recognition particle ribonucleoproteins. The identification of proteins that localise to the nucleolus will provide a crucial first step to many studies focused on understanding nucleolar function, dynamics and interaction with other sub-nuclear structures.

Many of the nucleolar proteins so far identified are involved in rRNA synthesis and processing and in ribosome assembly. Others are indicative of more diverse nucleolar function including numerous nucleic acid and nucleotide-binding proteins, the so-called DEAD box proteins (the tetrapeptide Asp-Glu-Ala-Asp), kinases, chaperones and translation factors (such as eIF5A, eIF4A1 and eTF1), supporting a role for nucleolar translational activity. Following the compilation of the new catalogue of nucleolar proteins by biochemical analysis, numerous surprising proteins have been found to be present, indicating novel functions of the nucleolus beyond ribosome biosynthesis.

The nucleolar proteomic study highlights the dynamic nature of the nucleolar proteome and shows that proteins can either associate with nucleoli transiently or can accumulate only under specific metabolic conditions. FLIP experiments have shown that, in transcriptionally active cells, PSP1 cycles between nucleoli and paraspeckles [5].

The current proteome list may include non-nucleolar, contaminant proteins – without a membrane, the nucleolus is notoriously difficult to separate from the nucleus and even the purest sample may contain molecules that accidentally stuck to it during the split. In addition, several dozen known ribosomal proteins are absent from the list. Almost certainly there are more nucleolar proteins to be found. Although there is much more to be done, an excellent start has been made. Of the remaining 90 or so proteins to be further investigated, it is realised that they participate in a range of functions. They include molecular chaperones, which prevent other proteins from sticking together, translation machinery, which coaxes mRNAs and ribosomes to form proteins, and DEAD-box proteins, which control the structure of RNA. The role of this wide range of proteins in disease is also only just becoming recognised. Treacher Collins syndrome, involving an autosomal dominant disorder of craniofacial development, was the first Mendelian disorder to be described of aberrant expression of a shuttling nucleolar phosphoprotein ('Treacle') resulting from mutation. Treacle is structurally related to Nopp140 but it is only now starting to be characterised [24]. Unlike Nopp140, it fails to co-localise to Cajal (coiled) bodies.

6.2.3 The role of the nucleolus in cell proliferation

NORs are the initiation sites for nucleologenesis where pre-nucleolar bodies fuse to form interphase nucleoli. An examination of the distribution of nucleolar proteins during the cell cycle has shown that nucleolin, fibrillarin and pre-RNAs are synthesised at the G_2/M phase and are readily recruited to UBF-associated NORs early in subsequent telophase before chromosome condensation. This is independent of RNA pol-I transcription and this may underlie nucleolar formation at the end of mitosis [25]. Mammalian nuclei contain three different RNA polymerases defined by their characteristic locations and drug sensitivities. Only RNA pol-I is found in the nucleolus – polymerases II and III are located in the nucleoplasm. The active forms of all three nuclear polymerases are concentrated in their own dedicated transcription 'factories', suggesting that different regions of the nucleus specialise in the transcription of different types of gene [26].

NORs are the sites of transcription of rRNA, the post-transcriptional modification of the RNA transcripts, and their assembly into functional ribosomes. The number of NORs expressed in a tissue has been related to the rate of cell proliferation, differentiation, and to neoplastic transformation. This has been used to demonstrate neoplastic potential and for evaluating the prognosis and aggression of malignant neoplasms. Models for studying the initiation of NOR activity have been devised by silver staining the NOR regions in human lymphocytes at various times after PHA activation [27]. An increase in NOR activity was seen as early as 4 hours after PHA treatment. As the size of the cell nucleus increased, the number and amount of silver staining increased as the cells progressed through the cycling phase [28]. Mechanisms governing both formation and maintenance of functional nucleoli seem to involve CDK activity [29] and the amount of two major AgNOR proteins, nucleolin and the protein B23, is cell cycle dependent, their amount being high in S/G_2 and low in G_1 [30, 31]. In the last couple of years, three interesting cell cycle regulators have been identified. Localisation of the protein phosphatase Cdc14 in the cell cycle during G_1, S phase and early mitosis, shows the molecule is sequestered in the nucleolus by Cfi1/Net1. After nuclear division is commenced, Cdc14 spreads into the nucleus and cytoplasm. It then dephosphorylates Cdh1 and Sic1 thereby promoting the inactivation of mitotic kinases leading to exit from mitosis [32]. Cdc14 – critical for exit from mitosis – is kept inactive in the nucleolus until the onset of anaphase, so preventing premature exit from mitosis. Mdm2 – an inhibitor of tumour suppressor protein sequestered in the nucleolus in response to activation of the oncoprotein Myc or replicative senescence, allowing p53 to become active – induces cell cycle arrest in response to DNA damage. Pch2 is a protein required for meiotic cell cycle progression in response to recombination and chromosome synapsis defects, and also localises to the nucleolus. This suggests a possible role for nucleolar proteins and the nucleolus in checkpoint signalling.

6.3 Techniques for studying nucleolar activity

A wide range of procedures is available for studying nucleolar structure and function. The 3D organisation of ribosomal genes has been studied by flow cytometry, confocal microscopy and electron microscopy to illustrate how they vary along with the architecture of the nucleolus during the G_1, S and G_2 phases of mitosis [33]. Different interpretations have been discussed for the location of transcription and processing of rRNA into ribonucleoproten (RNP) in relation to the ultrastructural components of the nucleolus [34].

6.3.1 Morphological procedures

Simple morphological staining methods, as described in standard histology texts, permit the visualisation of nucleoli under the LM, and these include general nuclear stains such as haematoxylin, the azure dyes, toluidine blue and Giemsa.

6.3.2 Procedures for preparing purified nucleolar isolates and investigating nucleolar function

Nucleoli are not bounded by membranes and, as the intact structure is not differentially extractable by salts, so the usual biochemical fractionation methods do not pertain. Instead, nucleoli must be detached from the nucleoplasm by extensive sonication, followed by sucrose centrifugation. The purity and integrity of the fractions can be confirmed by light microscopy and electron microscopy at each step of the procedure. Once a suitably pure fraction has been obtained, identification of the proteins can proceed. Various procedures have been described for isolating nucleoli, analysing individual nuclear components, and studying nuclear activity – with and without selective drug inhibition of nuclear function. The reader is recommended to research these in the established literature [27, 35].

Inducing apoptosis using short-term hypertonic stress leads to RNP rearrangement with the formation of heterogeneous ectopic RNP-derived structures (HERDS) which pass into the cytoplasm. Nucleolus-like bodies that resemble morphologically functional nucleoli may be extruded into the cytoplasm of apoptotic cells and are observed inside the cytoplasmic fragments blebbing out at the cell surface [36].

6.3.3 Use of GFP to study nucleolar dynamics

In the last 10 years, GFP has changed from a nearly unknown protein to a commonly used biological marker. Fusion of GFP to a protein does not seem to alter the function or location of that protein. The use of GFP is outside the scope of this chapter, but the subject has recently been excellently reviewed [37]. The dynamics of nucleolar reassembly have been studied for the first time in living cells using GFP fusions with the processing-related proteins fibrillarin, nucleolin and B23 [38]. Entry of specific proteins into the nucleolus was examined in relation to the timing of processing events showing that mitotically preserved processing complexes are implicated in regulating the distribution of components to reassembling daughter cell nucleoli.

6.3.4 Histochemical procedures based on the argyrophilic properties of NORs

Silver solutions have been used for staining cells and chromosomes for at least 100 years – since the days of the early silver stain neuropathologists. NORs have been defined as nucleolar components containing a set of argyrophilic proteins which are selectively stained by silver solutions, hence the term 'argyrophilic nucleolar organiser regions'. A number of studies carried out in different tumour types have demonstrated that malignant cells frequently present a greater AgNOR protein content than corresponding non-malignant cells. Moreover, in cancer tissues, AgNOR protein expression is strictly related to cell duplication rate. Consequently, AgNOR enumeration has become a focus of attention in cell proliferation studies over the last 30 years. The evolution of the AgNOR technique has been reviewed [39].

A reliable technique for staining human NORs was first described in 1976 [40] when the authors discovered that highly selective staining of NORs can be achieved after incubation in 50% silver nitrate without the need for a chemical developer. Prior to this time, methods already existed for NOR silver staining but these suffered from the major problems of excess background, non-uniformity of staining across a slide, inconsistent staining due to variations in staining time, and occasional total failure to achieve staining.

Early studies showed that during blast formation of human lymphocytes following PHA stimulation, silver-stained nucleolar regions increased both in size and in number with characteristic changes in shape [41]. After silver staining, the NORs can be easily identified as black dots exclusively localised throughout the nucleolar area. The interphase AgNORs are structural-functional units of the nucleolus in which all the components necessary for rRNA synthesis are located. Two argyrophilic proteins involved in rRNA transcription and processing, nucleolin and nucleophosmin, are largely responsible for stainability with silver [42].

During interphase, the NORs are located in the DFCs of the nucleoli. Loops of rDNA are responsible for the transcription of the rRNAs (except 5S rRNA) where the enabling NOR proteins are located (RNA pol-1, DNA topoisomerase-1, B23, C23 and fibrillarin). AgNOR relative area shows a strong correlation with the changes of the population doubling time induced by different cell culture temperatures. Although the correlation between AgNORs and the so-called proliferation markers (MIB-1/Ki-67, PCNA, p532) remains debatable, most agree that that AgNOR size or number is related to proliferative activity – the larger the AgNORs, the shorter the population doubling time. Combined MIB-1 and AgNOR staining and cytometry has been used to show that the quantity of AgNOR proteins in cycling cells, cell cycle time and the size of the ribogenesis machinery are co-regulated – measurements of AgNORs can thus be used as a static evaluation of the cell cycle duration in arbitrary units [43].

Lack of a standardised silver staining protocol has led to much misinterpretation of actual structures evaluated in individual studies. Indeed, the absolute AgNOR scores reported by different authors for the same types of tumours are sometimes scarcely comparable – even results produced in the same investigation may conflict. Problems commonly associated with earlier AgNOR demonstration techniques, incorporating developing reagents, have been addressed in detail [44, 45]:

1. In older protocols, development time can not be standardised because the ammoniacal silver and formalin developer solutions are unstable, causing under- or over-development of AgNORs.
2. Silver precipitate over the slide caused by too-rapid development can obscure the image and microscopic monitoring of the AgNOR development is usually required.
3. Uneven staining of AgNORs occurs across the slide making reliable counting difficult.
4. Incubation in aqueous silver nitrate alone is time consuming.
5. Expense can be a problem. The use of ammoniacal silver solutions is wasteful; solutions have a short shelf life and must be discarded after a few days.

In 1993, Lindner [46] examined various factors surrounding the range of persistent problems encountered with AgNOR demonstration, and the findings are summarised as follows:

1. Limited reliability and reproducibility.
2. The AgNOR procedure is extremely temperature dependent. Although increase in staining temperature reduces staining time, room temperature gives the best control over labelling.
3. Laboratory temperatures need careful monitoring as there is a 2 minute change in staining time for every degree Celsius change.
4. General background staining of tissues makes weakly stained NORs difficult to identify and NORs, in general, harder to resolve. Oxidation of sections, before silver staining using hydrogen peroxide or potassium dichromate, increases NOR staining but also raises the background level. Reduction with 1% potassium iodide, 1% sodium bisulphite or 1% DTT (Cleland's reagent) decreases non-specific silver deposition without affecting NOR demonstration. Pre-reduction treatment with alcoholic acetic acid has little effect although pre-incubation with glycine may improve AgNOR demonstration [47].
5. Staining of granules in the cytoplasm can be confused with NORs, especially using automated image analysis.
6. Black precipitates on the slide can confuse NOR analysis and such precipitates are almost certainly due to silver proteinate crystal formation. Using other protective protein colloids tends to compound the problem. The level of precipitate is variable depending on the gelatin source and it is worth examining different gelatins from different suppliers to see which produces the best results.

7. Preparations fade, often within days, due to residual chemicals in the section and acidic components of the mounting medium.

8. Variation in reagent concentrations is to be avoided. Lowering the silver concentration increases impregnation times markedly and lowering the formic acid concentration decreases staining. There are no advantages to raising the concentration of either reagent.

9. Variation in specimen fixation can drastically alter silver labelling results. Studies of the effects of different fixatives on the AgNOR procedure have shown that, in general, alcohol-based fixatives and conventional formalin fixation gives optimal results whereas mercurial and dichromate fixatives are highly detrimental [48, 49].

There have been many attempts to introduce a standardised approach to AgNOR quantitation. In order to achieve a definitive standardisation of the AgNOR method and produce comparable data in all laboratories, the 'International Committee on AgNOR Quantitation' was founded under the auspices of the European Society of Pathology, and during the first workshop 'AgNORs in Oncology' held in Berlin in 1993, guidelines for AgNOR protein evaluation were first established [50]. The following meeting in Regensburg in 1994 [51] and associated publications from committee members [52] examined the problems surrounding AgNOR quantification. Subsequent meetings in Reims (1995), Taormina (1996), Zaragoza (1997), Bologna (1998) and Innsbruck (1999) looked at AgNORs and rRNA transcription, the nucleolus in cell proliferation and the cell cycle, nucleologenesis and nucleolar proteins, explored the range of clinical application for AgNORs in cell proliferation and oncology, and reviewed technical problems with the intention of achieving reliable and reproducible results [53, 54]. These meetings have brought together the results of many research groups using powerful microscopical techniques and new combinations of molecular and structural research to uncover the details of the complex relationships between transcription, transcript processing and protein annexation in the formation of ribosomes in this large sub-domain of the eukaryote nucleus. Some centres have made dedicated studies of the use of the AgNOR evaluation and have published appraisals of the technology with attempts to suggest a standardised means of enumeration of NORs. Such reports have paid particular attention to rigorous technique and careful resolution of intra-nucleolar AgNOR dots [55, 56]. Standardised full and short (one step) AgNOR methods are described in Protocols 6.1 and 6.2 [57, 58].

Prior to the international AgNOR meetings, a series of European Nucleolar Workshops on basic nucleolar biology had also been held, and the first attempt to standardise the nomenclature associated with the nucleolus was made at the congress in Banyuls-sur-mer in 1983 [59].

Whereas most diagnostic pathology applications using AgNOR labelling have been with histological sections, there have been many reports evaluating its usefulness with cytological preparations, one of the original descriptions being for cell imprints of lymphoid tissue (Protocol 6.3). Absolute AgNOR counts in particular were found to be at least as accurate, if not more accurate, with single cell smears than in concomitant biopsy sections [60].

Treatment of sections at superheated steam temperatures before silver incubation leads to significantly better discrimination of single interphase AgNORs, counting subsequently either by eye or image analyser. Wet autoclave pretreatment is appropriate for routine use without substantial damage to sections or nuclear morphology – it enhances discrimination of AgNORs without evidence of any staining artefacts. This leads to improved staining quality which is homogeneously performed throughout the whole section and is apparently independent of tissue origin as well

Protocol 6.1: Basic AgNOR room temperature procedure for cryostat and paraffin wax sections (adapted from [39, 40, 44, 46, 57])

EQUIPMENT, MATERIALS AND REAGENTS

Light-proof humidified staining box.

5% sodium thiosulphate.

95% industrial methylated spirit (IMS).

Methanol:chloroform:glacial acetic acid fixative, 6:3:1 (by volume).

Carnoy's fixative: IMS:glacial acetic acid, 3:1 (by volume).

Graded alcohols: 50%, 70%, 90% and undiluted IMS.

Xylene.

Distrene Plasticizer Xylene (DPX).

Silver solution: dissolve 1 g gelatin in 100 ml 2% (v/v) formic acid and keep as stock solution at 4°C. Dissolve 5 g silver nitrate in 10 ml distilled water and keep as stock solution in a dark-glass bottle at room temperature. Immediately before use, mix 1 part stock gelatin solution with 2 parts stock silver nitrate solution.

Alternative silver solution [I]: 15% silver nitrate in 1% formic acid and 45 dl of 2% gelatin in 1% formic acid are equilibrated to 37°C for 1 hour and then mixed. Slides are incubated at 37°C.

Alternative silver solution [II]: Mix 2 volumes of 0.5% silver nitrate solution with one volume of 1 g ml^{-1} 20 000 polyethylene glycol (PEG) in 2% (v/v) formic acid.

Nuclear fast red. Dissolve 0.1 g nuclear fast red and 5 g aluminium sulphate in 100 ml hot distilled water. Allow to cool and filter.

Aniline blue. Dissolve 0.2 g water-soluble aniline blue powder in 100 ml 0.2% acetic acid. Filter before storing at room temperature.

Light green. Dissolve 0.2 g light green powder in 100 ml 0.2% acetic acid. Filter before storing at room temperature.

Methyl green. Dissolve 2 g methyl green powder in 100 ml distilled water. Filter before storing at room temperature.

Neutral red. Dissolve 1 g neutral red powder in 100 ml 0.1% acetic acid. Filter before storing at room temperature.

METHOD

1a. Cryostat sections. After drying, fix the sections for 2 minutes in 95% ethanol or in methanol:chloroform:glacial acetic acid. Post-fix in Carnoy's fixative for 30 minutes and rehydrate through graded alcohols to distilled water.

1b. Paraffin wax sections of alcohol-fixed tissues. De-wax sections in 3 changes of xylene, 2 minutes each change. Remove xylene from slides using 3 thorough changes of IMS. Post-fix in Carnoy's fixative for 30 minutes. Hydrate through graded alcohols to distilled water.

1c. Paraffin wax sections of conventional formalin-fixed tissues. De-wax sections in 3 changes of xylene, 2 minutes each change. Remove xylene from slides using 3 thorough changes of IMS. Follow Protocol 6.4 if required, and rinse in distilled water.

2. Transfer slides to the freshly prepared silver solution. Leave for 15 to 45 minutes in the dark. Reactions are performed at room temperature unless alternative silver solution II is used. Staining time should be reduced if nucleoplasmic staining is a problem. For preference, use the standard silver solution; try the alternative silver solutions if unsatisfactory results persist. It has been reported that background staining may be minimised by performing the reaction with the slide inverted [58]. To reduce variability between batches, it may be best to perform incubations in a thermostatically controlled environment at 37°C [54].

3. Rinse slides thoroughly in distilled water.

4. Post-treatment in 5% sodium thiosulphate for 3 minutes.

5. Rinse slides thoroughly in distilled water.

6. Optional toning of silver reaction (see Protocols 6.6 and 6.7).

7. Rinse slides thoroughly in distilled water.

8. Counterstain nuclear material using either nuclear fast red, aniline blue, light green, methyl green or neutral red for 15 seconds to 5 minutes according to required intensity.

9. Rinse slides thoroughly in running tap water.

10. Dehydrate slides thoroughly using 3 changes of IMS.

11. Clear slides thoroughly using 3 changes of xylene.

12. Mount sections in neutral synthetic mounting medium. Examine under LM. AgNORs appear as black dots, or purple-black/blue-black following toning.

Protocol 6.2: One step 70°C procedure for general purpose, rapid AgNOR demonstration [45]

EQUIPMENT, MATERIALS AND REAGENTS

Slide warmer.

5% sodium thiosulphate.

95% IMS.

Methanol:chloroform:glacial acetic acid fixative, 6:3:1 (by volume).

Carnoy's fixative: IMS:glacial acetic acid, 3:1 (by volume).

Graded alcohols: 50%, 70%, 90% and undiluted IMS.

Xylene.

DPX.

Colloidal developer solution: dissolve 2 g powdered gelatin in 100 ml deionised water containing 1 ml concentrated formic acid. Stir constantly for 10 minutes in order to dissolve the gelatin. Store in capped, amber-glass bottle. Solution is stable for 2 weeks.

Aqueous silver nitrate: dissolve 4 g silver nitrate in 8 ml deionised water. Store in capped, amber-glass bottle. Solution is stable over long periods.

METHOD

1. Prepare de-waxed paraffin sections or appropriately fixed cryostat sections/cytology slides.

2. Place 2 drops of the colloidal developer and 4 drops of the aqueous silver nitrate onto the preparation.

3. Mix the solutions and cover with a coverglass.

4. Place the slide on the surface of a slide warmer previously stabilised to 70°C. Within 30 seconds the silver-staining mixture will turn yellow, and within 2 minutes it will become golden-brown.

5. Remove the slide from the hotplate and remove the coverglass/silver mixture using running deionised water.

6. AgNORs are uniformly stained black against yellow nuclear material.

7. Scrupulous attention must be paid to slide and sample cleanliness. Any background deposition is likely to result from dust on the preparation.

Protocol 6.3: AgNOR procedure for cytological preparations (adapted from [60, 64])

EQUIPMENT, MATERIALS AND REAGENTS

Light-proof humidified staining box.

5% sodium thiosulphate.

95% IMS.

Methanol:chloroform:glacial acetic acid fixative, 6:3:1 (by volume).

Carnoy's fixative: IMS:glacial acetic acid, 3:1 (by volume).

Graded alcohols: 50%, 70%, 90% and undiluted IMS.

Xylene.

DPX.

Silver solution: Dissolve 1 g gelatin in 100 ml 2% (v/v) formic acid and keep as stock solution at 4°C. Dissolve 5 g silver nitrate in 10 ml distilled water and keep as stock solution in a dark-glass bottle at room temperature. Immediately before use, mix 1 part stock gelatin solution with 2 parts stock silver nitrate solution.

METHOD

1. Touch imprints, smears and cytospin preparations are air dried for at least 5 minutes.

2. Cell preparations are fixed in 100% ethanol for 5 minutes.

3. Cell preparations are rehydrated through graded 90%, 70%, 50% IMS baths.

4. Drop silver solution onto the slides and keep in a humidified chamber, protected from daylight, for 30 minutes at room temperature.

5. Wash slides with deionised water.

6. Stain for 30 seconds with 1% fast red, or examine without counterstain.

7. Dehydrate through 70%, 90% and 100% IMS.

8. Clear preparations with xylene and mount in DPX.

as of the duration of formalin fixation and of archival storage [61] (Protocol 6.4). It has also been reported that both fixation and staining can be improved by microwave fixation [62] (Protocol 6.5). It was suggested that microwave irradiation probably enhances fixation by controlled heat, whereas the increase in activity of the staining solution is a direct effect of the microwaves on the silver ions themselves. Post-microwave silver staining fixation in Farmer's solution has also been described as a suitable method for producing rapid, sharp AgNOR images [63] (Protocol 6.6).

Image cytometry is probably the method of choice for evaluation of AgNORs. Total area of AgNORs per cell is a basic indicator of AgNOR quantity. The coefficient of variation (CV) relative to the AgNOR protein area (CV = standard deviation of area per cell/mean area value) has been shown to be the most reproducible parameter of AgNOR protein quantity. This value should be determined in 100 cells using a 40× objective lens, when it is possible to see all cells of a given image in one focal plane. Counting of AgNORs by eye is the method of second choice and should be restricted to preparations exhibiting distinct silver-stained dots as definite nucleolar sub-structures [65]. The use of computerised image analysis systems for assessing AgNOR area represents a reliable means of comparing results obtained between different laboratories but standardisation of the staining procedure is of paramount importance [66–68].

A modification of the silver colloid technique for staining NORs in paraffin-embedded tissues has been described [69] which involves the application of a gold toning step with subsequent gold reduction following incubation of sections in the standard silver colloid solution. AgNORs in toned sections are more sharply delineated when compared to un-toned controls. In high-grade tumours the addition of the toning step results in significantly higher AgNOR counts due to the ability to discriminate more easily individual AgNORs in argyrophilic aggregates within the nucleus (Protocol 6.7). A straightforward and equally sensitive method for blue toning of AgNORs uses potassium hexacyanoferrate and ferric chloride [70] (Protocol 6.8). Since the introduction of photographic technology, toning solutions have been used to turn black and white silver images into a monochrome colour print. The process has been adapted for chromogenic enhancement of the sensitivity of AgNOR preparations [71] (Protocol 6.9). Combining conventional light studies of AgNOR labelling with fluorescence examination of DAPI labelling has permitted the simultaneous examination of NORs and DNA in the same nuclei [72] (Protocol 6.10). Frequency studies using the AgNOR-DAPI method have shown that different AgNOR shapes reflect both the number of ribosomal genes carried by each chromosome and the differential recruitment of active ribosomal genes in each NOR cluster, and that distribution of AgNOR shape among acrocentric chromosomes is non-random [73].

Combining a modified AgNOR procedure with the Feulgen reaction is useful for studying correlations between DNA activity and NORs in the same preparation. Cells are treated with partially deactivated Schiff's reagent, followed by Feulgen hydrolyis, silver staining and finally fresh Schiff's reagent. This sequence avoids interactions between the two components of the method that would be deleterious to the final result [74] (Protocol 6.11). There have been many combined applications of microscopical and histological procedures for examining AgNORs in relation to other aspects of cell structure and function. In some cases it is possible to combine protocols in this chapter with those found in other parts of the book. A method for the combined conventional light microscopy and FISH imaging of NORs and cellular rRNA has been described [75] which allows the correlation of specific NOR parameters to the cellular RNA content known to be related to cell growth. After combined specific labelling of DNA and AgNOR demonstration, it has been possible to study

Protocol 6.4: Wet autoclave pretreatment in AgNOR demonstration (adapted from [61])

EQUIPMENT, MATERIALS AND REAGENTS

Silane-coated glass slides.

Plastic Coplin jar.

Sodium citrate buffer: 0.01 M sodium citrate monohydrate pH 6.0.

Wet autoclave conditions or pressure cooker.

METHOD

1. Rinse suitably prepared preparations in distilled water.

2. Immerse preparations in citrate buffer and boil at 120°C for 20 minutes.

3. Cool down to room temperature.

4. Rinse in distilled water.

5. Proceed to silver staining in Protocol 6.1, 6.2 or 6.3.

Protocol 6.5: Microwave modification of the AgNOR procedure (adapted from [62])

EQUIPMENT, MATERIALS AND REAGENTS

Microwave processor with suitable temperature control probe.

Polystyrene block 2 cm high.

Fixative: 56% ethanol and 20% PEG in distilled water.

Absolute ethanol at 4°C.

Silver solution: Dissolve 1 g gelatin in 100 ml 2% (v/v) formic acid and keep as stock solution at 4°C. Dissolve 5 g silver nitrate in 10 ml distilled water and keep as stock solution in a dark-glass bottle at room temperature. Immediately before use, mix 1 part stock gelatin solution with 2 parts stock silver nitrate solution.

5% sodium thiosulphate.

METHOD

1. Roughly cuboidal, small (approximately 1 mm across) fragments of tissue are placed for fixation in 10 ml fixative.

2. The container is placed immediately in the centre of the microwave processor for irradiation.

3. Place two 250 ml water loads near the rear corners of the chamber.

4. The microwave power is set to 630 W and the cycle time to 2 seconds.

5. Irradiate for 3 minutes at 45°C, with continuous stirring by means of an air-bubbling device and using the microwave probe for temperature control.

6. Immediately following fixation, transfer samples to absolute ethanol at 4°C, and follow standard embedding protocols for the desired mode of sectioning.

7. Place polystyrene block near centre of the oven.

8. Take sections to distilled water. Dry the microscope slide thoroughly on the back and around the section.

9. Place 100 µl drop of freshly prepared silver solution over the section and transfer the slide to the top of the polystyrene block.

10. Irradiate as before, with the temperature control restricted to 38°C and with two pulses of 15 seconds spaced by a 30-second interval.

11. Rinse slides in distilled water, immerse in 5% sodium thiosulphate for 5 minutes, counterstain as required and mount in a neutral synthetic mounting medium.

Note: optimum microwave parameters should be investigated for different microwave units.

Protocol 6.6: Fixation – microwave modification of the AgNOR procedure (adapted from [63])

EQUIPMENT, MATERIALS AND REAGENTS

As for Protocol 6.5.

1% sodium thiosulphate in distilled water.

1% potassium hexacyanoferrate in distilled water.

METHOD

1. Place two 250 ml water loads near the rear corners of the oven chamber.

2. The microwave power is set to 630 W and the cycle time to 2 seconds.

3. Place polystyrene block near centre of the oven.

4. Take sections to distilled water. Dry the microscope slide thoroughly on the back and around the section.

5. Place 100 μl drop of freshly prepared silver solution over the section and transfer the slide to the top of the polystyrene block.

6. Irradiate with the temperature control restricted to 38°C and with two pulses of 15 seconds spaced by a 30-second interval.

7. Immerse the slides for 10 seconds in 1% Farmer's solution: mix 9 parts 1% sodium thiosulphate with 1 part 1% potassium hexacyanoferrate just before use. The duration for treatment should be adjusted according to the original staining intensity.

8. Rinse slides in distilled water, immerse in 5% sodium thiosulphate for 30 seconds, counterstain as required and mount in a neutral synthetic mounting medium.

Protocol 6.7: Gold chloride blue-black toning of AgNOR results (adapted from [69])

EQUIPMENT, MATERIALS AND REAGENTS

Stock solution: 2% gold chloride (chloroauric acid).

METHOD

1. Perform AgNOR demonstrations as described in any of Protocols 6.1–6.6, taking the preparations as far as distilled water following silver impregnation.

2. Flood the microscope slides with 0.05% gold chloride in distilled water, examining the preparations microscopically for sufficient blue-black colour change. According to the initial AgNOR staining this may take from a few seconds to several minutes.

3. Counterstain as required and mount in a neutral synthetic mounting medium.

Protocol 6.8: Potassium hexacyanoferrate (III) blue toning of AgNOR results (adapted from [70])

EQUIPMENT, MATERIALS AND REAGENTS

Toning solution made up in distilled water:
30 mmol l^{-1} ferric chloride;
11 mmol l^{-1} potassium hexacyanoferrate (III);
33 mmol l^{-1} oxalic acid.

METHOD

1. Perform AgNOR demonstrations as described in any of Protocols 6.1–6.6, taking the preparations as far as distilled water following silver impregnation.

2. Rinse slides in deionised water.

3. Tone preparations for 20–30 seconds.

4. Rinse slides in deionised water for 10 minutes.

5. Counterstain if required, dehydrate in ethanol, clear in xylene and mount in DPX.

Protocol 6.9: Chromogenic toning of AgNOR results (adapted from [71])

EQUIPMENT, MATERIALS AND REAGENTS

As for Protocols 6.1–6.6.

0.4% formalin.

0.6% sodium carbonate.

Rebromination solution: 10% potassium hexacyanoferrate III and 5% potassium bromide.

Cyan-blue (N-(1-hydroxy-2-naphtoyl)-2-(o-acetylaminophenyl) ethylamine).

Magenta red (1-(2,4,6-trichlorophenyl)-3-(p-nitroanilino)-2-pyrazolin-5-one).

Yellow dye (benzoylacetanilide).

Solution A: 5.0 ml 0.2 M NaOH in isopropanol containing 100 mg dye.

Solution B: 100 mg N^4n^4-diethyl-1,4-phenylendiamin (sulphate), 1.0 g Na_2CO_3, 50 mg KBr, 100 mg Na_2SO_4 dissolved in 45 ml distilled water.

METHOD

1. Perform AgNOR demonstrations as described in any of Protocols 6.1–6.6, taking the preparations as far as distilled water following silver impregnation.

2. Rinse slides in deionised water.

3. Treat with 0.4% formalin for 3 minutes.

4. Rinse slides in tap water.

5. Treat with 0.6% sodium carbonate for 3 minutes.

6. Rinse slides in tap water.

7. Treat with rebromination solution for 1 minute.

8. Rinse slides in distilled water.

9. Mix solutions A and B in a ratio of 1:10 and treat slides for a range of times between 30 seconds and 10 minutes.

10. 5% sodium thiosulphate for 5 minutes.

11. Counterstain if required, rinse in water, dehydrate in IMS, clear in xylene and mount in DPX.

Protocol 6.10: Combined DAPI and AgNOR demonstration of nuclear and nucleolar components (adapted from [72])

EQUIPMENT, MATERIALS AND REAGENTS

As for Protocols 6.1–6.6.

Dual illumination microscope with epifluorescence incorporating a DAPI filter and bright field diascopic illumination.

$1 \mu g \, ml^{-1}$ DAPI in distilled water.

PBS.

1% glycerol in PBS.

METHOD

1. Perform AgNOR demonstrations as described in any of Protocols 6.1–6.6, taking the preparations as far as distilled water following silver impregnation.

2. Counterstain with DAPI solution for 30 minutes at room temperature in the dark.

3. Rinse in PBS.

4. Mount preparations in glycerol-PBS and examine under combined epifluorescence and bright field illumination.

Protocol 6.11: Combined Feulgen and AgNOR demonstration of nuclear and nucleolar components (adapted from [74])

EQUIPMENT, MATERIALS AND REAGENTS

Silver solutions in Protocols 6.1 or 6.3, as appropriate.

Schiff's reagent – purchase from commercial source.

Partially deactivated Schiff's reagent.

5 M HCl.

37% sodium hydrogen sulphite in distilled water diluted 1:20 by volume in 0.005 M HCl.

METHOD

1. Fix preparations as described in Protocols 6.1 or 6.3, as appropriate.

2. Immerse preparations in partially deactivated Schiff's reagent in the dark at room temperature for 1 hour.

3. Rinse slides 3 times in distilled water.

4. Perform hydrolysis in 5 M HCl at 22°C for a range of test times from 5–60 minutes.

5. Halt hydrolysis by washing slides in distilled water at 4°C for 5 minutes.

6. Incubate in silver solution in a humidified chamber at 60°C for between 5–60 minutes.

7. Rinse thoroughly in distilled water.

8. Stain slides in fresh Schiff's reagent for 70 minutes in the dark at 22°C.

9. Rinse slides in sodium hydrogen sulphite solution – 3 successive baths for 2 minutes each.

10. Rinse well in distilled water, dehydrate in IMS, clear in xylene and mount in a neutral synthetic mounting medium.

the 3D organisation of silver-stained structures by confocal microscopy, showing that the argyrophilic components are organised as a twisted necklace structure within the interphase nuclei [76]. Laser scanning cytometry has been used to study nucleolin expression through phases of the cell cycle of mitogenically stimulated lymphocytes [77].

The potential qualitative and quantitative role of AgNOR demonstration in pathology has been discussed at length [78–80]. Double immunolabelling using Apase-Fast red fluorescence to reveal Ki-67/MIB-1 antigen on cycling cells and AgNOR staining has been described with flow cytometric analysis, making it possible to compare growth fraction and cycling speed of partially proliferating cell populations such as tumours [81]. Immunoperoxidase localisation of BrdU combined with AgNOR labelling of sections from the same biopsy tissue blocks has been used to provide kinetic and proliferation information in the study of tumour metastases [82].

The distribution and balance of AgNORs during interphase and mitosis related to the nucleolar ultrastructural regions has been demonstrated using combined light and electron microscopy [83]. It has been shown that particular types of bone marrow cell have characteristic AgNOR staining patterns and that AgNORs can be used in conjunction with conventional stains to provide more information on the different cell types seen in leukaemia samples [84]. In an attempt to further classify the role of AgNOR proteins in cancer diagnosis and prognosis, AgNOR staining profiles on western blots and under electron microscopy have been used to show the relative distribution of C23 and B23 and nucleolar phosphoprotein pp135 in tumour material [85]. Sequential silver staining and *in situ* hybridisation has been used to demonstrate the relative rDNA content of a NOR, determining its level of expression and the likelihood of it becoming active [86]. Thus AgNOR staining, although a powerful technique used alone and in its simplest or modified forms, is extremely versatile when used in conjunction with other modern microscopical procedures.

6.3.5 Immunocytochemical procedures for analysing nucleolar structure and function

Subsequent to the identification of such a host of nucleolar proteins, it was only to be expected that the immunocytological tracking of these molecules (see Protocols 6.12, 6.13, 6.14, 6.15 and 6.16), together with the application of molecular biology procedures to identify underlying gene activity, would become widely used. *In situ* and nitrocellulose filter staining of AgNOR proteins have led to the identification of some members of this family, which include nucleophosmin (B23), nucleolin (C23), RNA pol-I large subunit, and UBF [87]. Immunofluorescence and confocal microscopy studies of optical sections through mitotic cells from prophase to telophase were used to generate 3D images of nucleolar antigen distribution, showing that there is active targeting and transfer of nucleolar proteins away from the nucleolus to other specific nuclear sites [88]. This was the first indication that nucleoli participate in the protection and stablilisation of chromosomes with possible interaction with telomeres.

Immunohistochemical localisation of UBF (for which the genetic profile is now complete [91]) has been useful, especially in combination with AgNOR staining, in the histopathological diagnosis of malignancy [92] (*Figure 6.2*) and the cytological diagnosis of post-transplantation renal allograft dysfunction [93] (*Figure 6.3*). UBF and AgNOR labelling has also been used to identify a link between ribosomal biogenesis events and apoptosis in lymphoid neoplasia [94]. The structure of the human UBF gene has been shown to consist of 20 exons with intervening sequences, and spans approximately 15 kilobases of DNA [95].

Protocol 6.12: Immunoperoxidase procedure for intranuclear immunocytochemistry

EQUIPMENT, MATERIALS AND REAGENTS

Cold acetone (4°C).

Cytological fixative: methanol:acetone, 1:1 (by volume).

PBS pH 7.6.

0.5% copper (II) sulphate in 0.85% aqueous sodium chloride.

0.05% aqueous gold chloride prepared by diluting 2% yellow gold chloride (chloroauric acid) 1:40 with distilled water.

Horseradish peroxidase-conjugated secondary antibody. This should be directed at the appropriate species and immunoglobulin class according to the animal in which the primary antibody was raised.

Peroxidase substrate: immediately prior to use:
 (i) Add 80 ml PBS to a Coplin jar.
 (ii) Dissolve 65 mg 3,3'-diaminobenzidine tetrahydrochloride.
 (iii) Add 1% cobalt chloride at this stage if a blue-black result is required.
 (iv) Add 200 µl of hydrogen peroxide (20 volumes) to the Coplin jar.

IMS.

Xylene.

Neutral synthetic mounting medium.

METHOD

1a. Air dry cryostat sections for 1 hour and cytological preparations overnight.

1b. Fix cryostat sections for 10 minutes in cold acetone, and cytological preparations in methanol-acetone for 90 seconds at room temperature.

1c. De-wax paraffin sections using xylene and rinse thoroughly in IMS.

2. Rinse slides 3 times, 2 minutes each, in PBS.

3. Dry the slides carefully avoiding the cell preparation.

4. Apply the primary antibody over the cell preparation at an appropriate dilution in PBS and incubate in a humidified chamber at room temperature for 60 minutes.

5. Repeat steps 3 and 4.

6. Apply the horseradish peroxidase-conjugated secondary antibody, again appropriately diluted in PBS, over the cell preparation, and incubate in a humidified chamber at room temperature for 60 minutes.

7. Rinse extremely thoroughly in running tap water, followed by distilled water.

8. Incubate the slides in the freshly prepared substrate solution for 5 minutes at room temperature with occasional agitation.

9. Rinse slides well with distilled water.

10. If desired, and if cobalt chloride has not been added to the substrate, tone the brown peroxidase results using one of the following two metal chloride procedures. *Copper Sulphate (CuSO$_4$):* Immerse the preparations in the 0.5% copper sulphate solution for 2–10 minutes, checking microscopically. The colour of the original reaction will gradually darken. *Aqueous Gold Chloride (HAuCl$_4$):* Flood the preparations with 0.05% aqueous gold chloride solution and monitor reaction microscopically for adequate darkening.

11. Dehydrate in IMS, clear in xylene and mount in DPX.

Note: (i) Antibody dilutions should be identified by Latin square dilution experiments where a range of dilutions of the primary antibody (suggested range from 1:25 to 1:500) should be tested in a chequer-board fashion on positive control sections against a range of dilutions of the labelled secondary antibody (suggested range from 1:25 to 1:100). (ii) Appropriate controls should be incorporated into every immunocytochemistry run. Negative controls may involve omission of the primary antibody or employ a matched isotype control. Positive controls include cell preparations known to contain the target antigen. For UBF, cytocentrifuge slides of activated peripheral blood lymphocytes are ideal. (iii) Poor labelling of intranuclear antigens may be due to poor reagent access across the cell membrane and/or the nuclear envelope. Permeabilisation techniques may be helpful in such instances (Protocols 6.13–6.16). (iv) Incubation with the primary antibody overnight at 4°C may improve antigen detection.

Protocol 6.13: Saponin permeabilisation for intracellular immunocytochemistry (adapted from [89])

EQUIPMENT, MATERIALS AND REAGENTS

As for Protocol 6.12.

0.1% saponin in PBS.

METHOD

1. Follow Protocol 6.12, except 0.1% saponin in PBS replaces conventional PBS in all washing and antisera dilution buffers.

2. Following incubation with the peroxidase conjugate, wash thoroughly in running tap water, followed by distilled water prior to incubation with the substrate.

Protocol 6.14: Nonidet permeabilisation for intracellular immunocytochemistry (from [90])

EQUIPMENT, MATERIALS AND REAGENTS

Tris-buffered saline pH 7.2 (TBS).

1% paraformaldehyde in TBS.

Lysolecithin.

Cold methanol at 4°C.

0.1% Nonidet in TBS.

METHOD

1. Dissolve 2.5 mg lysolecithin in 100 ml 1% paraformaldehyde in TBS.

2. Incubate slides for 2 minutes at room temperature.

3. Transfer slides directly to cold methanol for 10 minutes.

4. Transfer slides directly to 0.1% Nonidet in TBS for 5 minutes at room temperature.

5. Transfer to TBS pH 7.6 for 5 minutes at room temperature.

6. Continue with immunoperoxidase procedure from step 2 of Protocol 6.12.

Protocol 6.15: Acetone permeabilisation for intracellular immunocytochemistry (from [92, 94])

EQUIPMENT, MATERIALS AND REAGENTS

Acetone.

Poly-L-lysine or silane-coated slides.

37°C incubator.

METHOD

1. Sections or cells are prepared onto poly-L-lysine or silane-coated slides.

2. Preparations are fixed in acetone for 1–3 hours at 4°C.

3. Proceed to step 2 of Protocol 6.12.

Protocol 6.16: Temperature permeabilisation for intracellular immunocytochemistry (from [92, 94])

EQUIPMENT, MATERIALS AND REAGENTS

As for Protocol 6.12.

37°C incubator.

METHOD

As for Protocol 6.12 except:

Step 4. Apply the primary antibody over the cell preparation at an appropriate dilution (which may be considerably greater than for room temperature labelling) in PBS and incubate for 45 minutes in a humidified chamber stabilised at 37°C in an incubator.

Step 6. Apply the horseradish peroxidase-conjugated secondary antibody (again at an appropriate dilution in PBS which may be considerably greater than for room temperature labelling) over the cell preparation, and incubate for 45 minutes in a humidity chamber stabilised at 37°C in an incubator.

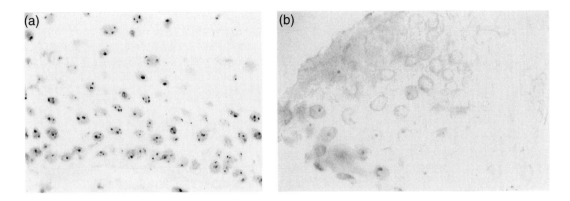

Figure 6.2

Cervical epithelium labelled (a) with AgNOR, and (b) with UBF immunoperoxidase. AgNOR counts are increased in parabasal cells and remain in superficial layers, whereas UBF is quickly lost in epithelial cells as they differentiate, migrate to superficial layers and cease differentiating. Magnification ×800.

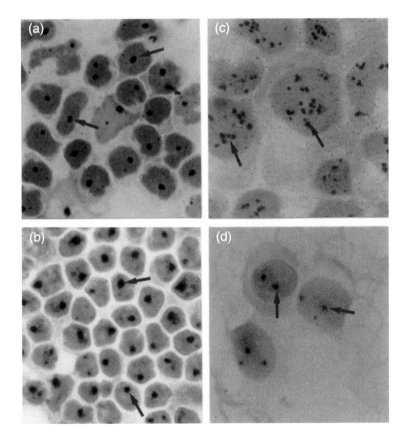

Figure 6.3

Photomicrographs comparing cytocentrifuge preparations of resting human peripheral blood lymphocytes (following (a) AgNOR demonstration and (b) UBF immunoperoxidase labelling) with 24 hour PHA-stimulated lymphocytes (following (c) AgNOR demonstration and (d) UBF immunoperoxidase labelling). The numbers of AgNOR and UBF dots were similar in resting cells but UBF counts were significantly lower than AgNORs in activated cells [93]. Magnification ×800.

Immunohistochemistry provides a reliable benchmark against which to investigate the AgNOR procedure. Although high electron charge density and the presence of phosphate moieties are important, aspartic acid residues in acidic regions of B23 and C23 are primarily involved in, and probably responsible for, silver staining [96]. There is also evidence that aspartic acid residues in acidic domains of RNA pol-I large subunit and UBF, an activator for RNA pol-I, are responsible for silver staining of these two proteins; adjacent aspartate residues in UBF are absolutely essential for silver staining [97]. UBF immunolabelling is more sensitive than localisation of the RNA pol-I large subunit. Chemical studies on the silver staining of nucleoli have indicated that AgNOR labelling need not necessarily be caused by a special silver affinity protein and that carboxyl groups alone could attract silver [98].

The group of auxiliary proteins required for the initiation of transcription, including RNA pol-I and its transcription factor, UBF, have received a lot of recent attention. UBF binds specifically to the rDNA promotor to form a stable pre-initiation complex. Two molecules, UBF1 and UBF2, have been sequenced in the hamster and were shown to be highly homologous with other mammalian UBF molecules, including human [99]. Regulation of transcription of the UBF gene by RNA pol-II may represent a pathway through which cells modulate transcription by RNA pol-I sites [100]. The existence of different forms of UBF may have important ramifications for transcription by RNA pol-I [101]. Interestingly, UBF is twenty times more sensitive to staining with the AgNOR procedure than RNA pol-I [97] and at least two aspartic acid residues are required for such efficient labelling. There is evidence that basal transcription factors such as UBF are inactivated during mitosis and reactivated by dephosphorylation at the exit from mitosis and during G_1 progression [102].

Double immunolabelling can be used to study the fate of different nucleolar components during cell division and apoptosis [103]. RNA pol-I disappears as long as UBF is associated with fibrogranular threaded bodies. In contrast, fibrillarin, B23 and C23 remain detectable in the granular material present amid micronuclei of late apoptotic cells. Immunoblotting has shown that UBF is proteolytically degraded whereas fibrillarin, B23 and C23 are not, which may explain the presence of anti-nucleolar antibodies in various pathological disorders.

6.4 References

1. Thiry, M. and Goessens, G. (1996) *The Nucleolus During the Cell Cycle*. R.G. Landes Company: Austin, Texas.
2. Olsen, M.O.J., Dundr, M. and Szebeni, A. (2000) The nucleolus: an old factory with unexpected capabilities. *Trends Cell Biol.* **10**: 189–196.
3. Scheer, U. and Benavente, R. (1990) Functional and dynamic aspects of the mammalian nucleolus. *BioEssays* **12**: 14–21.
4. Shaw, P.J. and Jordan, E.G. (1995) The nucleolus. *Ann. Rev. Cell. Dev. Biol.* **11**: 93–121.
5. Chen, D. and Huang, S. (2001) Nucleolar components involved in ribosome biogenesis cycle between the nucleolus and nucleoplasm in interphase cells. *J. Cell Biol.* **153**: 169–176.
6. Maggio, R. (1966) Progress report on the characterization of nucleoli from guinea pig liver. *Natl Cancer Inst. Monogr.* **23**: 213–222.
7. Scheer, U. and Hock, R. (1999) Structure and function of the nucleolus. *Curr. Opin. Cell Biol.* **11**: 385–390.
8. Leary, D.J. and Huang, S. (2001) Regulation of ribosome biogenesis within the nucleolus. *FEBS Lett.* **509**: 145–150.
9. Carmo-Fonseca, M., Mendes-Soares, L. and Campos, I. (2000) To be or not to be in the nucleolus. *Nat. Cell Biol.* **2**: E107–E112.
10. Schwarzacher, H.G. and Mosgoeller, W. (2000) Ribosome biogenesis in man: current views on nucleolar structures and function. *Cytogenet. Cell Genet.* **91**: 243–252.

11. Utama, B., Kennedy, D., Ru, K. and Mattick, J.S. (2002) Isolation and characterization of a new nucleolar protein, Nrap, that is conserved from yeast to humans. *Genes Cells* **7**: 115–132.

12. Mosgoeller, W., Schofer, C., Steiner, M., Sylvester, J.E. and Hozak, P. (2001) Arrangement of ribosomal genes in nucleolar domains revealed by detection of 'Christmas tree' components. *Histochem. Cell Biol.* **116**: 495–505.

13. Hernandez-Verdun, D. (1991) The nucleolus today. *J. Cell Sci.* **99**: 465–471.

14. Pederson, T. (1998) The plurifunctional nucleolus. *Nucl. Acids Res.* **26**: 3871–3876.

15. Huang, S. (2000) Review: perinucleolar structures. *J. Struct. Biol.* **129**: 233–240.

16. Jacob, S.T. (1995) Regulation of ribosomal gene transcription. *Biochem. J.* **306**: 617–626.

17. Bosman, F.T. (1999) The nuclear matrix pathology. *Virchows Arch.* **435**: 391–399.

18. Pandy, A. and Mann, M. (2000) Proteomics to study genes and genomes. *Nature* **405**: 837–846.

19. Yanagida, M., Shimamoto, A., Nishikawa, K., Furuichi, Y., Isobe, T. and Takahashi, N. (2001) Isolation and proteomic characterization of the major proteins of the nucleolin-binding ribonucleoprotein complexes. *Proteomics* **1**: 1390–1404.

20. Andersen, J.S., Lyon, C.E., Fox, A.H., Leung, A.K., Lam, Y.W., Steen, H., Mann, M. and Lamond, A.I. (2002) Directed proteomic analysis of the human nucleolus. *Curr. Biol.* **12**: 1–11.

21. Pederson, T. (2002) Proteomics of the nucleolus: more proteins, more functions? *Trends Biochem. Sci.* **27**: 111–112.

22. Fox, A.H., Lam, Y.W., Leung, A.K., Lyon, C.E., Andersen, J., Mann, M. and Lamond, A.I. (2002) Paraspeckles: a novel nuclear domain. *Curr. Biol.* **12**: 13–25.

23. Ospina, J.K. and Matera, A.G. (2002) Proteomics: the nucleolus weighs in. *Curr. Biol.* **12**: R29–R31.

24. Emes, R.D. and Ponting, C.P. (2001) A new sequence motif linking lissencephaly, Treacher Collins and oral-facial-digital type 1 syndromes, microtubule dynamics and cell migration. *Hum. Mol. Genet.* **10**: 2813–2820.

25. Dousset, T., Wang, C., Verheggen, C., Chen, D., Hernandez-Verdun, D. and Huang, S. (2000) Initiation of nucleolar assembly is independent of RNA polymerase I transcription. *Mol. Biol. Cell* **11**: 2705–2717.

26. Pombo, A., Jackson, D.A., Hollinshead, M., Wang, Z., Roeder, R.G. and Cook, P.R. (1999) Regional specialization in human nuclei: visualization of discrete sites of transcription by RNA polymerase III. *EMBO J.* **18**: 2241–2253.

27. Ochs, R.L. and Smetana, K. (1989) Fibrillar center distribution in nucleoli of PHA–stimulated human lymphocytes. *Exp. Cell Res.* **184**: 552–557.

28. Arrighi, F.E., Lau, Y.F. and Spallone, A. (1980) Nucleolar activity in differentiated cells after stimulation. *Cytogenet. Cell Genet.* **26**: 244–250.

29. Sirri, V., Hernandez-Verdun, D. and Roussel, P. (2000) Cyclin-dependent kinases govern formation and maintenance of the nucleolus. *J. Cell Biol.* **156**: 969–981.

30. Sirri, V., Roussel, P., Gendron, M-C. and Hernandez-Verdun, D. (1997) Amount of two major AgNOR proteins nucleolin and protein B23 is cell-cycle dependent. *Cytometry* **28**: 147–156.

31. Sirri, V., Roussel, P. and Hernandez-Verdun, D. (2000) The AgNOR proteins: qualitative and quantitative changes during the cell cycle. *Micron* **31**: 121–126.

32. Visintin, R. and Amon, A. (2000) The nucleolus: the magician's hat for cell cycle tricks. *Curr. Opin. Cell Biol.* **12**: 372–377.

33. Jordan, E.G. (1991) Interpreting nucleolar structure: where are the transcribing genes? *J. Cell Sci.* **98**: 437–442.

34. Junera, H.R., Masson, C., Geraud, G. and Hernandez-Verdun, D. (1995) The three-dimensional organization of ribosomal genes and the architecture of the nucleoli vary G_1, S and G_2 phases. *J. Cell Sci.* **108**: 3427–3441.

35. Busch, H. and Smetana, K. (1970) *The Nucleolus*. Academic Press, New York.

36. Pellicciari, C., Bottone, M.G., Scovassi, A.I., Martin, T.E. and Biggiogera, M. (2000) Rearrangement of nuclear ribonucleoproteins and extrusion of nucleolus-like bodies during apoptosis induced by hypertonic stress. *Eur. J. Histochem.* **44**: 247–254.

37. Zimmer, M. (2002) Green fluorescent protein (GFP): applications, structure and related photophysical behaviour. *Chem. Rev.* **102**: 759–781.

38. Dundr, M., Misteli, T. and Olson, M.O. (2000) The dynamics of post-mitotic reassembly of the nucleolus. *J. Cell Biol.* **150**: 433–446.

39. Trerè, D. (2000) AgNOR staining and quantification. *Micron* **31**: 127–131.

40. Bloom, S.E. and Goodpasture, C. (1976) An improved technique for selective silver staining of nucleolar organizer regions in human chromosomes. *Hum. Genet.* **28**: 199–206.

41. Field, D.H., Fitzgerald, P.H. and Sin, F.Y. (1984) Nucleolar silver-staining patterns related to cell cycle phase and cell generation of PHA-stimulated human lymphocytes. *Cytobios* **41**: 23–33.

42. Derenzini, M. (2000) The AgNORs. *Micron* **31**: 117–120.

43. Canet, V, Montmasson, M-P., Usson, Y., Giroud, F. and Brugal, G. (2001) Correlation between silver-stained nucleolar organizer region area and cell cycle time. *Cytometry* **43**: 110–116.

44. Goodpasture, C. and Bloom, S.E. (1975) Visualisation of nucleolar organizer regions in mammalian chromosomes using silver staining. *Chromosoma* **53**: 37–50.

45. Howell, W.M. and Black, D.A. (1980) Controlled silver staining of nucleolus organizer regions with a protective colloidal developer: a 1-step method. *Experientia* **36**: 1014.

46. Lindner, L.E. (1993) Improvements in the silver-staining technique for nucleolar organizer regions (AgNOR). *J. Histochem. Cytochem.* **41**: 439–445.

47. Cromie, C.J., Benbow, E.W., Stoddart, R.W. and McMahon, R.F.T. (1988) Pre-incubation with glycine solution aids the demonstration of nucleolar organizer-associated proteins. *Histochem. J.* **20**: 722–724.

48. Smith, P.J., Skilbeck, N., Harrison, A. and Crocker, J. (1988) A study of the effects of different fixatives on the AgNOR technique. *J. Pathol.* **155**: 109–112.

49. Rowlands, D.C., Ayres, J.G. and Crocker, J. (1993) The effect of different fixatives and length of fixation time on subsequent AgNOR staining for frozen and paraffin-embedded tissue sections. *Histochem. J.* **25**: 123–132.

50. Committee for AgNOR quantitation within the European Society for Pathology (1994) AgNORs in oncology. Proceedings of a workshop. Berlin, 1–3 October 1993. *Zentralbl. Pathol.* **140**: 1–108.

51. Aubele, M., Biesterfeld, S., Derenzini, M., Hufnagl, P., Martin, H., Ofner, D., Ploton, D. and Ruschoff, J. (1994) Guidelines of AgNOR quantitation – Committee on AgNOR Quantitation within the European Society of Pathology. *Zentralbl. Pathol.* **140**: 107–108.

52. Trere, D., Migaldi, M. and Trentini, G.P. (1995) Higher reproducibility of morphometric analysis over the counting method for interphase AgNOR quantification. *Anal. Cell Pathol.* **8**: 57–65.

53. Committee for AgNOR quantitation within the European Society for Pathology (1998) Cell Proliferation in oncology. Proceedings of the 6th international workshop on applications of AgNORs in pathology. University of Bologna, Italy.

54. Tuccari, G., Guiffre, G., Öfner, D. and Rüschoff, J. (2000) Standardized use of the AgNOR technique. *J. Oral Pathol. Med.* **29**: 526–527.

55. Crocker, J., Boldy, D.A.R. and Egan, M.J. (1989) How should we count AgNORs? Proposals for a standardised approach. *J. Pathol.* **158**: 185–188.

56. Crocker, J. (1995) The trials and tribulations of interphase AgNORs. *J. Pathol.* **175**: 367–368.

57. Plotton, D., Menager, M., Jeannesson, P., Himber, G., Pigeon, F. and Adnet, J. (1986) Improvement in the staining and visualization of the argyrophilic proteins of the nucleolar organizer regions at the optical level. *Histochem. J.* **18**: 5–14.

58. Coghill, G., Grant, A., Orrell, J.M., Jankowski, J. and Evans, T.A. (1990) Improved silver staining of nucleolar organiser regions in paraffin wax sections using an inverted incubation technique. *J. Clin. Pathol.* **43**: 1029–1031.

59. Jordan, E.G. (1984) Nuclear nomenclature. *J. Cell Sci.* **67**: 217–220.

60. Boldy, D.A.R., Crocker, J. and Ayres, J.G. (1989) Applications of the AgNOR method to cell imprints of lymphoid tissue. *J. Pathol.* **157**: 75–79.

61. Öfner, D., Bankfalvi, A., Rieheman, K., Bier, B., Böcker, W. and Schmid, K.W. (1994) Wet autoclave pre-treatment of silver-stained nucleolar organizer region-associated proteins in routinely formalin-fixed and paraffin-embedded tissues. *Mod. Pathol.* **7**: 946–950.

62. Medina, F-J. and Marco, A.C.R. (1995) Microwave irradiation improvements in the silver staining of the nucleolar organiser (Ag-NOR) technique. *Histochem. Cell Biol.* **103**: 403–413.

63. Li, Q., Hacker, G.W., Danscher, G., Sonnleitner-Wittauer, U. and Grimelius, L. (1995) Argyrophilic nucleolar organizer regions. A revised version of AgNOR staining technique. *Histochem. Cell Biol.* **104**: 145–150.

64. Cia, E.M., Trevisan, M. and Metze, K. (1999) Argyrophilic nucleolar organizer region (AgNOR) technique: a helpful tool for differential diagnosis in urinary cytology. *Cytopathology* **10**: 582–584.

65. Öfner, D., Aubele, M., Biestefield, S., Derenzini, M., Gimenez-Mas, J.A., Hufnagl, P., Ploton, D., Trerè, D. and Rüschoff, J. (1995) Guidelines of AgNOR quantification – first update. *Virchows Arch.* **426**: 341.

66. Öfner, D., Hittmair, A., Maier, H., Riedmann, B., Rumer, A., Lucciarini, P., Offner, F., Mikuz, G. and Schmid, K.W. (1994) Sequential quantification of AgNOR area and number during silver staining by means of an image analysing system. *Zentralbl. Pathol.* **140**: 37–40.

67. Martin, H., Beil, M., Hufnagl, P., Wolf, G. and Korek, G. (1991) Computer-assisted image analysis of nucleolar organizer regions (NORs): a pilot study of astrocytomas and glioblastomas. *Acta Histochem.* **90**: 189–196.

68. Öfner, D. and Schmid, K.W. (1996) Standardized AgNOR analysis: its usefulness in surgical oncology. *Histochem. Cell Biol.* **106**: 193–196.

69. Delahunt, B., Avallone, F.A., Ribas, J.L. and Mostofi, F.K. (1991) Gold toning improves the visualization of nucleolar organizer regions in paraffin embedded tissues. *Biotech. Histochem.* **66**: 316–320.

70. Yekeler, H., Erel, Ö., Yumbul, A.Z., Doymaz, M.Z., Dogan, Ö., Özercan, M.R. and Iplikçi, A. (1995) A sensitive staining method for NORs. *J. Pathol.* **175**: 449–452.

71. Fritz, P., Hoenes, J., Multhaupt, H., Schenk, J., Mischlinski, A., Klein, C., Tuczek, H.V. and Wolf, M. (1990) Application of the chromogenic reaction to conventional silver staining, the Ag-NOR staining and the silver-intensified immunogold technique. *Acta Histochem.* **38**: 239–246.

72. Bilinksi, S.M. and Bilinska, B. (1996) A new version of the Ag-NOR technique. A combination with DAPI staining. *Histochem. J.* **28**: 651–656.

73. Héliot, L., Mongelard, F., Klein, C., O'Donohue, M.F., Chassery, J.M., Robert-Nicoud, M. and Usson, Y. (2000) Non-random distribution of metaphase AgNOR staining patterns on human acrocentric chromosomes. *J. Histochem. Cytochem.* **48**: 13–20.

74. Foucrier, J., Rigaut, J.P. and Pechinot, D. (1990) A combined AgNOR-Feulgen staining technique. *J. Histochem. Cytochem.* **38**: 1591–1597.

75. Pajor, L. and Honeyman, T.W. (1995) Combined light and fluorescent microscopical imaging of nucleolar organizer regions and cellular RNA as detected by fluorescence in situ hybridization. *Cytometry* **19**: 171–176.

76. Ploton, D., Gilbert, N., Ménager, M., Kaplan, H. and Adnet, J-J. (1994) Three-dimensional co-localisation of nucleolar argyrophilic components and DNA in cell nuclei by confocal microscopy. *J. Histochem. Cytochem.* **42**: 137–148.

77. Gorczyca, W., Smolewski, P., Grabarek, J., Ardelt, B., Ita, M., Melamed, M.R. and Darzynkiewicz, Z. (2001) Morphometry of nucleoli and expression of nucleolin analyzed by laser scanning cytometry in mitogenically stimulated lymphocytes. *Cytometry* **45**: 206–213.

78. Egan, M.J. and Crocker, J. (1992) Nucleolar organiser regions in pathology. *Br. J. Cancer* **65**: 1–7.

79. Öfner, D. (2000) In situ standardised AgNOR analysis: a simplified method for routine use to determine prognosis and chemotherapy efficiency in colorectal adenocarcinoma. *Micron* **31**: 161–164.

80. Pich, A., Chiusa, L. and Margaria, E. (2000) Prognostic relevance of AgNORs in tumor pathology. *Micron* **31**: 133–134.

81. Jacquet, B., Canet, V., Giroud, F., Montmasson, M-P. and Brugal, G. (2001). Quantitation of AgNORs by flow versus image cytometry. *J. Histochem. Cytochem.* **49**: 433–438.

82. Mourad, W.A., Connelly, J.H., Sembera, D.L., Atkinson, E.N. and Bruner, J.M. (1993) The correlation of two argyrophilic nucleolar organizer region counting methods with bromodeoxyuridine labelling index: a study of metastatic tumors of the brain. *Hum. Pathol.* **24**: 206–210.

83. Ploton, D., Thiry, M., Ménager, M., Lepoint, A., Adnet, J-J. and Goessens, G. (1987) Behaviour of nucleolus during mitosis. A comparative ultrastructural study of various cancer cell lines using the AgNOR staining procedure. *Chromosoma* **95**: 95–107.

84. Nikicicz, E.P. and Norback, D.H. (1990) Argyrophilic nucleolar organiser region (AgNOR) staining in normal bone marrow cells. *J. Clin. Pathol.* **43**: 723–727.

85. Vandalaer, M., Thiry, M. and Goessens, G. (1999) AgNOR proteins from morphologically intact isolated nucleoli. *Life Sci.* **64**: 2039–2047.

86. Zurita, F., Jiménez, R., Díaz de la Guardia, R. and Burgos, M. (1998) The relative rDNA content of a NOR determines its level of expression and its probability of becoming active. A sequential silver staining and in situ hybridization study. *Chromosome Res.* **7**: 563–570.

87. Roussel, P. and Hernandez–Verdun, D. (1994) Identification of AgNOR proteins, markers of proliferation related to ribosomal gene activity. *Exp. Cell Res.* **214**: 465–472.

88. Gautier, T., Robert-Nicoud, M., Guilly, M.N. and Hernandez-Verdun, D. (1992) Relocation of nucleolar proteins around chromosomes at mitosis. A study by confocal laser scanning microscopy. *J. Cell Sci.* **102**: 729–737.

89. Behringer, D.M., Sunderer, B., Andersson, U., Kresin, V., Mertelsmann, R. and Lindemann, A. (1996) Simultaneous detection of cytokine and immunophenotype at the single cell level by immunoenzymatic double staining. *Histochem. J.* **28**: 461–466.

90. Hupin, J., Cantinieaux, B. and Fondu, P. (1993) Applications of the monoclonal antibody PC10 assessment of proliferative grade in cell smears. *Am. J. Clin. Pathol.* **99**: 673–676.

91. Bolivar, J., Iglesias, C., Ortiz, M., Goenechea, L.G., Grenett, H., Torres-Montaner, A., Valdivia, M.M. (1999) The DNA binding domains of UBF represent major human autoepitopes conserved in vertebrate species. *Cell Molec. Biol.* **45**: 277–284.

92. Torres-Montaner, A., Bolivar, J., Ortiz, M. and Valdivia, M.M. (1998) Immunohistochemical detection of ribosomal transcription factor UBF: diagnostic value in malignant specimens. *J. Pathol.* **184**: 77–82.

93. Hughes, D.A., Weerasinghe, S., Torres-Montaner, A. and Welsh, K.I. (in preparation) Comparison of argyrophilic nucleolar organiser region (AgNOR) and ribosomal transcription factor (UBF) localisation in cells aspirated from human renal allografts.

94. Torres-Montaner, A., Bolivar, J., Astola, A., Gimenez-Mas, J.A., Brieva, J.A. and Valdivia, M.M. (2000) Immunohistochemical detection of ribosomal transcription factor UBF and AgNOR staining identify apoptotic events in neoplastic cells of Hodgkin's disease and in other lymphoid cells. *J. Histochem. Cytochem.* **48**: 1521–1530.

95. Dühr, S., Torres-Montaner, A., Astola, A., García-Cozar, F.J., Pendón, C., Bolívar, J. and Valdivia, M.M. (2001) Molecular analysis of the 5' region of human ribosomal transcription factor UBF. *DNA Sequence* **12**: 267–272.

96. Valdez, B.C., Henning, D., Le, T.V. and Busch, H. (1995) Specific aspartic-acid rich sequences are responsible for silver staining of nucleolar proteins. *Biochem. Biophys. Res. Comm.* **207**: 485–491.

97. Valdez, B.C., Henning, D., Liangjin, Z. and Stetler, D.A. (1998) Silver (AgNOR) staining of nucleolar transcription factor UBF requires adjacent aspartic acid residues. *J. Histotechnol.* **21**: 13–18.

98. Kling, H., Lepore, L., Krone, W., Olert, J. and Sawatzki, G. (1980) Chemical studies on silver staining of nucleoli. *Experientia* **36**: 249–251.

99. Bolivar, J., Goenechea, L.G., Grenett, H., Pendon, C. and Valdivia, M.M. (1996) Cloning and sequencing of the genes encoding the hamster transcription factors UBF1 and UBF2. *Gene* **176**: 257–258.

100. Glibetic, M., Taylor, L., Larson, D., Hannan R., Sells, B. and Rothblum, L.I. (1995) The RNA polymerase I transcription factor UBF is the product of a primary response gene. *J. Biol. Chem.* **270**: 4209–4212.

101. O'Mahony, D.J. and Rothblum, L.I. (1991) Identification of the two forms of the RNA polymerase I transcription factor UBF. *Proc. Natl Acad. Sci. U.S.A.* **88**: 3180–3184.

102. Klein, J. and Grummt, I. (1999) Cell cycle-dependent regulation of RNA polymerase I transcription: the nucleolar transcription factor UBF is inactive in mitosis and early G1. *Proc. Natl Acad. Sci.* **96**: 6096–6101.

103. Martelli, A.M., Robuffo, I., Bortul, R., Ochs, R.L., Luchetti, F., Cocco, L., Zweyer, M., Bareggi, R. and Falcieri, E. (2000) Behaviour of nucleolar proteins during the course of apoptosis in camptothecin-treated HL60 cells. *J. Cell Biochem.* **78**: 264–277.

Membrane Alterations in Dying Cells

7

Hugo van Genderen, Heidi Kenis, Ewald Dumont, Waander van Heerde, Leo Hofstra and Chris P. Reutelingsperger

Contents

7.1 Introduction

Programmed cell death (PCD) is an essential part of both normal embryonic development and tissue homeostasis in the adult organism. Failure to invoke appropriate PCD may result in malformations, autoimmune disease or cancer, whereas increased cell death occurs in acute pathologies such as infection by toxin-producing micro-organisms or ischaemia/reperfusion damage (infarction), as well as in chronic diseases such as immunodeficiency or neurodegenerative disorders. Instead of being a cause of disease, PCD may also be employed as a target in the clinic to treat diseases. For instance, many anti-cancer therapies including chemotherapy and radiotherapy are effective because they induce PCD of the tumour cell. This understanding has led

Cell Proliferation and Apoptosis, David Hughes and Huseyin Mehmet (Eds)
© 2003 BIOS Scientific Publishers Ltd, Oxford

to the exploration of novel therapeutic strategies, which target parts of the molecular machinery of PCD.

On the basis of morphology and biochemistry, PCD can be divided into four forms: apoptosis, apoptosis-like PCD, necrosis-like PCD and necrosis [1]. The first three forms have in common that the programme of cell death culminates in a structural change of the plasma membrane resulting in the exposure of so-called 'eat me' signals to the environment. Phagocytes recognise these signals and respond to them by removing the dying cell through engulfment [2]. Triggering the phagocytic response is considered a pivotal part of the PCD concept because it ensures that dying cells are removed in a timely fashion from tissues before they induce an inflammatory response. The nature of these 'eat me' signals have been the subject of many investigations, although phosphatidylserine has been identified as a ubiquitous recognition signal of PCD *in vitro* and *in vivo* [3]. This has initiated the development of a variety of methodologies to measure PCD *in vitro* and *in vivo* in animals and humans. Other plasma membrane features of PCD include loose packaging of the phospholipid bilayer, an altered carbohydrate composition and a typical cell adhesion molecule profile. These features appear to be less ubiquitous than phosphatidylserine exposure and have been, therefore, a less attractive target for measuring PCD.

7.2 Apoptosis

The most abundant form of PCD is apoptosis. It is characterised by cell shrinkage, nuclear fragmentation and membrane blebbing. These morphological features appear to result from an energy-dependent process with a high degree of regulation at the biochemical level [4]. Caspases, which constitute a family of cysteine-dependent aspartate-directed proteases, play an important role in the biochemistry of apoptosis. They reside in the cytosol and intracellular organelles like mitochondria and ER as dormant zymogens. They are activated by proteolysis and then trigger a caspase cascade, which generates a proteolytic burst within the cell. Caspase activation can start from receptor/ligand-associated signal transduction (extrinsic activation) or from mitochondrial-dependent apoptosome formation (intrinsic activation).

7.2.1 The extrinsic activation of caspases

In the receptor-mediated pathway, apoptosis can be extrinsically imposed on cells through the interaction of the so-called death receptors with their ligand, such as Fas with Fas-Ligand, TRAIL receptor with TRAIL, TNF-R1 or R2 with TNF [5]. Ligand-induced clustering of death receptors results in the recruitment of initiator caspases, such as pro-caspases-8 and -10, to the intracellular domains of the clustered receptors. These pro-caspases become proteolytically active most likely by auto-activation events. Active caspase-8 can then proteolytically activate downstream effector caspases such as pro-caspase-3. Cells with sub-optimal pro-caspase-8 activation may engage Bid, a pro-apoptotic member of the Bcl-2 family, to involve mitochondria, which deploy the intrinsic activation of caspases [6].

7.2.2 The intrinsic activation of caspases

Many apoptotic signals, including the generation of tBid by active caspase-8, increased cytosolic calcium and pH shift, impinge on mitochondria, which respond by releasing proteins such as cytochrome c into the cytosol. Once in the cytosol, cytochrome c interacts with Apaf-1, pro-caspase-9 and ATP to form a complex called an 'apoptosome' [7]. Pro-caspase-9 can now undergo intrinsic activation, probably by auto-activation events. Active caspase-9 then cleaves pro-caspase-3 into its active form, which is considered the major executioner of apoptosis.

7.2.3   The execution of apoptosis

As discussed above, the execution of apoptosis occurs through the activation of 'effector' or 'executioner' caspases, such as caspase-3. These effector caspases cleave and inactivate vital cellular proteins and trigger the activation of caspase-activated DNAse (CAD) which mediates oligonucleosomal DNA fragmentation [8]. Caspase-3 cleaves ROCK-1, a Rho-activated serine/threonine kinase responsible for membrane blebbing and for the relocalisation of fragmented DNA into apoptotic bodies [9]. To date, unidentified pathways cause a change in plasma membrane architecture, resulting in molecular changes that are recognised by phagocytes. A major part of this change involves the rearrangement of the phospholipid asymmetry of the plasma membrane.

7.3   Phospholipid asymmetry of the plasma membrane

The plasma membrane of healthy cells is characterised by an asymmetric distribution of the various phospholipid species over the two membrane leaflets. The inner leaflet, facing the cytosol, contains phosphatidylserine, phosphatidylinositol, phosphatidylethanolamine and phosphatidylcholine. The outer leaflet, facing the environment, harbours phosphatidylethanolamine (but to a lesser extent than the inner leaflet), phosphatidylcholine and sphingomyelin (SM). This asymmetry is generated and maintained by energy-dependent processes, which rely on the activity of a socalled aminophospholipid translocase (APT) (*Figure 7.1*). APT is defined by its ability to transport phosphatidylethanolamine and phosphatidylserine from the outer to

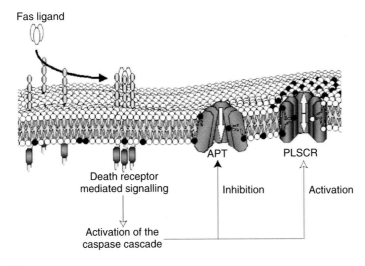

Figure 7.1

Schematic presentation of the phospholipid asymmetry of the plasma membrane. Phosphatidylserine is predominantly located in the inner leaflet due to the action of the APT. Activation of the caspase cascade, for example through Fas-ligation, results in the inhibition of APT and the activation of the phospholipid scramblase (PLSCR). Consequently, phosphatidylserine is presented in the outer leaflet of the plasma membrane.

the inner leaflet, but has not been identified at the molecular level thus far. Potential candidates include members of a new sub-family of P-type ATPases.

From an evolutionary perspective, asymmetry of the plasma membrane is probably an old and general feature of eukaryotic cells, but it is not a static feature. Under specific circumstances, for example an activated apoptotic programme, cells are able to change the asymmetry by presenting phosphatidylserine in the outer leaflet of the plasma membrane.

7.3.1 Phosphatidylserine exposure during apoptosis

Apoptosis generates signals that act on the plasma membrane by inhibiting APT and activating a phospholipid scramblase (PLSCR), an enzyme that resides in or close to the bilayer (*Figure 7.1*). The combination of these actions results in the appearance of phosphatidylserine in the outer plasma membrane leaflet while the integrity of the plasma membrane remains largely uncompromised. The nature of the PLSCR is still not fully understood. It has been described unambiguously only by biological activity. PLSCR transports bi-directionally all phospholipid species from the inner to the outer leaflet and *vice versa* in an ATP-independent manner. Candidate proteins with PLSCR activity are members of a recently discovered gene family, which has accordingly been called the PLSCR gene family.

It is still unclear how the apoptotic machinery inhibits APT and activates PLSCR. It seems that active caspase-3 does not act directly on these molecular complexes but likely via intermediates that may include the flux of calcium ions across the plasma membrane, activated protein kinase Cδ, and the ABC1-transporter. The Rho-1 kinase pathway, previously thought to be involved, does not harbour candidates since inhibition of this pathway prevents membrane blebbing but not phosphatidylserine exposure.

The pathways responsible for phosphatidylserine exposure are not exclusively activated by caspases. For example, cathepsin B is capable of inducing phosphatidylserine exposure in a caspase-independent fashion if it is translocated from the lysosome into the cytosol. Such translocation may arise from caspase-independent processes in apoptosis-like PCD.

The translocation of phosphatidylserine to the outer leaflet is one of the plasma membrane changes that occur during PCD. Other changes include looser packing of plasma membrane phospholipids. It is suggested that this is the consequence of scrambling the phospholipid species over the two leaflets of the plasma membrane. While restructuring of the plasma membrane is occurring its integrity and barrier function are kept intact. Gradually, however, the apoptotic process changes the plasma membrane from a non-permeable barrier into a leaky fence, which first allows small and later large molecules to traverse across the plasma membrane. *In vivo*, the leakage of larger molecules across the plasma membrane is unlikely to happen under normal circumstances since phagocytes recognise dying cells by changes in plasma membrane architecture and engulfment occurs before the plasma membrane integrity is compromised.

7.3.2 Phagocyte recognition of apoptotic cells

Removal of the majority of apoptotic cells is performed by professional phagocytes such as macrophages and dendritic cells, although other cell types including fibroblasts, epithelial cells and endothelial venule cells are also capable of phagocytosing apoptotic cells. Recognition and engulfment are mediated by interactions between phagocytic receptors and a variety of ligands on the surface of the dying cell. These ligands are either a part of the apoptotic plasma membrane (e.g. phosphatidylserine for almost all apoptotic cells and ICAM-3 for apoptotic leukocytes) or appear at the surface from opsonisation as a consequence of altered plasma membrane architecture. The opsonins include members of the complement system C1q and iC3b, the glycoprotein β2-GPI, thrombospondin, pentraxin C-reactive protein (CRP) and serum amyloid protein (SAP). The latter two belong to a family of phospholipid binding proteins with unknown function. The binding sites for the opsonins remain unidentified, although phosphatidylserine appears to function as a binding site for a number of them (C1q, iC3b and β2-GPI). The loose packing of the plasma membrane phospholipids may also expose binding sites to the opsonins.

7.4   Structure and function of Annexin A5

To date, surface-exposed phosphatidylserine is the most widely used 'eat me' signal for measuring apoptosis. This 'eat me' signal can be visualised with conjugates of Annexin A5, a phosphatidylserine binding protein [10]. Annexin A5 is a member of a family of proteins called the annexins. The annexins share structural and functional features and the primary structure is organised in an N-terminal tail, which is unique for each annexin, and a C-terminal core, which consists of a canonically repeated conserved domain containing the endonexin loop [10]. The annexins bind to phospholipids in the presence of calcium ions, which bind to the endonexin loop sequences thereby creating phospholipid interaction sites. Annexin A5 contains four homologous domains in its C-terminal core. In the presence of calcium ions Annexin A5 binds to phosphatidylserine-containing membranes with an affinity constant of more than 10^9 M^{-1}. Bound Annexin A5 dissociates from the phosphatidylserine-containing membrane if the calcium ions are removed, for example by chelation with EDTA. At a synthetic phospholipid surface Annexin A5 organises in a 2D lattice of a monolayer of Annexin A5 molecules in building blocks of trimers (*Figure 7.2*). The Annexin A5 trimer is formed by reversible protein–protein interactions and the individual Annexin A5 molecules interact with the phosphatidylserine-containing phospholipid surface. It has been proposed that Annexin A5 contains a binding pocket specific for the phosphoserine group of phosphatidylserine. The phospholipid binding property of Annexin A5 is stable in a pH range of 5.5–10 and at temperatures below 50°C. Annexin A5 thus appears to be a stable probe to measure apoptosis *in vitro* as well as *in vivo*.

7.4.1   *In vitro* measurement of phosphatidylserine exposure during apoptosis

Conjugates of Annexin A5 with fluorophores (e.g. FITC, Alexa's, phycoerythrin, and Cy's) offer versatility to measure phosphatidylserine exposure of cells by flow cytometry (FCM) and confocal scanning laser microscopy (CSLM) (Protocols 7.1 and 7.2). These measurements require that cells are mixed with Annexin A5 conjugates in the presence of calcium ions. Following a short incubation period phosphatidylserine exposure can be measured immediately without the necessity of washing. Using green fluorescent annexin in combination with red fluorescent PI one can distinguish viable, apoptotic and late-apoptotic/necrotic cells with FCM (*Plate 6a*). The exposure of phosphatidylserine starts early after the onset of execution of apoptosis but can precede the morphological alterations and the loss of plasma membrane integrity. At all stages of cell death (early apoptotic cells (*Plate 6b*), blebbed apoptotic cells, late-apoptotic/necrotic cells (*Plate 6c*) and apoptotic bodies) cells have phosphatidylserine exposed on their surface.

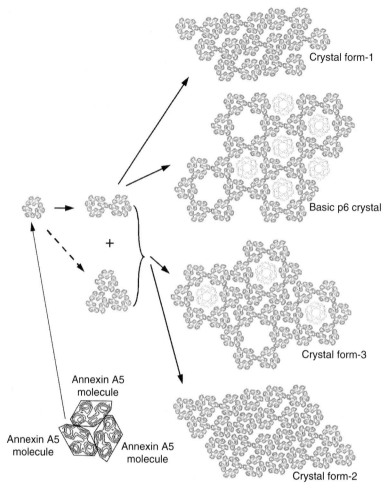

Crystal form-1

Basic p6 crystal

Crystal form-3

Annexin A5
molecule

Annexin A5
molecule

Annexin A5
molecule

Crystal form-2

Figure 7.2

Two-dimensional (2D) crystallization of Annexin A5 at the phospholipid surface.
Annexin A5 forms trimers at the phospholipid surface by protein–protein interactions.
These trimers organise in different 2D crystals as was measured by atomic force
microscopy (adapted from [11]).

Phosphatidylserine exposition can also be visualised at the EM level using
Annexin A5-biotin. The cells treated with biotinylated Annexin A5 have to be fixed
with a solution containing calcium ions. Cellular bound Annexin A5 can be visualised by the avidin/streptavidin-peroxidase system. Using this EM technology,
exposure of phosphatidylserine can be correlated with intracellular processes at the
ultrastructural level, including chromatin margination and the release of ribosomes
from the ER.

7.4.2 *In vivo* measurement of phosphatidylserine exposure during apoptosis in animal models

Although plasma membrane phospholipid asymmetry is considered a general feature of eukaryotic cells, data supporting this came essentially from *in vitro* experiments. By injecting biotinylated Annexin A5 into the living mouse embryo [12]

Protocol 7.1: Flow cytometric analysis of phosphatidylserine exposure and plasma membrane permeability of Jurkat cells, which are triggered to execute apoptosis by anti-Fas antibody

EQUIPMENT, MATERIALS AND REAGENTS

Jurkat cells in RPMI 1640, 10% newborn calf serum with a cell count of 10^6 cells ml^{-1}.

200 $\mu g\,ml^{-1}$ anti-Fas antibody.

Binding buffer: 25 mM Hepes/NaOH pH 7.4, 140 mM NaCl, 2.5 mM $CaCl_2$.

25 $\mu g\,ml^{-1}$ Annexin A5-FITC.

250 $\mu g\,ml^{-1}$ PI.

Ice.

CO_2 incubator.

Flow cytometer.

METHODS

1. Add 20 ng ml^{-1} anti-Fas antibody to the Jurkat cells and incubate at 37°C in CO_2 incubator.

2. Take at determined time points a 50 μl sample from the cell suspension.

3. Dilute the sample with 440 μl binding buffer and add 5 μl Annexin A5-FITC and 5 μl PI.

4. Incubate 5–15 minutes at ambient temperature or on ice depending on the research question.

5. Analyse the reaction mixture by flow cytometry. *Plate 6a* illustrates an analysis performed with Jurkat cells, which were treated 4 hours with the anti-Fas antibody.

Protocol 7.2: Confocal scanning laser microscopic analysis of phosphatidylserine exposure and plasma membrane permeability of apoptotic Jurkat cells

EQUIPMENT, MATERIALS AND REAGENTS

Jurkat cells in M199, 10% newborn calf serum with a cell count of 10^6 cells ml^{-1}.

200 μg ml^{-1} anti-Fas antibody.

25 μg ml^{-1} Annexin A5-FITC.

250 μg ml^{-1} PI.

Eight-well glass slides.

Coverslips.

CO_2 incubator.

Confocal scanning laser microscope.

METHODS

1. Wash 1×10^6 cells in ice-cold PBS pH 7.2.

2. Add 20 ng ml^{-1} anti-Fas antibody to the Jurkat cells and incubate at 37°C in CO_2 incubator.

3. Take at determined time points a 100 μl sample from the cell suspension.

4. Add 1 μl Annexin A5-FITC and 1 μl PI.

5. Incubate 5 minutes at ambient temperature.

6. Place a drop of the reaction mixture on a well of the glass slide to cover the whole well with the fluid.

7. Cover the well with a coverslip.

8. Allow the cells to settle, which takes about 2–5 minutes, and analyse by CSLM using dual wavelength excitation at 485 nm (Annexin A5-FITC) and 568 nm (PI). *Plate 6* shows merged CSLM images of an early apoptotic cell with surface-exposed phosphatidylserine (*Plate 6b*) and a late apoptotic cell with surface-exposed phosphatidylserine, a leaky plasma membrane and fragmented chromatin (*Plate 6c*).

(Protocol 7.3) we have found that cell membrane asymmetry also exists *in vivo* in the complex environment of the tissue. Our studies demonstrated that *in vivo*, cells change this asymmetry to expose phosphatidylserine at the plasma membrane surface during apoptosis (*Plate 7*). Other important conclusions from these studies were that the Annexin A5 molecule rapidly distributes through the extracellular compartments of the organism and that the calcium concentration in the extracellular fluids of the mouse embryo is sufficient to support binding of Annexin A5 to the apoptotic cells, suggesting that phosphatidylserine exposure could be measured in whole embryos.

In contrast to the embryo, apoptosis in the healthy adult organism is hardly detectable, likely due to the fact that small numbers of apoptotic cells are rapidly cleared in the adult. However, pathologies can change this balance dramatically giving rise to a prolonged existence of apoptotic cells in diseased tissues. For example, ischaemia/reperfusion of the heart induces apoptosis in cardiomyocytes in the area at risk. These cells are not removed immediately by phagocytes so they exist long enough to allow detection. The apoptotic cardiomyocytes expose phosphatidylserine at their surface as can be visualised with biotinylated Annexin A5 in a mouse model of ischaemia/reperfusion injury of the heart (*Plate 7*). This was achieved by injecting biotinylated Annexin A5 intravenously into the mouse prior to ischaemia/reperfusion of the heart [13] (Protocol 7.4). The results of this study show that the extracellular compartments of the adult mouse also contain sufficient calcium levels to support binding of Annexin A5 to apoptotic cells.

One drawback of the methodology to assess phosphatidylserine exposure *in vivo* with biotinylated Annexin A5 is that the tissues require fixation and sectioning. Fluorescent Annexin A5 in combination with vital fluorescence microscopy circumvents this requirement and allows the real time measurement of phosphatidylserine exposure in a living mouse model. This technology gathers spatio-temporal information about phosphatidylserine exposure during apoptosis of cells in their natural environment. Application of this technology to a mouse model of cardiac ischaemia/reperfusion injury reveals that cardiomyocytes in the area at risk exhibit a kinetically invariant process of phosphatidylserine exposure that can be prevented by pan-caspase inhibitors [14].

7.4.3   *In vivo* measurement of phosphatidylserine exposure in human disease

The promising results of measuring phosphatidylserine exposure *in vitro* and in animal models have catalysed the development of non-invasive apoptosis imaging technology in humans based on Annexin A5. For this purpose Annexin A5 is labelled with the radionuclide technetium (Tc-99m) and injected intravenously into patients. Tc-99m-Annexin A5 accumulates at sites where cell death occurs and this can be visualised by Single Photon Emission Computed Tomography (SPECT) using gamma cameras (Protocol 7.5). Proof of principle for the Annexin A5 imaging protocol was obtained with patients suffering from an acute myocardial infarction and one patient with an intracardiac tumour. Immunohistochemical analysis of tumour tissue of the latter patient demonstrated that Annexin A5 bound to the plasma membrane of tumour cells, which had activated caspase-3. Hence, phosphatidylserine exposure occurs also in human cells during apoptosis.

The Annexin A5 imaging protocol is currently only applicable to patients with apoptotic lesions in the extremities and above the diaphragm, such as those with acute myocardial infarction or those with lung or breast cancer. One consideration is that Tc-99m-Annexin A5 is rapidly cleared by the liver and kidney, which increases the background signal in the abdominal region to levels that render imaging unreliable.

Protocol 7.3: Light microscopic analysis of apoptotic cells in mouse embryonic tissue using Annexin A5-biotin

EQUIPMENT, MATERIALS AND REAGENTS

Pregnant FVB mice, from days 11–13 post coitus (plug = day 0).

Ether.

Hepes buffer: 20 mM Hepes/NaOH pH 7.4, 132 mM NaCl, 2.5 mM CaCl$_2$, 6 mM KCl, 1 mM MgSO$_4$, 1.2 mM K$_2$HPO4, 5.5 mM glucose, 0.5% (w/v) BSA.

500 μg ml^{-1} Annexin A5-biotin.

37% formaldehyde solution.

Paraffin.

ABC Elite kit (Vector Laboratories).

Haematoxylin.

Hamilton syringe pipetting system with glass needles (tip diameter ~20 μm).

Surgical microscope.

Microtome.

Approval of the local animal committee to conduct these experiments.

METHODS

1. Euthanise the pregnant mice by cervical dislocation after ether anesthesia.

2. Dissect the uteri and remove the embryos.

3. Immerse the embryos in Hepes buffer warmed to 37°C.

4. Inject 3 μl Annexin A5-biotin solution through the ventricle of the heart using the surgical microscope and the Hamilton syringe pipetting system.

5. Keep the embryos for 30 minutes in the warmed Hepes buffer. Only embryos that show heart activity after the 30 minutes incubation period are further processed.

6. Fix the embryos overnight in 4% formaldehyde in Hepes buffer at 4°C.

7. Embed the fixed embryos in paraffin according to standard protocols.

8. Section the fixed embryo.

9. De-wax the sections and block endogenous peroxidase according to standard procedures.

10. Stain Annexin A5-biotin in the section with the ABC Elite kit and counterstain with haematoxylin. *Plates 7a* and *7b* show sections of embryos stained by this procedure.

Protocol 7.4: Light microscopic analysis of apoptotic cells in the ischaemia/reperfusion injured heart of the adult mouse using Annexin A5-biotin

EQUIPMENT, MATERIALS AND REAGENTS

2-month-old male Swiss mice.

Pentobarbital.

$1.8\,\mathrm{mg\,ml^{-1}}$ Annexin A5-biotin.

Hepes buffer: 25 mM Hepes/NaOH pH 7.4, 140 mM NaCl, 2.5 mM CaCl$_2$.

37% formaldehyde solution.

Paraffin.

ABC Elite kit.

Haematoxylin.

Approval of the local animal committee to conduct these experiments.

METHODS

1. Anaesthetise the mice with pentobarbital ($100\,\mathrm{mg\,kg^{-1}}$ i.p.) and intubate the trachea perorally with a stainless steel tube.

2. Ventilate the animals mechanically with room air.

3. Expose the heart by thoracotomy.

4. Apply ischaemia and subsequently reperfusion of the heart as described by Dumont et al. [13].

5. Inject 50 µl Annexin A5-Biotin into the carotid artery during reperfusion 30 minutes before excision of the heart.

6. Excise the heart and fix in 4% formaldehyde in Hepes buffer overnight at 4°C.

7. Embed the fixed heart in paraffin according to standard protocols.

8. Section the fixed heart perpendicular to its long axis.

9. De-wax the sections and block endogenous peroxidase according to standard procedures.

10. Stain Annexin A5-biotin in the section with the ABC Elite kit and counterstain with haematoxylin. *Plates 7c* and *7d* show sections of hearts stained by this procedure. More examples are described by Dumont *et al.* [13]. They also describe a method to measure the plasma membrane integrity by staining for intracellular IgG.

Protocol 7.5: Non-invasive analysis of cell death in patients using technetium-labelled Annexin A5 and SPECT

EQUIPMENT, MATERIALS AND REAGENTS

Kit to label Annexin A5 with technetium for clinical use.

MultiSPECT2 dual-head gamma camera.

Protocol approved by the medical ethical committee.

Informed consent of the patient.

METHODS

1. Label 1 mg Annexin A5 with technetium (Tc-99m) according to the Apomate protocol.

2. Inject Tc-99m-labelled Annexin A5 intravenously into the patient, who gave informed consent for the study, according to the approved protocol.

3. Blood pool activity of Tc-99m-labelled Annexin A5 generates a high background. The optimal time for SPECT analysis is between 10 and 20 hours after injection of Tc-99m-labelled Annexin A5.

4. The liver and kidneys clear Tc-99m-labelled Annexin A5. The tissue to be analysed by SPECT should therefore preferably be above the diaphragm, like the heart, and in the extremities. This will affect the choice of the patient populations.

5. Perform scintigraphy with the gamma camera with low energy, high resolution collimators, and an energy peak of 140 keV and a window of 20%.

6. Examples of SPECT analysis of cell death in the hearts of acute myocardial infarction patients are described by Hofstra *et al.* [15].

7. An example of SPECT analysis of cell death in a patient with an intracardiac tumour is presented by Hofstra *et al.* [16].

7.5   Detection of plasma membrane permeability

At later stages following phosphatidylserine exposure, apoptosis is often associated with a looser packing or unpacking of the phospholipids of the plasma membrane allowing lipophilic dyes such as fluorescent merocyanine 540 (MC 540) to penetrate the membrane more readily. MC 540 has an excitation and an emission maximum around 550 nm and 580 nm, respectively. It can be excited by the argon line of 488 nm generated by the argon-ion laser. The increased uptake of MC 540 by apoptotic cells can therefore be measured by techniques such as FCM and CSLM. MC 540 not only stains apoptotic cells but also necrotic cells (including those undergoing secondary necrosis) with reduced plasma membrane integrity. Like the exposure of phosphatidylserine, the unpacking of the plasma membrane phospholipids during apoptosis appears to precede the loss of plasma membrane integrity. However, phosphatidylserine exposure and unpacking of plasma membrane phospholipids can occur independently from each other.

Plasma membrane integrity can be monitored by low molecular weight fluorescent dyes such as 7-aminoactinomycin D (7-AAD) and PI. These dyes are non-vital DNA intercalating agents, which cannot enter the cell if the plasma membrane integrity is intact. Once the plasma membrane is compromised both dyes enter the cell and intercalate in the DNA double strands, where they gain fluorescent properties. 7-AAD and PI have an excitation maximum of 535 nm and 550 nm and an emission maximum of 650 nm and 620 nm, respectively. These dyes can be excited by the 488 nm argon laser and can therefore be detected by both FCM and CSLM (*Plate 6*). The spectral properties of MC 540, 7-AAD and PI make them difficult to combine for dual-colour analyses. MC 540, 7-AAD and PI are predominantly used to measure cell death *in vitro* and this type of probe has only limited, if any, use *in vivo*, mainly because they are highly toxic.

7.6   References

1. Leist, M. and Jaattela, M. (2001) Four deaths and a funeral: from caspases to alternative mechanisms. *Nat. Rev. Mol. Cell Biol.* **2**: 589–598.
2. Savill, J. and Fadok, V. (2000) Corpse clearance defines the meaning of cell death. *Nature* **407**: 784–788.
3. Schlegel, R.A. and Williamson, P. (2001) Phosphatidylserine, a death knell. *Cell Death Differ.* **8**: 551–563.
4. Hengartner, M.O. (2000) The biochemistry of apoptosis. *Nature* **407**: 770–776.
5. Wajant, H. (2002) The Fas signaling pathway: more than a paradigm. *Science* **296**: 1635–1636.
6. Scaffidi, C., Fulda, S., Srinivasan, A., Friesen, C., Li, F., Tomaselli, K.J., Debatin, K.M., Krammer, P.H. and Peter, M.E. (1998) Two CD95 (APO-1/Fas) signaling pathways. *EMBO J.* **17**: 1675–1687.
7. Liu, X., Kim, C.N., Yang, J., Jemmerson, R. and Wang, X. (1996) Induction of apoptotic program in cell-free extracts: requirement for dATP and cytochrome c. *Cell* **86**: 147–157.
8. Sakahira, H., Enari, M. and Nagata, S. (1998) Cleavage of CAD inhibitor in CAD activation and DNA degradation during apoptosis. *Nature* **391**: 96–99.
9. Coleman, M.L., Sahai, E.A., Yeo, M., Bosch, M., Dewar, A. and Olson, M.F. (2001) Membrane blebbing during apoptosis results from caspase-mediated activation of ROCK I. *Nat. Cell Biol.* **3**: 339–345.
10. Reutelingsperger, C.P.M. (2001) Annexins: key regulators of haemostasis, thrombosis and apoptosis. *Thromb. Haemost.* **86**: 413–419.

11. Oling, F., Bergsma-Schutter, W. and Brisson, A. (2001) Trimers, dimers of trimers, and trimers of trimers are common building blocks of annexin A5 two-dimensional crystals. *J. Struct. Biol.* **133**: 55–63.

12. van den Eijnde, S.M., Lips, J., Boshart, L., Vermeij-Keers, C., Marani, E., Reutelingsperger, C.P. and De Zeeuw, C.I. (1999) Spatiotemporal distribution of dying neurons during early mouse development. *Eur. J. Neurosci.* **11**: 712–724.

13. Dumont, E.A., Hofstra, L., van Heerde, W.L., van den Eijnde, S., Doevendans, P.A., DeMuinck, E., Daemen, M.A., Smits, J.F., Frederik, P., Wellens, H.J., Daemen, M.J. and Reutelingsperger, C.P. (2000) Cardiomyocyte death induced by myocardial ischemia and reperfusion: measurement with recombinant human annexin-V in a mouse model. *Circulation* **102**: 1564–1568.

14. Dumont, E.A., Reutelingsperger, C.P., Smits, J.F., Daemen, M.J., Doevendans, P.A., Wellens, H.J. and Hofstra, L. (2001) Real-time imaging of apoptotic cell-membrane changes at the single-cell level in the beating murine heart. *Nat. Med.* **7**: 1352–1355.

15. Hofstra, L., Liem, I.H., Dumont, E.A., Boersma, H.H., van Heerde, W.L., Doevendans, P.A., De Muinck, E., Wellens, H.J., Kemerink, G.J., Reutelingsperger, C.P. and Heidendal, G.A. (2000) Visualization of cell death in vivo in patients with acute myocardial infarction. *Lancet* **356**: 209–212.

16. Hofstra, L., Dumont, E.A., Thimister, P.W., Heidendal, G.A., DeBruine, A.P., Elenbaas, T.W., Boersma, H.H., van Heerde, W.L. and Reutelingsperger, C.P. (2001) In vivo detection of apoptosis in an intracardiac tumor. *JAMA* **285**: 1841–1842.

Morphological Changes in Dying Cells

8

Katharina D'Herde, Sylvie Mussche and
Karin Roberg

Contents

8.1 Introduction

8.1.1 Defining modalities of cell death: the dichotomic view is no longer valid

The concept of 'programmed cell death' was initially introduced by Lockshin to describe cell death that occurred in defined developmental niches and at predictable times during development, emphasising that cells are programmed to die during the normal development of the organism [1]. The first evidence that a genetic program

Cell Proliferation and Apoptosis, David Hughes and Huseyin Mehmet (Eds)
© 2003 BIOS Scientific Publishers Ltd, Oxford

existed for physiological cell death came much later from studying development in the nematode *Caenorhabditis elegans* [2–4] and the observation that the genes involved had mammalian counterparts [5–7].

Kerr, Wyllie and Currie [8] described in a landmark paper the morphological features of cell death during development and tissue homeostasis characterised by nuclear and cytoplasmic shrinkage, which they termed 'apoptosis' (derived from the Greek word meaning 'falling off', as of autumn leaves from a tree). Before the term apoptosis was coined this mode of cell death was referred to as shrinkage necrosis.

Initially, a dichotomous view was generally accepted in that all programmed cell deaths were of apoptotic morphology, while accidental, non-physiological passive cell death displayed features of what was called cell necrosis (from the Greek word for 'dying') or defined by Majno and Joris [9] as oncosis, from the Greek word *ónkos*, which means swelling. It must be emphasised that this classification of cell death has made a crucial contribution, promoting the research of programmed cell death and providing clear-cut morphological clues to predict the presence or absence of regulatory mechanisms behind cell death.

However, recent reports indicate that necrotic cell death also occurs during normal cell physiology and development [10, 11] and *in vitro* findings revealed that this necrotic-like cell death is also regulated by a cellular intrinsic death programme [12]. From the literature, at least two other types besides classical apoptosis should be recognised: type 2 physiological cell death or autophagic degeneration, and type 3 physiological cell death or non-lysosomal disintegration [13–17]. Thus, the terms apoptosis and programmed cell death should no longer be used as synonyms. Moreover, necrotic cell death may also need to be considered as a heterogeneous entity containing both active and passive cell deaths.

8.1.2 Morphological changes in apoptotic and necrotic cells

In apoptotic cells morphological changes are prominent in the nuclear compartment with initially marked chromatin condensation and margination accompanied by convolution of the nuclear and cellular outlines and nuclear fragmentation into discrete fragments. Surface protuberances separate from the plasma membrane and then re-seal, converting the cell into a number of membrane-bound apoptotic bodies, that may or may not contain nuclear fragments. Due to early cell shrinkage, the apoptotic cell also loses its normal cell contacts and specialised surface elements, such as microvilli and cell–cell junctions, well before the cellular budding and fragmentation take place. The process of nuclear and cytoplasmic fragmentation is restricted in cells with a high nucleocytoplasmic ratio, such as thymocytes. Depending on the tissue and pathology studied, apoptotic bodies have been defined in the past by many different names including civatte bodies, sunburn cells, colloid bodies and councilman bodies. Significantly, apoptotic bodies generally do not provoke inflammation and are phagocytosed by adjacent healthy cells or by professional phagocytes. In cell culture systems however, apoptotic bodies undergo secondary necrosis, although cytoplasmic organelles initially remain relatively well preserved. In contrast, in necrotic cell death the most prominent feature is an early disruption of the plasma membrane, with spilling of the contents into the surrounding tissue, provoking inflammation. Swelling of necrotic cells and disintegration of the cytoplasmic organelles are apparent at both the LM and EM level. Mitochondria become swollen and often contain granular densities, while ribosomes detach from their associated membranes. In the nucleus, chromatin condenses in ill-defined patches, as opposed to the sharply circumscribed large chromatin masses often abutting against the nuclear membrane or occupying the entire area within the nuclear membrane in apoptotic cells.

At late stages of cell necrosis the chromatin disappears, an event termed karyolysis by pathologists. As opposed to the swelling in necrotic cells, there is a marked shrinkage in the apoptotic cells due to the extrusion of ions and the contraction of the cytoskeleton. In some apoptosis models (ovary, prostate, pituitary, chondrocytes, neurons) very dark cells, distinct from classical apoptotic cells, can be identified. These cells show electron-dense cytoplasm and the nucleus retains nucleolar structure, while dilated ER and conservation of intercellular junctions are other distinguishing features. While these very dark cells – also called type B cells – were previously suspected to be fixation artefacts, increasing numbers of studies report their presence in association with apoptosis [18]. In general, apoptotic nuclei have their chromatin condensed into simple geometrical masses, while other subtypes of programmed cell death have a less compact or complete chromatin condensation pattern [12].

8.1.3   Focusing on cytoplasmic events during apoptosis

Mitochondria

Research on the morphology of apoptosis was initially focused on events that occur in the nucleus, such as chromatin condensation and nuclear fragmentation. However, in the mid-1990s, specialised cysteine proteases (caspases) were discovered and found to be responsible for apoptotic morphology, hence scientists became more interested in cytoplasmic events that take place well before the nuclear changes seen during the degradation phase of apoptosis. Today, the mitochondrion is referred to as the centre of death control, and release of multiple death-triggering proteins from this organelle seems to be essential for the execution phase of apoptosis (for review see [19]).

Among these proteins, cytochrome c, AIF and SMAC/DIABLO are probably the most important, although mitochondria also contain pro-caspases-3, -2 and -9 that are released in the cytosol during apoptosis [20, 21]. In so-called type 1 cells responding to Fas/Apo1/CD 95 cross-linking, the mitochondrial compartment is bypassed and the binding of FasL to its receptor results in generation of a large amount of caspase-8 at the death-inducing signalling complex (DISC), leading to efficient activation of downstream effector caspases when no cytochrome c is being released. In addition to liberation of pro-apoptotic factors and activation of the caspase cascade, a typical early characteristic of apoptosis is mitochondrial membrane permeabilisation followed by dissipation of the mitochondrial transmembrane potential ($\Delta\psi_m$). This potential is due to asymmetric distribution of protons and other ions on either side of the inner mitochondrial membrane, which gives rise to a chemical (pH) and an electrical gradient, both of which are essential for many mitochondrial functions (e.g. protein translocation and ATP synthesis). Disruption of $\Delta\psi_m$ (indicating permeability transition pore opening) has been detected in many cell systems undergoing apoptosis (for review see [22]), but the exact relationship (e.g. cause or consequence) between $\Delta\psi_m$ and the release of apoptogenic factors is still a matter of debate. Other related issues include the mechanism by which cytochrome c traverses the mitochondrial outer membrane and how the Bcl-2 family members may regulate this process.

Morphological approaches involving the mitochondrial compartment have been successful in showing that all mitochondria of an individual cell do not react simultaneously to a given apoptotic stimulus and that sub-populations can be distinguished with respect to cytochrome c release and dissipation of $\Delta\psi_m$ (see Sections 8.3 and 8.5).

Lysosomes

Lysosomal destabilisation also plays an important role in apoptosis initiated by a variety of stimuli, including oxidative stress, growth factor withdrawal, activation of

Fas and exposure to α-tocopheryl succinate [23–25]. Besides caspases, lysosomal proteases such as cathepsins D, B and L have been shown to act as positive mediators of apoptosis in different cell systems [26–28]. Most evidence linking cathepsins with apoptosis has been obtained in experiments using cathepsin D and cathepsin B. Deiss *et al.* [26] suggested that cathepsin D is involved in apoptosis, because they noted that inhibition of this enzyme with pepstatin A blocked the death of cells exposed to TNF-α, interferon-γ, or Fas/APO-1. Furthermore, it has recently been reported that the lysosomal cysteine protease cathepsin B causes mitochondrial release of cytochrome c and activates pro-caspases during apoptosis induced by TNF-α/actinomycin D [27].

Using immunocytochemistry and electron microscopy, we have observed that treatment of cardiomyocytes with naphthazarin (a redox-cycling quinone responsible for intracellular generation of superoxide radicals and H_2O_2) causes early translocation of cathepsin D from lysosomal structures to the cytosol, resulting in apoptotic cell death [28]. Furthermore, we found that the release of cathepsin D precedes the release of cytochrome c and the loss of $\Delta\psi_m$ [29, 30]. Significantly, pretreatment of fibroblasts and cardiomyocytes with pepstatin A, an inhibitor of cathepsin D, prevents both the liberation of cytochrome c from mitochondria and the decrease in $\Delta\psi_m$, even after 48 hours of exposure to naphthazarin [29–31]. Thus, lysosomal destabilisation and active cytosolic cathepsin D seem to be involved in mediating the apoptotic signal to the mitochondria in this system.

8.2 Determination of apoptosis by light microscopy

Phase contrast microscopy or Nomarski interference microscopy of cultured (unstained) cells can reveal the characteristic early rounding-up and blebbing of the cell membrane of apoptotic cells and the formation of apoptotic bodies. Examination of stained cells by light microscopy reveals cell shrinkage, nuclear condensation and the formation of apoptotic bodies. These may or may not contain condensed chromatin, since numerous studies now indicate that classical oligonucleosomal DNA fragmentation does not occur in all apoptotic cell death. Thus, morphological assessment of apoptosis by routine light microscopical procedures has regained an important place in the investigation of cell death, while TUNEL (terminal deoxy-nucleotidyl transferase-mediated desoxyuridinetriphosphate nick end-labelling) staining of nuclei with fragmented DNA (see Section 8.3), which was widely accepted as a major light microscopic detection technique for apoptotic cells in the early 1990s, should be interpreted with caution.

Microscopic examination of living cells is facilitated by the ability of early apoptotic cells to exclude vital dyes such as Trypan blue consistent with the integrity of the plasma membrane during the initial stages of apoptosis. This feature offers the possibility to combine cell-permeable and cell-impermeable DNA-binding dyes to distinguish viable and early apoptotic cells from necrotic and late apoptotic cells (see Section 8.3).

Methyl green is recommended as a light microscopical technique for nuclear counterstaining in combination with a number of immunocytochemical or *in situ* end-labelling techniques (Protocol 8.1; *Plate 8a*). Besides classical haematoxylin and eosin staining (Protocol 8.2) to analyse both nuclear and cytoplasmic changes in dying cells, clear images may also be obtained with Feulgen staining (Protocol 8.3), a selective non-fluorescent DNA dye (*Plate 8b*).

Protocol 8.1: Methyl green nuclear staining

EQUIPMENT, MATERIALS AND REAGENTS

1% methyl green solution: dissolve 1 g methyl green in 100 ml distilled water. Methyl green samples are often mixtures of this dye with methyl violet. Before use, methyl green should be freed from the violet component by dissolving methyl green in equal parts of distilled water and chloroform (100 ml distilled water + 100 ml chloroform). After two or three days' standing, the aqueous supernatant is removed for use, containing pure methyl green.

METHOD

1. Add 25 ml ethanol.

2. Stain sections in methyl green solution for 1–10 minutes.

3. Dehydrate the slides through a graded series of ethanol (70%, 96%, 100%, 100%, 100%).

4. Clear in xylene and mount with non-aqueous mounting medium (e.g. DPX).

5. Check the slides with a microscope. If the colouration is not satisfactory follow the opposite way through the graded series of ethanol and re-stain.

6. Allow slides to dry overnight prior to microscopic examination.

Protocol 8.2: Haematoxylin and eosin staining (H&E)

EQUIPMENT, MATERIALS AND REAGENTS

Harris' haematoxylin solution:

Solution A: dissolve 2 g haematoxylin in 20 ml ethanol.

Solution B: dissolve 40 g ammonium aluminium sulphate in 400 ml distilled water and heat. After 24 hours, mix solutions A and B and heat to boiling. Let the solution cool and add 1 g mercuric oxide. Heat again to dark purple and cool down quickly. Filter.

1% aqueous eosin yellow.

METHOD

1. Incubate 2–10 minutes in Harris' haematoxylin solution.

2. Rinse in running water (5 minutes).

3. Differentiate quickly in 1% HCl (3 seconds).

4. Rinse in running water (5–10 minutes). If the blue colouration is too dark (this can be checked quickly using a microscope), the stain can be lightened by replacing the slides in 1% HCl for a few seconds. Rinse the slide in running water and re-examine.

5. Transfer to 1% eosin yellow for 1–2 minutes.

6. Rinse in running water.

7. Dehydrate the slides through a graded series of ethanol (70%, 96%, 100%, 100%, 100%).

8. Clear in xylene and mount with non-aqueous mounting medium (e.g. DPX).

9. Allow slides to dry overnight prior to microscopic examination.

10. View cytospins, adherent cell cultures, paraffin slides or whole mounts by light microscopy using a 40 × objective. An estimation of the percentage apoptotic and necrotic cells is given by calculating the respective indices. Apoptosis (necrosis) is scored in ten randomly selected microscopic fields and expressed as the number of apoptotic (necrotic) cells per number of total nuclei counted in the same microscopic field. Small groups of apoptotic bodies are counted as remnants of one apoptotic cell. Apoptotic cells: blebbing, chromatin condensation and margination giving rise to the appearance of densely basophilic masses, reduction of cell volume, formation of apoptotic bodies with eosinophilic cytoplasm. Necrotic cells: chromatin flocculation into ill-defined masses giving rise to the appearance of hyperchromatic nuclei (pyknosis) followed by karyolysis with loss of all basophilia in the nucleus, nuclear swelling, cellular swelling and vacuolation, cell membrane lysis, resulting in the appearance of ghost cells.

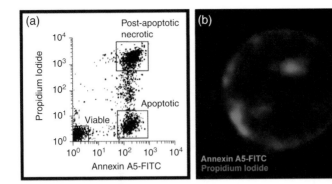
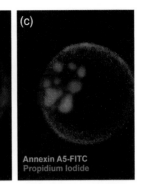

Plate 6

Measurement of phosphatidylserine exposure and integrity of the plasma membrane with Annexin A5-FITC (green) and PI (red), respectively. (a) Dual colour FCM analysis of anti-Fas-treated Jurkat cells. Three populations are distinguishable: the Annexin A5-FITC-negative and PI-negative cells (viable cells); the Annexin A5-FITC-positive and PI-negative cells (apoptotic cells); and the Annexin A5-FITC-positive and PI-positive cells (post-apoptotic necrotic cells). The apoptotic cells (b) and the post-apoptotic necrotic cells (c) can be visualised by CSLM. Note that the post-apoptotic necrotic cell with a permeable plasma membrane has multiple fragmented chromatin (c).

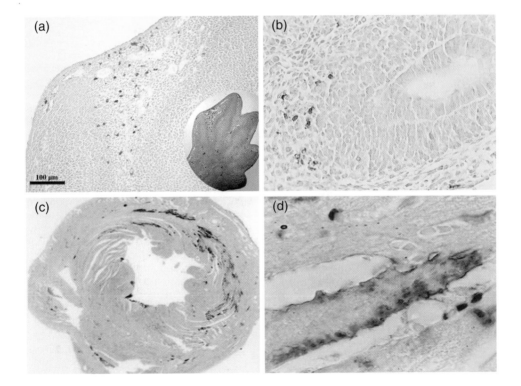

Plate 7

Visualisation of phosphatidylserine exposure during apoptosis *in vivo*. Annexin A5-biotin was injected into the beating heart of a live mouse embryo (a, b) and into the tail vein of an adult mouse just before ischaemia/reperfusion injury to the heart (c, d). The tissues were then fixed, sectioned and stained with an ABC Elite kit for the peroxidase-based detection of annexin A5-biotin, which stains brown. (a) Limb of a E13 embryo. (b) Metanephrogenic blastema of E11 embryo. (c) Cross section of the mouse heart with ischaemia/reperfusion injury. The area at risk is in the top right area. (d) Enlarged detail of (c). Panels (a) and (b) courtesy of Drs. Chr. Vermeij-Keers and S.M. van den Eijnde.

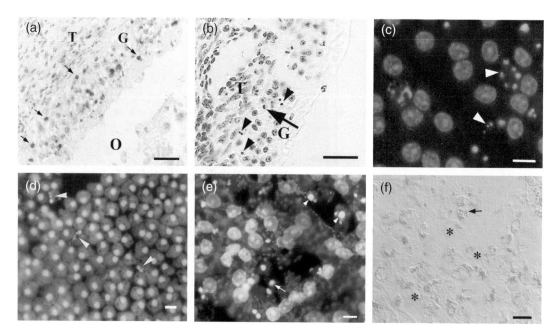

Plate 8

(a) TUNEL-stained paraffin section of atretic quail granulosa (G) with adjacent thecal layers (T) and oocyte (O). Incorporation of digoxigenin-11-dUTP by terminal deoxynucleotidyl transferase, and direct immunoperoxidase detection of the digoxigenin-labelled DNA via DAB counterstained by methyl green. Numerous apoptotic bodies show brown reaction product in their cytoplasm, while the condensed green chromatin masses have not incorporated dUTP (arrows). Bar: 20 μm (b) Granulosa of a starvation-induced atretic follicle (Feulgen staining). Note the numerous Feulgen-positive apoptotic chromatin masses, some of them lying in clusters, and the apoptotic nucleus with a crescent-shaped mass under its nuclear membrane (arrow). Bar: 20 μm. (c) Epifluorescence image: granulosa explant, whole mount, numerous apoptotic condensed chromatin masses (arrowheads), DAPI staining. Bar: 10 μm. (d) Epifluorescence image: apoptotic granulosa explant vitally stained with acridine orange. Arrowheads mark three fragmented apoptotic nuclei. Bar: 10 μm. (e) Apoptotic granulosa explant stained with CM-H$_2$TMRos. Persistence of polarised mitochondria surrounding apoptotic chromatin masses (arrowheads). Other apoptotic nuclei show no stained granules (arrow). Bar: 10 μm. (f) Nomarski interference contrast: cytochrome c localisation (brown) in living granulosa explant cultured for 72 h in the absence of follicle stimulating hormone (FSH); normal nuclei are faintly stained. In one apoptotic cell with condensed chromatin (methyl green positive masses) several reactive mitochondria are seen (arrow), while other apoptotic bodies are devoid of stained mitochondria. Bar: 10 μm. ((a) and (b) reprinted from Biochem. Cell Biol., d'Herde et al., 1994 © with permission from National Research Council Canada, (d) reprinted from Apoptosis, D'Herde and Leybaert, 1998 © with permission from Kluwer Academic Publishers, (e) reprinted from J. Histochem. Cytochem., Krysko et al., 2001 © with permission from the Histochemical Society, (f) reprinted from Cell Death Differ., D'Herde et al., 2000 © with permission from Nature Publishing Group.)

Protocol 8.3: Feulgen staining

EQUIPMENT, MATERIALS AND REAGENTS

Schiff's reagent: dissolve 1 g of basic fuchsin in 200 ml of boiling distilled water. Shake to dissolve and cool to 52°C. Filter and add to the filtrate 20 ml of 1 N HCl. Cool to 25°C and add 1 g sodium metabisulphite ($Na_2S_2O_5$). Keep this solution in the dark for 14–24 hours. Add 2 g of activated charcoal and shake for 1 minute. Filter. Store in the dark at 4°C. Allow to reach room temperature before use.

Bisulphite solution: 5 ml 10% $Na_2S_2O_5$, 5 ml 1 N HCl, water to 100 ml.

METHOD

1. Treat sections with 5 N HCl at room temperature for the optimum time of hydrolysis (10–40 minutes).

2. Rinse in distilled water.

3. Transfer to Schiff's reagent for 30–60 minutes.

4. Rinse in three changes of freshly prepared bisulphite solution.

5. Rinse in running water (15 minutes).

6. Dehydrate through a graded series of ethanol (70%, 96%, 100%, 100%, 100%).

7. Clear in xylene and mount in non-aqueous mounting medium (e.g. DPX).

8.3 Determination of apoptosis by fluorescence microscopy

Fluorescence microscopy constitutes an invaluable tool in cell death studies. In the vast array of techniques, the use of DNA-specific probes provides an excellent means to detect the characteristic chromatin condensation and nuclear fragmentation in apoptosis as well as nuclear changes in necrotic cells.

8.3.1 DAPI

DAPI is a DNA-specific dye, which forms a fluorescent complex by attaching in the minor groove of consecutive (3 to 4 base pairs) A-T-rich sequences of DNA. The DNA specificity is confirmed by the observation that DAPI fluorescence is sensitive to DNAse but not to RNAse. Single-stranded DNA and both single- and double-stranded RNA form complexes with DAPI, most probably by intercalation, but the fluorescence is much weaker than the fluorescence of the DAPI/double-stranded (ds) DNA complex. Typically, the DAPI/RNA complex exhibits a longer wavelength fluorescence emission maximum than the DAPI/dsDNA complex (± 500 nm versus ± 460 nm) and a quantum yield that is only about 20% of the DNA value. DAPI staining has been applied to flow cytometry, chromosome staining, DNA visualisation and quantitation in histochemistry and biochemistry (for review see [32]).

Hoechst 33258 is another DNA-specific dye with similar properties and binding characteristics to DAPI, but its emission is quenched by halogenated DNA (e.g. BrdU incorporated in DNA). The bright blue DAPI-stained condensed chromatin of apoptotic cells differs greatly from that in healthy and necrotic cells (*Plate 8c*). Following DNA fragmentation, oligonucleosomes diffusing outside the nucleus are initially retained in the cytoplasm due to the integrity of the plasma membrane. While this cytoplasmic DNA can be stained by DAPI, rinsing with PBS prior to DAPI staining eliminates the blue hue in the cytoplasm, since the oligonucleosomes are not irreversibly immobilised by fixation.

Single necrotic nuclei in cell cultures are sometimes relatively unremarkable. Their nuclear membrane becomes wrinkled and chromatin condenses into ill-defined small masses. At later stages the chromatin staining is lost giving the appearance of ghost nuclei. However, the characteristic cytoplasmic swelling in necrosis due to osmotic shock is not accompanied by nuclear swelling in all model systems. An absolute feature to discriminate between necrotic and healthy (or early apoptotic) cells is the physical integrity of the plasma membrane. The intact membrane of live cells excludes a variety of dyes such as propidium iodide (PI). Necrotic cells instead take up the dye, which readily crosses the damaged plasma membrane, binds to DNA and dsRNA and fluoresces intensely. A useful assay based on this principle combines DAPI, which can penetrate through the intact plasma membrane to stain DNA, and PI which is excluded by the intact membrane. This co-staining technique combining DAPI and PI thus provides an excellent assay to discriminate between apoptotic, necrotic and live cells. Excitation of both PI and DAPI can be achieved; compared with live cells, DAPI fluorescence will be suppressed in dead cells together with a more intensive PI stain, while apoptotic nuclei can be recognised by their morphology. Under conditions where cytoplasmic oligonucleosomes are lost (detergent treatment, ethanol fixation), early apoptotic cells show reduced staining with DAPI compared to healthy cells (measured by flow cytometry). Late apoptotic cells have increased staining levels with PI due to loss of plasma membrane function. The DAPI protocol proposed here (Protocol 8.4) allows easy and simple determination of apoptotic and necrotic cells and is suitable for most specimens, for example fixed single

Protocol 8.4: DAPI staining

EQUIPMENT, MATERIALS AND REAGENTS

PBS pH 7.4.

4% PFA in PBS.

DAPI stock solution (1 mg ml^{-1} in distilled water).

PI stock solution (1 mg ml^{-1} in distilled water). Stock solutions of DAPI and PI should be prepared in distilled water since at relatively high concentrations these dyes tend to precipitate in PBS.

METHOD 8.4A   DAPI ALONE

1. Rinse cells 3 times with PBS.

2. Fix in 4% PFA in PBS for 20 minutes (not necessary for fixed material e.g. de-waxed slides).

3. Rinse 3 times with PBS.

4. Stain for 3 minutes with DAPI stock solution diluted 1:1000 in PBS.

5. Rinse 3 times with PBS.

6. Mount in an aqueous anti-fade fluorescence mounting.

7. Examine cells by fluorescence microscopy.

METHOD 8.4B   COMBINATION DAPI AND PI

1. Rinse cells 3 times with PBS.

2. PI staining: stock PI solution diluted 1:50 in PBS for 30 minutes on ice.

3. As from step 2 in Method 8.4a.

cells or cytospins, whole mounts of tissue explants and paraffin sections. In contrast, the combined DAPI-PI protocol can be performed starting with living cells only.

8.3.2 Acridine orange

Among acridine dyes having different affinity to various cell constituents, 3,6-dimethylaminoacridine or acridine orange (AO) is the most popular. AO can penetrate viable as well as non-viable cells, staining the nucleus (DNA) green and the cytoplasm (RNA) red-orange. AO can be used to stain cell nucleic acids for morphological and/or quantitative analysis, to distinguish DNA from RNA and to measure the extent of DNA denaturation. At low concentrations, as a vital stain, AO is useful for analysing nuclear morphology (*Plate 8d*). Like DAPI it allows the detection of condensed chromatin masses in apoptotic cells but offers the further possibility of combination with other probes that require living cells, such as intracellular free $(Ca^{2+})_i$ measurements and $\Delta\psi_m$ probing ([33]; Protocol 8.5).

8.3.3 Terminal deoxynucleotidyl transferase-mediated dUTP nick end-labelling (TUNEL)

During the degradation stage of apoptosis, high molecular weight and internucleosomal DNA cleavage occurs [34–38]. The TUNEL method developed by Gavrieli *et al.* [39] has enabled *in situ* visualisation of DNA fragmentation at the single-cell level. DNA fragments can be labelled enzymatically with modified nucleotides (X-dUTP, where the tag X = fluorescein, biotin or digoxigenin) and this is followed by the synthesis of a labelled polydeoxynucleotide. Terminal deoxynucleotidyl transferase (TdT) is used as the labelling enzyme, which is able to label both recessed and blunt ends of double-stranded DNA breaks. This contrasts with the use of DNA polymerase I as the labelling enzyme, which catalyses the template-dependent addition of nucleotides at recessed 3'-OH termini only, not protruding or blunt ends. Theoretically, this latter reaction can detect not only apoptotic non-random DNA fragmentation, but also single-strand nicks existing in normal viable cells [40]. Use of fluorescein-dUTP as the label allows the detection of the incorporated nucleotides directly with a fluorescence microscope or flow cytometer and results in general low-background staining. The immunocomplexes where DNA is labelled with biotin-dUTP or digoxigenin-dUTP can be detected by staining with an avidin-conjugated peroxidase or anti-digoxigenin-conjugated peroxidase, respectively.

Permeabilisation of the plasma membrane allows exogenous enzymes to enter the cell. To avoid loss of low molecular mass DNA from the permeabilised cells, fixation with cross-linking agents such as formaldehyde or paraformaldehyde is necessary. Gorczyca *et al.* [41] reported that the low molecular mass DNA is cross-linked to cellular constituents. Another key step in the TUNEL technique, requiring fine tuning for each cell type, is the pretreatment of the specimen in order to make the DNA breaks accessible to the labelling enzymes. For this purpose, besides conventional proteinase K pretreatment, microwave irradiation can also be used (Protocol 8.6).

As mentioned previously, the TUNEL assay should not be used as the only method to investigate apoptosis, as oligonucleosomal DNA fragmentation is not an obligatory event for all apoptotic cells [42, 43]. Moreover, the TUNEL assay fails to discriminate between apoptosis, necrosis and autolytic cell death [44]. To overcome this lack of specificity, a method has been described for simultaneously detecting necrosis and apoptosis using a combination of the TUNEL technique and Trypan blue staining [45]. In a few model systems where detectable oligonucleosomal DNA fragmentation precedes chromatin condensation, the TUNEL technique is more sensitive then classical histochemical staining.

Protocol 8.5: AO staining

EQUIPMENT, MATERIALS AND REAGENTS

PBS pH 6.5.

Stock solution of $100\,\mu g\,ml^{-1}$ acridine orange in PBS.

Fura-2-AM.

Normal Buffered Solution with 2.5 mM probenecid.

Pluronic.F-127.

Fluorescence microscope with photomicrography facility.

METHOD 8.5A

1. Cells are superfused for 30 seconds in PBS at pH 6.5 containing $4\,\mu g\,ml^{-1}$ AO.

METHOD 8.5B   DOUBLE STAINING PROTOCOL (E.G. WITH Ca^{2+} MEASUREMENT (FURA-2-AM))

1. Incubate cells in $5\,\mu M$ fura-2-AM + 0.01% Pluronic in Normal Buffered Solution with probenecid for 90 minutes at room temperature.

2. Rinse with Normal Buffered Solution + probenecid.

3. De-esterification of fura-2-AM: 15–30 minutes in Normal Buffered Solution + probenecid at 37°C.

4. Take $(Ca^{2+})_i$ image.

5. Superfuse with AO as in Method 8.5a and take image of exactly the same region.

Protocol 8.6: TUNEL labelling

Many commercial TUNEL kits are available for use with tissue sections, adherent cell cultures, cytospins and cell smears and they provide an extensive description of the procedure involved. One key step to improve the pretreatment step, necessary to make the DNA breaks accessible to the TdT, is pretreatment of the slides with microwave irradiation instead of proteinase K.

EQUIPMENT, MATERIALS AND REAGENTS

Commercial TUNEL kit.

0.1 M citrate buffer at pH 6.0.

METHOD

1. Submerse the slides in a plastic jar containing 200 ml of 0.1 M citrate buffer at pH 6.0.

2. Put the jar in a microwave oven and apply 750 W microwave irradiation for 1 minute.

3. Cool the slides quickly by adding immediately 80 ml of distilled water at room temperature.

4. Transfer the slides to PBS at room temperature and continue the procedure as indicated by the manufacturer.

8.3.4   Staining of mitochondrial transmembrane potential

An important issue in studies focusing on the mitochondrial control of cell death is the precise timing of dissipation of $\Delta\psi_m$ in relation to other key steps such as caspase activation or phosphatidylserine exposure. Many different classes of non-fixable probes to measure $\Delta\psi_m$ are commercially available, but some (e.g. $DiOC_6$ and rhodamine 123) are fraught with potential artefacts [46]. JC-1, a dual-emission potential-sensitive probe (5,5′,6,6′-tetrachloro-1,1′,3,3′-tetraethylbenzimidazolocarbocyanine iodide), seems the most sensitive and reliable non-fixable probe, but necessitates (due to its two different emission wavelengths) a dual-laser flow cytometer or confocal microscope [47]. Divergent data in this field may also be due to the fact that flow cytometric studies utilise data from mitochondrial populations. The visualisation of $\Delta\psi_m$ at the single organelle indicates a gradual onset of depolarisation of mitochondria during apoptosis induction [48, 49]. A double-staining protocol has even revealed the presence of polarised mitochondria in apoptotic cells at the degradation stage [46]. For this approach a chloromethyl rosamine-derived probe, CM-H_2TMRos, was used which becomes fluorescent only when it is oxidised in the cell. Once the probe accumulates in the mitochondria, the chloromethyl group can react with accessible thiol groups of peptides and proteins to form an aldehyde-fixable conjugate [50, 51] (Protocol 8.7; *Plate 8e*).

Protocol 8.7: Mitochondrial staining by CM-H₂TMRos

EQUIPMENT, MATERIALS AND REAGENTS

CM-H₂TMRos.

DMSO.

Phenol red-free culture medium.

Dulbecco's PBS pH 7.4.

4% PFA in PBS buffer.

Acetone.

Anti-fade fluorescence mounting medium.

Valinomycin.

Carbonyl cyanide p-trifluoromethoxyphenylhydrazone (FCCP).

METHOD 8.7A   STANDARD PROCEDURE

1. Prepare dye stock solution: CM-H₂TMRos is dissolved in DMSO to give a 1 mM stock solution.

2. Prepare staining medium: prepare immediately before use by adding the dye stock solution to phenol red-free culture medium to obtain a dye concentration of 200 nM.

3. Living cells are incubated with the dye in phenol red-free medium for 20–60 minutes at 37°C, shaking in the dark.

4. Rinse cells 3 × 5 minutes in PBS at 37°C.

5. Fix cells in freshly prepared 4% PFA in PBS at 37°C for 15 minutes.

6. Rinse cells 3 × 5 minutes in PBS and permeabilise cells by incubation in ice-cold acetone.

7. Rinse cells 3 × 5 minutes in PBS followed by a rinse in distilled water.

8. Mount the samples with an anti-fade fluorescence mounting medium.

9. Examine samples on a fluorescence microscope equipped with epifluorescence optics and suitable filters for DAPI and TRITC detection.

METHOD 8.7B   DOUBLE STAINING (CM-H₂TMRos AND DAPI)

Perform DAPI staining (Protocol 8.4) after step 7 of Protocol 8.7a.

METHOD 8.7C   CONTROLS

Controls whereby mitochondria are depolarised prior to CM-H$_2$TMRos staining should be included. Due to the binding of the probe after oxidation in respiring mitochondria, conventional protocols with uncoupling after dye loading do not work [46].

1. Valinomycin and FCCP are dissolved in 95% ethanol to obtain stock solutions of 10 mM. Store at −20°C.

2. Dilute the stock solutions in culture medium immediately before adding to the cells to obtain a final concentration of 1 μM and 50 μM, respectively.

3. Treat the cells with the mitochondrial uncouplers: valinomycin (5 minutes at 37°C) or FCCP (15 minutes at 37°C).

8.4 Determination of apoptosis by electron microscopy

At present, quantification of apoptosis is achieved by using a combination of various methods to analyse individual features such as caspase activation, phosphatidyl-serine exposure and DNA fragmentation. However, many of these techniques are accompanied by one or more of the following methodological problems. For example, apoptotic bodies can be erroneously identified as individual apoptotic cells, it may be impossible to distinguish between truly apoptotic cells and viable, non-apoptotic cells that are engulfing apoptotic bodies, and finally apoptotic cells do not always exhibit all of the classical features of apoptosis. Apoptosis and necrosis were originally defined on the basis of morphological criteria and it is still necessary to combine the findings of light or electron microscopy with results obtained using other methods. In their seminal publication, Kerr and co-workers [8] used electron microscopy as the main technique to describe apoptosis. Compared to light microscopy, electron microscopy is time consuming and requires more expensive equipment, but it provides much more detailed information about cellular morphology and is therefore the most accurate method to distinguish apoptosis and necrosis in cell cultures and in tissues (Protocol 8.8).

Most early experiments were performed on different types of mononuclear blood cells displaying fragmented nuclei and apoptotic bodies. Compared to other types of cells (e.g. fibroblasts and epithelial cells), a mononuclear leukocyte has a large nucleus and reduced cytoplasm, and Collins and colleagues [52] have found that DNA fragmentation does not always appear in cells that contain more cytoplasm. Thus, nuclei of healthy Jurkat and U937 cells are irregular in shape with many gulfs and protrusions and in the very early stages of apoptosis, the nuclei round up and the chromatin starts to aggregate in regularly shaped clumps. This is followed by fragmentation through budding, which is an actin-dependent process, similar to the formation of apoptotic bodies (*Figure 8.1a*). Conversely, in normal fibroblasts the oval-shaped nuclei have few, if any, indentations or protrusions and early in apoptosis they adopt a very irregular appearance while the chromatin begins to condense in geometrically more complex masses (*Figure 8.1b*). Later in the apoptotic process, fibroblast nuclei also exhibit the typical condensation and fragmentation pattern as classically described. In contrast, in necrotic cells, marginal clumping of loosely textured nuclear chromatin is apparent besides general disorganisation of the cytoplasm (*Figure 8.1c*). For these reasons, electron microscopy remains the 'gold standard' for determining the mode of cell death.

Protocol 8.8: Preparation of cells growing in Petri dishes for transmission electron microscopy

EQUIPMENT, MATERIALS AND REAGENTS

1 M sodium cacodylate buffer (stock solution pH 7.4, 2000 mOsm).

0.15 M sodium cacodylate buffer (Nacac).

1 M sucrose.

2% glutaraldehyde in 0.1 M Nacac-HCl buffer (pH 7.4, 510 mOsm).

1% osmium tetroxide in 0.15 M Nacac (1 part 4% OsO_4 in distilled water + 3 parts 0.2 M Nacac).

2% uranyl acetate in 50% ethanol.

Epon 812.

Reynold's lead citrate.

METHOD

1. Fix cells growing in a Petri dish by treatment with 2% glutaraldehyde in 0.1 M sucrose-Nacac-HCl buffer (1:1) for 5 minutes and subsequently pure fixative for 2 hours. Cells fixed in this manner can be stored at 4°C for a short time.

2. Rinse 4 × 5 minutes in 0.15 M Nacac.

3. Post-fix the cells in 1% OsO_4 in 0.15 M Nacac for 1 h at room temperature.

4. Rinse 4 × 5 minutes in 0.15 M Nacac.

5. Dehydrate in 50% ethanol (2 × 5 minutes).

6. Block stain the cells overnight with 2% uranyl acetate in 50% ethanol at room temperature.

7. Continue the dehydration in a graded series of ethanol (70%, 85%, 95%, 100%; 10 minutes each).

8. Infiltrate with 100% ethanol:Epon 812 (1:1) for 30 minutes.

9. Cover the cells in the Petri dish with a thin layer (1–2 mm) of Epon 812 and allow to stand for at least 2 hours at room temperature. Thereafter, place inverted Beem capsules containing polymerised Epon upside down at a straight angle to the cell layer.

10. Polymerise for at least 48 hours at 60°C.

11. Remove the Epon layer and the attached Beem capsules from the Petri dish. This procedure captures the cells on the surface on the Epon layer.

12. Remove the Beem capsules with tweezers and cut ultra-thin (60 nm) sections with a diamond knife. Collect the sections on Formvar-coated copper 100-mesh grids and counterstain for 2 minutes with Reynold's lead citrate. Wash three times in double-distilled water and allow to air dry.

13. Examine the sections by transmission electron microscopy. For tissues it can be necessary to cut semi-thin sections (2 μm) and to stain the pre-heated sections (70°C) with a drop of toluidine blue solution (0.1% in distilled water, pH 11.0) in order to select an interesting zone of the tissue block for ultra-thin sectioning.

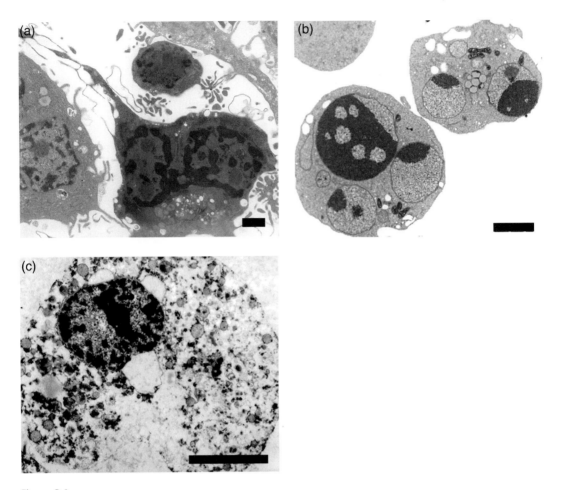

Figure 8.1

Electron micrographs of (a) human foreskin fibroblasts exposed to naphthazarine for 8 hours, and (b) Jurkat cells exposed to Fas ligands for 8 hours. (a) Ultrastructural features revealed by electron microscopy during apoptosis in this cell type are condensation of the cytoplasm, deformation of the nuclear contour with chromatin condensation into geometrically complex masses and retention of the nucleolar structure, dilated endoplasmic reticulum and formation of apoptotic bodies all, or not, containing nuclear fragments. (b) In Jurkat cells the more pathognomonic ultrastructural features of apoptosis are present: condensation of chromatin into regular circle of crescent-shaped masses, often abutting against the nuclear membrane and fragmentation of the nucleus into two or more micronuclei with intact nuclear membranes. (c) Electron micrograph of a necrotic cell with ill-defined condensed chromatin masses, general swelling and disorganisation of cytoplasm and loss of plasma membrane integrity. Bars: 2 μm. ((a) reprinted from Free Radic. Biol. Med. (27) pp 1228–1237, Roberg *et al.*, © 1999 with permission from Elsevier Science; (c) reprinted from Reprod. Nutr. Dev. (36) pp 175–189, D'Herde *et al.*, 1996 © with permission from Elsevier Science.)

8.5   Location of cytochrome c during apoptosis

Cytochrome c is a nuclear DNA-encoded protein that is an essential component of the respiratory chain in mitochondria. It is present in the intermembrane space and on the surface of the inner membrane. It was recently shown that translocation of cytochrome c from mitochondria to the cytosol is a fundamental step in apoptosis. Once in the cytosol, cytochrome c becomes part of a multi-protein complex called the apoptosome. This complex comprises Apaf-1 (apoptosis protein-activating factor 1), pro-caspase-9 and dATP or ATP [53]. In 1996, Liu and co-workers [54] showed that cytochrome c from a cytosolic fraction could induce apoptotic changes in isolated nuclei and it was later found that apoptosis was induced in a variety of cell types following microinjection of cytochrome c into the cytosol [55, 56]. In addition, numerous studies of intact cells have confirmed that release of cytochrome c is an early event in apoptosis induced by a variety of agents such as Ara-C [57], UVB and staurosporine [58].

8.5.1 Immunofluorescence

Translocation of cytochrome c during apoptosis has been shown mainly by western blotting or immunofluorescence techniques and only a few investigators have used immunocytochemistry and electron microscopy to determine the exact location of this haemoprotein. These methods were recently compared in a study of cytochrome c release during oxidative stress in cardiomyocytes and it was found that immunofluorescence is less sensitive than immunoelectron microscopy and western blotting [30]. That notwithstanding, for detection of cytochrome c, immunofluorescence offers certain advantages as it is easier to perform and can provide information about single cells or whole populations (*Figure 8.2*). Western blotting, on the other hand, provides information only about the cell population, but it is a quantitative and more sensitive technique [30] (Protocol 8.9).

8.5.2 Immunoelectron microscopy

Immunocytochemistry has become a powerful technique for localising proteins using the EM. Furthermore, introduction of ultra-small gold conjugates combined with silver enhancement has led to a breakthrough in microscopic detection of intracellular antigens based on pre-embedding procedures (i.e. immunocytochemical labelling

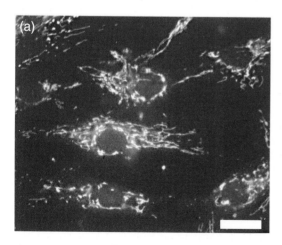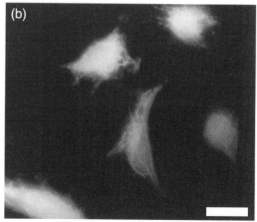

Figure 8.2

Immunofluorescence detection of cytochrome c in human foreskin fibroblasts. The micrographs show (a) control cells and (b) cells exposed to 0.5 μM of naphthazarin for 2 hours. Relocalisation of cytochrome c can be seen as diffuse staining of the treated cells while control cells show cytochrome c fluorescence in a threadlike network of mitochondria. Bars: 30 μm. (Reprinted from Free Radic. Biol. Med. (27) pp 1228–1237, Roberg *et al.*, © 1999 with permission from Elsevier Science.)

Protocol 8.9: Immunofluorescence detection of cytochrome c in cultured cells

EQUIPMENT, MATERIALS AND REAGENTS

Dulbecco's PBS pH 7.4.

4% PFA in PBS.

Incubation buffer: 5% FCS and 0.1% saponin in PBS.

Appropriately titrated monoclonal mouse-anti-human cytochrome c antibody.

Appropriately titrated mouse anti-IgG-Texas Red conjugate.

Anti-fade fluorescence mounting medium.

METHOD

1. Fix cells growing on a coverslip (diameter 22 mm) in 4% PFA in PBS for 20 minutes at 4°C.

2. Rinse the coverslip 2 × 5 minutes in PBS.

3. Pre-block and permeabilise by incubating the cells in 0.1% saponin and 5% FCS in PBS (incubation buffer) for 20 minutes at room temperature.

4. Incubate with monoclonal mouse-anti-human cytochrome c (diluted in incubation buffer) overnight at 4°C. Negative controls should be incubated without anti-cytochrome c antibodies, or with the same amount of mouse IgG negative control as anti-cytochrome c antibodies.

5. Rinse the coverslip 2 × 5 minutes in incubation buffer.

6. Incubate with mouse anti-IgG-Texas Red conjugate (diluted in incubation buffer) for 1 hour at room temperature in the dark.

7. Rinse the coverslip 2 × 5 minutes in incubation buffer.

8. Rinse in PBS and distilled water and mount the coverslip upside down on a slide with an aqueous mountant with an anti-fade additive.

9. Examine the cells in a fluorescence photomicroscope, using green exciting light and a red barrier filter.

Note: for immunofluorescence detection of cathepsin D, Protocol 8.9 can be adapted replacing the primary antibody with a polyclonal rabbit anti-human cathepsin D antibody (diluted 1:100) and the secondary antibody with a goat anti-rabbit IgG-Texas Red conjugate (diluted 1:200).

before embedding for electron microscopy). The availability of these probes offers exciting new possibilities to study intracellular antigens at the ultrastructural level [59] and the procedure is particularly propitious because it does not require sophisticated equipment and can therefore be carried out routinely in any immunocytochemical laboratory.

Penetration of the antibody–gold complex is the most important aspect of pre-embedding labelling of intracellular antigens. There are several factors that affect penetration including the condition of the specimen (e.g. fixation and degree of cross-linking), the characteristics of the reagents (Fab fragment, intact immunoglobulin and size of gold particles) and the incubation conditions (e.g. ion concentration, pH, temperature and time). Fixation for immunocytochemistry is often carried out with freshly buffered 2–4% PFA combined with a low concentration (0.05–0.5%) of glutaraldehyde [60]. Time, temperature and concentration determine the degree of cross-linking and the density of the matrix after fixation. Extremely mild permeabilising agents, such as very low concentrations of sodium borohydride or saponin, must be used in pre-embedding techniques to avoid negative effects on morphology. Treatment with detergents (e.g. Triton-X) enables antibodies to reach cytosolic antigens but tends to disrupt membranes and morphologically distort cultured cells. Moreover, the time required for labelling depends on the density of the matrix and, for satisfactory results, it is essential that there is a balance between the times and concentrations used for fixation, permeabilisation and labelling [61].

Immunocytochemistry at the EM level is the best method to obtain information about the exact location of cytochrome c in single cells. When studying cells grown in Petri dishes, pre-embedding is the most suitable method, because labelling of the antigen, silver enhancement and embedding can be carried out directly in the culture dish without any trypsination or scraping of the cells. In an investigation using pre-embedding to examine the location of cytochrome c in apoptotic cardio-myocytes, most of the particles labelling cytochrome c were found on the outer mitochondrial membrane in control cells [31], thus it seems likely that, with this technique, the primary antibody can only penetrate the outer membrane. Nonetheless, specific labelling of cytochrome c, with a negligible background, is achievable with this immunoelectron microscopy method (*Figure 8.3a*) and it is straightforward to monitor translocation of cytochrome c from mitochondria to the cytosol (*Figure 8.3b*) [30]. There are two main advantages of using immunoelectron microscopy to detect cytochrome c translocation during apoptosis: first, the method is very sensitive, providing information about the exact location of the protein, and second it is possible to study differences between individual cells and even between individual mitochondria in the same cell (Protocols 8.10 and 8.11).

Protocol 8.10: Pre-embedding immunogold labelling of intracellular antigens (cytochrome c) using ultra-small gold probes

EQUIPMENT, MATERIALS AND REAGENTS

1 M sodium cacodylate buffer (stock solution pH 7.4, 2000 mOsm).

4% PFA.

0.05% glutaraldehyde in 0.15 M sodium cacodylate HCl buffer (pH 7.6).

2% glutaraldehyde in PBS (pH 7.4).

Permeabilisation buffer: 0.05% saponin or 0.05% sodium borohydrate in 0.1% glycine in PBS.

Incubation buffer: 0.8% BSA, 0.1% gelatin and 20 mM NaN$_3$ in PBS (pH 7.6). This buffer should be filtered through a Millipore filter (0.22 μm) and stored at 4°C.

Monoclonal mouse-anti-human cytochrome c antibody, clone 6H2.B4 (dilution 1:50).

Mouse anti-IgG antibody tagged with 0.8 nm gold particles (dilution 1:100).

METHOD

1. Fix cells growing in a Petri dish (35 mm) with 4% PFA and 0.05% glutaraldehyde in 0.15 M sodium cacodylate HCl buffer (pH 7.6) for 20 minutes at 4°C.

2. Rinse 2 × 5 minutes in PBS.

3. Permeabilise the cells by incubating with 0.05% sodium borohydrate or 0.05% saponin in 0.1% glycine for 10 minutes at room temperature.

4. Incubate for at least 30 minutes in incubation buffer at room temperature.

5. Incubate with a monoclonal mouse anti-human cytochrome c antibody (diluted 1:50 in incubation buffer) overnight at 4°C.

6. Rinse for at least 4 hours in incubation buffer at 4°C.

7. Incubate the cells with mouse anti-IgG antibody tagged with 0.8 nm gold particles (diluted 1:50 in incubation buffer) overnight at 4°C.

8. Rinse 4 × 30 minutes in incubation buffer.

9. Rinse in PBS for 1 hour and then fix the cells in 2% glutaraldehyde for 10 minutes and thereafter rinse again in PBS.

10. Proceed to Protocol 8.11 for silver enhancement, which is necessary to visualise the ultra-small gold particles in the EM.

Protocol 8.11: Silver enhancement of gold labelling

EQUIPMENT, MATERIALS AND REAGENTS

200 mM Hepes buffer (pH adjusted to 6.8 with NaOH).

20 mM Hepes buffer containing 280 mM sucrose (pH 6.8).

Developer solution: 6 parts gum arabic (500 g l^{-1} in deionised water), 1 part Hepes buffer stock, 1.5 parts hydroquinone (0.52 M in deionised water) and 1.5 parts silver lactate (37 mM in deionised water). The silver lactate should be protected from light and added immediately before use.

METHOD

1. Rinse the cells 3 × 2 minutes in 20 mM Hepes (pH 6.8).

2. Incubate in developer solution (pH 6.8) for 6 minutes at 26°C in the dark.

3. Rinse quickly in deionised water.

4. Rinse 2 × 5 minutes in 50% ethanol.

5. Continue the dehydration in a graded series of ethanol (70%, 85%, 95%, 100%; 10 minutes each).

6. Go to step 8 in Protocol 8.7 for further preparation for transmission electron microscopy.

For immunoelectron microscopy detection of cathepsin D in cultured cells, Protocols 8.10 and 8.11 should be followed, with the primary antibody replaced by a polyclonal rabbit anti-human cathepsin D antibody (diluted 1:100) and the secondary antibody replaced by a goat anti-rabbit IgG tagged with 0.8 nm gold particles (diluted 1:100).

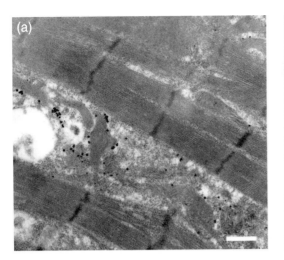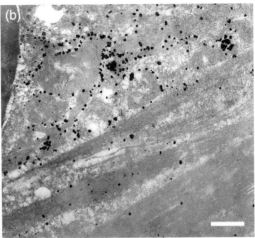

Figure 8.3

Electron micrographs of rat cardiomyocytes subjected to immunocytochemistry of cytochrome c using antibodies tagged with ultra-small gold particles and subsequent silver enhancement. The micrographs show (a) a control cell and (b) a cell treated with 1.5 μM naphthazarin for 3 hours. Silver-enhanced gold particles are seen on the outer mitochondrial membrane in both control cells and treated cells, but only in the treated cells are silver-gold particles spread throughout the cytosol. Bar = 1 μm. (Reprinted from Lab. Invest., Roberg, © 2001 with permission from United States and Canadian Academy of Pathology, Inc.)

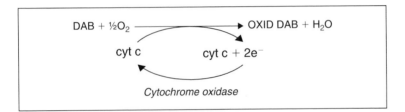

Figure 8.4

This reaction localises cytochrome c and is dependent on cytochrome oxidase activity [63, 64].

8.5.3   Histochemical localisation of cytochrome c at the EM level

Cytochrome c can also be localised in respiring mitochondria using a histochemical method based on the oxidation and precipitation of diaminobenzidine (see *Figure 8.4*; Protocol 8.12). This technique requires cytochrome oxidase activity to oxidise the reduced cytochrome c resulting in the generation of H_2O [62]. Because of the need for both cytochrome c and cytochrome oxidase, loss of either enzyme can theoretically cause loss of staining. However, cytochrome oxidase activity does not decrease during apoptosis and the absence of staining in the mitochondria is due to the leakage of cytochrome c. The advantage of this technique is that it can localise cytochrome c in the intermembrane and intracristal spaces of individual mitochondria in intact living cells. In contrast to the immunoelectron microscopy technique it cannot visualise cytoplasmic cytochrome c due to the requirement of the co-presence of cytochrome oxidase activity. This approach has shown that during apoptosis two sub-populations

Protocol 8.12: Localisation of cytochrome c in respiring mitochondria at the LM and EM level

EQUIPMENT, MATERIALS AND REAGENTS

Incubation medium: culture medium containing 20 mg DAB (3-3′ diaminobenzidine-tetrahydrochloride dihydrate) per 10 ml and 250 μl bovine catalase.

Fixative: 4% formaldehyde in 0.1 M sodium-cacodylate (Nacac) buffer containing 1% $CaCl_2$ (pH 4.4).

Rinsing solution: 13% sucrose in double-distilled water.

1% osmium tetroxide in 0.1 M Nacac buffer (1 part 4% OsO_4 in distilled water + 3 parts 0.134 M Nacac) with 1.5% $K_3Fe(CN)_6$.

Reynold's lead citrate.

Uranyl acetate stock solution: 7.5% in double-distilled water.

Uranyl acetate working solution: the stock solution is diluted (1:1) with 100% ethanol before use.

METHOD

1. Living cells are incubated in freshly prepared culture medium with addition of DAB at 37°C during 40 to 90 minutes at pH 7.4. Check the pH after addition of the DAB. The required incubation time is different for each cell type; the intensity of the staining (oxidised DAB = brown reaction product) can be monitored at light microscopical level.

2. Controls are incubated in DAB medium containing 10 M KCN as a cytochrome oxidase inhibitor.

3. Rinse 3 × 5 minutes in culture medium after the incubations (steps 1 or 2).

4. Fix in 4% formaldehyde solution buffered with Nacac containing 1% $CaCl_2$.

5. Rinse 3 × 5 minutes in 13% sucrose solution.

6. Post-fix the cells overnight at 4°C in osmium tetroxide solution.

7. Rinse 3 × 5 minutes in double-distilled water.

8. Dehydrate in a graded series of ethanol.

9. Continue with steps 8 to 11 as described in the general Protocol 8.7 for transmission electron microscopy.

10. Ultra-thin sections are contrasted with uranyl acetate (20 min) and Reynold's lead citrate (5 min). For light microscopical evaluation, nuclei are counterstained with methyl green. After step 4 (fixation) the cells are rinsed 3 × 5 minutes in double-distilled water and counterstained with methyl green following Protocol 8.3.

of mitochondria can be distinguished with respect to cytochrome c content and morphological appearance (i.e. orthodox versus condensed configuration). In short, apoptotic cells with condensed and fragmented chromatin still contain a few respiring mitochondria (*Plate 8f*).

8.6   Detection and localisation of lysosomal protease activity during apoptosis

Lysosomes are bound by a single membrane and contain hydrolases that operate optimally at acidic pH. Consequently, these structural compartments are the major site of intracellular degradation of cytoplasmic components such as organelles, long-lived proteins and other macromolecules that are transported to lysosomes either from other parts of the cell (autophagy) or from outside the cell after internalisation by endocytosis or crinophagy (for review see [65]). Lysosomal enzymes are glycoproteins and it has been estimated that they degrade about 90% of all long-lived proteins and a proportionately large fraction of the short-lived proteins in lysosomes [66]. Cathepsin D is one of the most abundant lysosomal enzymes in mammalian cells [67]. In addition to digesting peptides, purified cathepsin D can activate the precursor of transforming growth factor-β, degrade the basement membrane [68] and activate other proteases such as procathepsin B.

8.6.1   Immunofluorescence

Immunofluorescence and immunocytochemical techniques at the LM level have revealed translocation of cathepsins D, B and L early during apoptosis in a variety of cell types [27, 29, 69]. For example, exposure of cardiomyocytes to oxidative stress causes early translocation of cathepsin D from lysosomal structures to the cytosol [28], with lysosomal destabilisation detected as diffuse immunofluorescence staining of cathepsin D in apoptotic cells. For immunofluorescence detection of cathepsin D, Protocol 8.9 can be adapted, replacing the primary antibody with a polyclonal rabbit anti-human cathepsin D antibody (diluted 1:100) and the secondary antibody with a goat anti-rabbit immunoglobulin G (IgG) Texas Red conjugate (diluted 1:200).

8.6.2   Immunoelectron microscopy

Post-embedding immunogold techniques have been useful in the localisation of lysosomal cathepsins in cultured cells and tissues. For example, such methods have been employed by Ishii and colleagues [70] to determine the presence of cathepsins B, H and L in Epon sections of bronchoalveolar epithelial cells, and by Raczek and co-workers [71] to detect cathepsin D in the lysosomes of human epididymal epithelial cells embedded in Lowicryl K4M. We have also developed a pre-embedding immunogold technique to detect cathepsin D in different types of cultured cell [72]. Use of this method in combination with ultra-small gold probes and subsequent silver enhancement offers sensitive detection of lysosomal enzymes and morphological preservation that is sufficient to determine the localisation of the protein (*Figure 8.5*). For immunoelectron microscopy detection of cathepsin D in cultured cells, Protocols 8.10 and 8.11 should be followed, with the primary antibody replaced by a polyclonal rabbit anti-human cathepsin D antibody (diluted 1:100), and the secondary antibody replaced by a goat anti-rabbit immunoglobulin G (IgG) tagged with 0.8 nm gold particles (diluted 1:100).

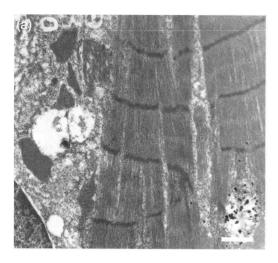 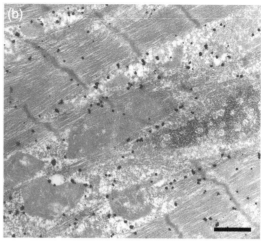

Figure 8.5

Electron micrographs of rat cardiomyocytes subjected to immunocytochemistry of cathepsin D using antibodies tagged with ultra-small gold particles and subsequent silver enhancement. Silver-enhanced gold particles can be seen in lysosome-like structures in control cells (a), but they are spread throughout the cytosol in cells treated with 1.5 μM naphthazarin for 3 hours (b). Bar = 1 μm. (Reprinted from Lab. Invest., Roberg, 2001 © with permission from United States and Canadian Academy of Pathology, Inc.)

8.7　References

1. Lockshin, R.A. and Willams, C.M. (1964) Programmed cell death. II. Endocrine potentiation of the breakdown of the intersegmental muscles of silk moths. *J. Insect Physiol.* **10**: 643–649.
2. Ellis, H.M. and Horvitz, H.R. (1986) Genetic control of programmed cell death in the nematode *C. elegans. Cell* **44**: 817–829.
3. Horvitz, H.R., Chalfie, M., Trent, C., Sulston, J.E. and Evans, P.D. (1982) Serotonin and octopamine in the nematode *Caenorhabditis elegans. Science* **16**: 1012–1014.
4. Sulston, J.E. and Horvitz, H.R. (1977) Post-embryonic cell lineages of the nematode, *Caenorhabditis elegans. Dev. Biol.* **56**: 110–156.
5. Hengartner, M.O. and Horvitz, H.R. (1994) Programmed cell death in *Caenorhabditis elegans. Curr. Opin. Genet. Dev.* **4**: 581–586.
6. Yuan, J., Shaham, S., Ledoux, S., Ellis, H.M. and Horvitz, H.R. (1993) The *C. elegans* cell death gene ced-3 encodes a protein similar to mammalian interleukin-1 beta-converting enzyme. *Cell* **75**: 641–652.
7. Zou, H., Henzel, W.J., Liu, X., Lutschg, A. and Wang, X. (1997) Apaf-1, a human protein homologous to *C. elegans* CED-4, participates in cytochrome c-dependent activation of caspase-3. *Cell* **90**: 405–413.
8. Kerr, J.F., Wyllie, A.H. and Currie, A.R. (1972) Apoptosis: a basic biological phenomenon with wide-ranging implications in tissue kinetics. *Br. J. Cancer* **26**: 239–257.
9. Majno, G. and Joris, I. (1995) Apoptosis, oncosis and necrosis. An overview of cell death. *Am. J. Pathol.* **146**: 3–15.
10. Wyllie, A.H. (1981) Cell death: a new classification separating apoptosis from necrosis. In: *Cell Death in Biology and Pathology* (eds I.D. Bowen and R.A. Lockshin). Chapman and Hall, London, pp. 9–34.
11. Chautan, M., Chazal, G., Cecconi, F., Gruss, P. and Golstein, P. (1999) Interdigital cell death can occur through a necrotic and caspase-independent pathway. *Curr. Biol.* **9**: 967–970.

12. Kitanaka, C. and Kuchino, Y. (1999) Caspase-independent programmed cell death with necrotic morphology. *Cell Death Differ.* **6**: 508–515.

13. Leist, M. and Jäättelä, M. (2001) Four deaths and a funeral: from caspases to alternative mechanisms. *Nat. Rev. Mol. Cell Biol.* **2**: 589–598.

14. Bursch, W. (2001) The autophagosomal-lysosomal compartment in programmed cell death. *Cell Death Differ.* **8**: 569–581.

15. Clarke, P.G.H. (1990) Developmental cell death: morphological diversity and multiple mechanisms. *Anat. Embryol.* **181**: 195–213.

16. D'Herde K., De Prest B. and Roels F. (1996) Subtypes of active cell death in the granulosa of ovarian atretic follicles in the quail (*Coturnic coturnix japonica*). *Reprod. Nutr. Dev.* **36**: 175–189.

17. Schweichel, J.U. and Merker, H.J. (1973) The morphology of various types of cell death in prenatal tissues. *Teratology* **7**: 253–266.

18. Zakeri, Z., Bursch, W., Tenniswood, M. and Lockshin, R.A. (1995) Cell death: programmed apoptosis, necrosis, or other? *Cell Death Differ.* **2**: 83–92.

19. Green, D.R. and Reed, J.C. (1998) Mitochondria and apoptosis. *Science* **281**: 1309–1312.

20. Chai, J., Du, C., Wu, J.W., Kvin, S., Wang, X. and Shi, Y. (2000) Structural and biochemical basis of apoptotic activation by Smac/DIABLO. *Nature* **406**: 855–862.

21. Mancini, M., Nicholson, D.W., Roy, S., Thornberry, N.A., Peterson, E.P., Casciola-Rosen, L. A. and Rosen, A. (1998) The caspase-3 precursor has a cytosolic and mitochondrial distribution: implications for apoptotic signaling. *J. Cell Biol.* **140**: 1485–1495.

22. Kroemer, G., Zamzami, N. and Susin, S.A. (1997) Mitochondrial control of apoptosis. *Immunol. Today* **18**: 44–51.

23. Brunk, U.T., Zhang, H., Roberg, K. and Öllinger, K. (1995) Lethal hydrogen peroxide toxicity involves lysosomal iron-catalyzed reactions with membrane damage. *Redox Rep.* **1**: 267–277.

24. Brunk, U.T. and Svensson, I. (1999) Oxidative stress, growth factor starvation and Fas activation may all cause apoptosis through lysosomal leak. *Redox Rep.* **4**: 3–11.

25. Neuzil, J., Svensson, I., Weber, T., Weber, C. and Brunk, U.T. (1999) Alpha-tocopheryl succinate-induced apoptosis in Jurkat T cells involves caspase-3 activation and both lysosomal and mitochondrial destabilization. *FEBS Lett.* **445**: 295–300.

26. Deiss, L.P., Galinka, H., Barissi, H., Cohen, O. and Kimichi, A. (1996) Cathepsin D protease mediates programmed cell death induced by interferon-gamma, Fas/APO-1 and TNF-alpha. *EMBO J.* **15**: 3861–3870.

27. Guicciardi, M.E., Deussing, J., Miyoshi, H., Bronk, S.F., Svingen, P.A., Peters, C., Kaufmann, S.H. and Gores, G.J. (2000) Cathepsin B contributes to TNF-alpha-mediated hepatocyte apoptosis by promoting mitochondrial release of cytochrome c. *J. Clin. Invest.* **106**: 1127–1137.

28. Roberg, K. and Öllinger, K. (1998) Oxidative stress causes relocation of the lysosomal enzyme cathepsin D with ensuing apoptosis in neonatal rat cardiomyocytes. *Am. J. Pathol.* **152**: 1151–1156.

29. Roberg, K., Johansson, U. and Öllinger, K. (1999) Lysosomal release of cathepsin D precedes relocation of cytochrome c and loss of mitochondrial transmembrane potential during apoptosis induced by oxidative stress. *Free Radic. Biol. Med.* **27**: 1228–1237.

30. Roberg, K. (2001) Relocalization of cathepsin D and cytochrome c early in apoptosis revealed by immuno-electron microscopy. *Lab. Invest.* **81**: 149–158.

31. Öllinger, K. (2000) Inhibition of cathepsin D prevents free-radical-induced apoptosis in rat cardiomyocytes. *Arch. Biochem. Biophys.* **373**: 346–351.

32. Kapuscinski, J. (1995) DAPI: a DNA-specific fluorescent probe. *Biotech. Histochem.* **70**: 220–233.

33. D'Herde, K. and Leybaert, L. (1997) Intracellular free calcium related to apoptotic cell death in quail granulosa cell sheets kept in serum-free culture. *Cell Death Differ.* **4**: 59–65.

34. Arends, M.J., Morris, R.G. and Wyllie, A.H. (1990) Apoptosis. The role of the endonuclease. *Am. J. Pathol.* **136**: 593–608.

35. Cohen, J.J. and Duke, R.C. (1984) Glucocorticoid activation of a calcium-dependent endonuclease in thymocyte nuclei leads to cell death. *J. Immunol.* **132**: 38–42.

36. Compton, M.M. (1992) A biochemical hallmark of apoptosis: Internucleosomal degradation of the genome. *Cancer Metastasis Rev.* **11**: 105–119.

37. Kokileva, L. (1998) Disassembly of genome of higher eukaryotes: pulse-field gel electrophoretic study of initial stages of chromatin and DNA degradation in rat liver and thymus nuclei by VM-26 and selected proteases. *Comp. Biochem. Physiol. Biochem. Mol. Biol.* **121**: 145–151.

38. Wyllie, A.H. (1980) Glucocorticoid-induced thymocyte apoptosis is associated with endogenous endonuclease activation. *Nature* **284**: 555–556.

39. Gavrieli, Y., Sherman, Y. and Ben–Sasson, S.A. (1992) Identification of programmed cell death in situ via specific labelling of nuclear DNA fragmentation. *J. Cell Biol.* **119**: 493–501.

40. Wijsman, J.H., Jonker, R.R., Keijzer, R., van de Velde, C.J., Cornelisse, C.J. and van Dierendonck, J.H. (1993) A new method to detect apoptosis in paraffin sections: in situ end-labelling of fragmented DNA. *J. Histochem. Cytochem.* **41**: 7–12.

41. Gorczyca, W., Bigman, K., Mittelman, A., Ahmed, T., Gong, J., Melamed, M.R. and Darzynkiewicz, Z. (1993) Induction of DNA strand breaks associated with apoptosis during treatment of leukemias. *Leukemia* **7**: 659–670.

42. D'Herde K., De Pestel G. and Roels F. (1994) In situ end labelling of fragmented DNA in induced ovarian atresia. *Biochem. Cell Biol.* **72**: 573–579.

43. Schulze-Osthoff, K., Walczak, H., Dröge, W. and Krammer, P.H. (1994) Cell nucleus and DNA fragmentation are not required for apoptosis. *J. Cell Biol.* **127**: 15–20.

44. Grasl-Kraupp, B., Ruttkay-Nedecky, B., Koudelka, H., Bukowska, K., Bursch, W. and Schulte-Hermann R. (1995) In situ detection of fragmented DNA (TUNEL assay) fails to discriminate among apoptosis, hepatology, necrosis and autolytic cell death: A cautionary note. *Hepatology* **21**: 1465–1468.

45. Perry, S.W., Epstein, L.G. and Gelbard, H.A. (1997) Simultaneous in situ detection of apoptosis and necrosis in monolayer cultures by TUNEL and trypan blue staining. *BioTechniques* **22**: 1102–1106.

46. Krysko, D.V., Roels, F., Leybaert, L., D'Herde, K. (2001) Mitochondrial transmembrane potential changes support the concept of mitochondrial heterogeneity during apoptosis. *J. Histochem. Cytochem.* **49**: 1277–1284.

47. Salvioli, S., Dobrucki, J., Moretti, L., Troiano, L., Fernandez, M.G., Pinti M., Pedrazzi, J., Franceschi, C. and Cossarizza, A. (2000) Mitochondrial heterogeneity during staurosporine-induced apoptosis in HJ60 cells: analysis at the single cell and single organelle level. *Cytometry* **40**: 189–197.

48. Bradham, C.A., Qian, T., Streetz, K., Trautwein, C., Brenner, D.A. and Lemasters, J.J. (1998) The mitochondrial permeability transition is required for tumor necrosis factor alpha-mediated apoptosis and cytochrome c release. *Mol. Cell. Biol.* **18**: 6353–6364.

49. Heiskanen, K.M., Bhat, M.B., Wang, H.W., Ma, J. and Nieminen, A.L. (1999) Mitochondrial depolarization accompanies cytochrome c release during apoptosis in PC6 cells. *J. Biol. Chem.* **274**: 5654–5658.

50. Macho, A., Decaudin, D., Castedo, M., Hirsch, T., Susin, S.A., Zamzami, N. and Kroemer, G. (1996) Chloromethyl-X-Rosamine is an aldehyde-fixable potential-sensitive fluorochrome for the detection of early apoptosis. *Cytometry* **25**: 333–340.

51. Poot, M., Zhang, Y.Z., Kramer, J.A., Wells, K.S., Jones, L.J., Hanzel, D.K., Lugade, A.G., Singer, V.L. and Haugland, R.P. (1996) Analysis of mitochondrial morphology and function with novel fixable fluorescent stains. *J. Histochem. Cytochem.* **44**: 1363–1372.

52. Collins, J.A., Schandi, C.A., Young, K.K., Vesely, J. and Willingham, M.C. (1997) Major DNA fragmentation is a late event in apoptosis. *J. Histochem. Cytochem.* **45**: 923–934.

53. Cai, J., Yang, J. and Jones, D.P. (1998). Mitochondrial control of apoptosis: the role of cytochrome c. *Biochim. Biophys. Acta* **1366**: 139–149.

54. Liu, X., Kim, C.N., Yang, J., Jemmerson, R. and Wang, R. (1996) Induction of apoptotic program in cell-free extracts: requirement for dATP and cytochrome c. *Cell* **86**: 147–157.

55. Li, F., Srinivasan, A., Wang, Y., Armstrong, R.C., Tomaselli, K. J. and Fritz, L.C. (1997) Cell-specific induction of apoptosis by microinjection of cytochrome c. Bcl-xL has activity independent of cytochrome c release. *J. Cell Biol.* **272**: 30299–30305.

56. Zhivotovsky, B., Orrenius, S., Brustugun, O.T. and Doskeland, S.O. (1998) Injected cytochrome c induces apoptosis. *Nature* **391**: 449–450.

57. Kim, C.N., Wang, X., Huang, Y., Ibrado, A.M., Liu, L., Fang, G. and Bhalla, K. (1997) Overexpression of Bcl-X(L) inhibits Ara-C-induced mitochondrial loss of cytochrome c and other perturbations that activate the molecular cascade of apoptosis. *Cancer Res.* **57**: 3115–3120.

58. Bossy-Wetzel, E., Newmeyer, D.D. and Green, D.R. (1998) Mitochondrial cytochrome c release in apoptosis occurs upstream of DEVD-specific caspase activation and independently of mitochondrial transmembrane depolarization. *EMBO J.* **17**: 37–49.

59. Lah, J.J., Hayes, D.M. and Burry, R.W. (1990) A neutral pH silver development method for the visualization of 1-nanometer gold particles in pre-embedding electron microscopic immunocytochemistry. *J. Histochem. Cytochem.* **38**: 503–508.

60. Tokuyasu, K.T. (1984) Immuno-cryoultramicrotomy. In: *Immunolabelling for Electron Microscopy* (eds J.M. Polak and I.M. Varndell). Elsevier Science, Amsterdam, New York, Oxford, pp. 71–80.

61. Willingham, M.C. (1983) An alternative fixation-processing method for pre-embedding ultrastructural immunocytochemistry of cytoplasmic antigens: the GBS (glutaraldehyde-borohydride-saponin) procedure. *J. Histochem. Cytochem.* **31**: 791–798.

62. D'Herde K., De Prest B., Mussche, S., Schotte, P., Van Coster, R. and Roels, F. (2000) Ultrastructural localization of cytochrome c in apoptosis demonstrates mitochondrial heterogeneity. *Cell Death Differ.* **7**: 331–337.

63. Seligman, A.M., Karnovsky, M.J., Wasserkrug, H.L. and Hanker, J.S. (1968) Nondroplet ultrastructural demonstration of cytochrome oxidase activity with a polymerizing osmiophilic reagent, diaminobenzidine (DAB). *J. Cell Biol.* **38**: 1–14.

64. Roels, F. (1974) Cytochrome c and cytochrome oxidase in diaminobenzidine staining of mitochondria. *J. Histochem. Cytochem.* **22**: 442–444.

65. Cuervo, A.M. and Dice, J.F. (2000) When lysosomes get old. *Exp. Gerontol.* **35**: 119–131.

66. Ahlberg, J., Berkenstam, A., Henell, F. and Glaumann, H. (1985) Degradation of short and long lived proteins in isolated rat liver lysosomes. Effects of pH, temperature and proteolytic inhibitors. *J. Biol. Chem.* **260**: 5847–5854.

67. Barrett, A.J. (1977) Human cathepsin D. *Adv. Exp. Med. Biol.* **95**: 291–300.

68. Briozzo, P., Morisset, M., Capony, F., Rougeot, C. and Rochefort, H. (1988) In vitro degradation of extracellular matrix with Mr 52,000 cathepsin D secreted by breast cancer cells. *Cancer Res.* **48**: 3688–3692.

69. Kågedal, K., Johansson, U. and Öllinger, K. (2001) The lysosomal protease cathepsin D mediates apoptosis induced by oxidative stress. *FASEB J.* **15**: 1592–1594.

70. Ishii, Y., Hashizume, Y., Watanabe, T., Waguri, S., Sato, N., Yamamoto, M., Hasegawa, S., Kominiami, E. and Uchiyama, Y. (1991) Cysteine proteinases in broncho-alveolar epithelial cells and lavage fluid of rat lung. *J. Histochem. Cytochem.* **39**: 461–468.

71. Raczek, S., Yeung, C. H., Hasilik, A., Robenek, L., Hertle, L., Schulze, H. and Cooper, T.G. (1995) Immunocytochemical localization of some lysosomal hydrolases, their presence in luminal fluid and their directional secretion by human epididymal cells in culture. *Cell Tissue Res.* **280**: 415–425.

72. Roberg, K. and Öllinger, K. (1998b) A pre-embedding technique for immunocytochemical visualization of cathepsin D in cultured cells subjected to oxidative stress. *J. Histochem. Cytochem.* **46**: 411–418.

Caspase Detection and Analysis

9

*Emma M. Creagh, Colin Adrain and
Seamus J. Martin*

Contents

9.1 Introduction

The caspases (cysteinyl aspartate-specific proteases) are a group of proteolytic enzymes that were first implicated in programmed cell death through analysis of cell death-defective mutants of the nematode worm *Caenorhabditis elegans*. The *C. elegans* caspase, CED-3, is essential for all developmental-related programmed cell deaths in

Cell Proliferation and Apoptosis, David Hughes and Huseyin Mehmet (Eds)
© 2003 BIOS Scientific Publishers Ltd, Oxford

the nematode [1]. A large family of mammalian CED-3-related proteases (the caspases) has now been identified and many of these play important roles in apoptosis.

Caspases are typically expressed as latent pro-enzymes and become activated in response to divergent pro-apoptotic signals. Active caspases target an array (probably hundreds) of cellular proteins for degradation and this results in the controlled disassembly of the cell (apoptosis). Synthetic, viral and cellular inhibitors of caspases can effectively inhibit the morphological and biochemical features of apoptosis, irrespective of the particular initiating stimulus [2]. This evidence, combined with analysis of caspase-deficient mice or cell lines [3], suggests an absolute requirement for caspases during most, if not all, forms of apoptosis in mammals [4]. Other, caspase-independent, cell death pathways may also exist but these do not seem to produce the same morphological or biochemical endpoints.

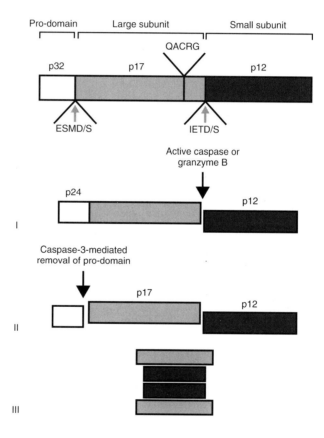

Figure 9.1

Activation of caspase-3 during apoptosis. Diagram illustrating the two-step mechanism involved in the apoptosis-associated maturation of caspase-3. Inactive pro-caspase-3 is a ~32 kDa molecule composed of large and small subunits and a small N-terminal pro-domain. Indicated are the active site QACRG motif within the large subunit, and the tetrapeptide motifs flanking the large subunit that represent the sites of proteolytic processing. The cleavage site after the P1 aspartate residue is indicated by the grey arrows. Numbers refer to the approximate molecular weight in kDa of the subunits or cleavage products. Activation of the molecule is initiated by cleavage between the large and small subunits (I) by an active caspase molecule or by a serine protease called granzyme B. Active caspase-3 then facilitates removal of its own pro-domain as indicated (II). Finally, the mature enzyme assumes a tetrameric structure, composed of a core of two small subunits, flanked on the periphery by two large subunits (III).

Latent (inactive) caspases are activated through proteolytic processing at internal aspartic acid residues. The latter processing event separates the pro-caspase into large and small subunits which heterodimerise to form the active enzyme (*Figure 9.1*). As the primary specificity of caspases is aspartic acid residues, this implies that caspases can be activated either through autoproteolysis or by other proteases with the same specificity (including other caspases or the cytotoxic T lymphocyte protease granzyme B).

The caspase family can be broadly divided into two functional sub-families; those that are activated during apoptosis (caspases-2, -3, -6, -7, -8, -9 and -10) and those that are involved in cytokine processing in the immune response (caspases-1, -4, -5 and -11). In the context of apoptosis, much evidence suggests that caspases are activated in hierarchical cascades [4, 5]. These caspase cascades consist of initiator caspases (caspases-8, -9 and -10) and effector caspases (caspases-3, -6 and -7), as well as their inhibitors and activators (reviewed in [4] and [5]).

Initiator caspases are typically activated through autoproteolysis, which is induced by binding to specific activators. Structurally, initiator caspases tend to have long N-terminal pro-domains with motifs such as CARDs or DEDs, that are also present in molecules that promote their aggregation, such as Apaf-1 and Fas-associated death domain (FADD). Each initiator caspase is activated in response to a particular subset of cytotoxic stimuli. For example, caspase-9 is activated in response to divergent cellular stresses such as cytotoxic drugs, heat shock and ionising radiation. These cellular stresses trigger the release of mitochondrial cytochrome c into the cytosol, thereby stimulating Apaf-1 oligomerisation and the subsequent recruitment and activation of caspase-9 [5, 6]. Another well-established initiator caspase, caspase-8, becomes activated through formation of a complex with FADD/TRADD (TNF receptor associated death domain) following stimulation of death receptors such as FasR and TNFR [7, 8].

Activated initiator caspases propagate death signals by direct proteolytic activation of downstream effector caspases (caspases-3, -6 and -7). The effector caspases then orchestrate the direct dismantling of cellular structures, disruption of cellular metabolism, inactivation of cell death inhibitory proteins and the activation of additional destructive enzymes [4, 5, 9]. This chapter details methods that are widely used to detect caspase activation/activity either directly, through assessment of the processing status or activity of the enzyme itself, or indirectly, by monitoring the cleavage of caspase substrates.

9.2 Direct detection of caspase activation status by SDS-PAGE/immunoblot analysis

As discussed above, caspases are constitutively present in cells as inactive precursors that require limited proteolysis to become activated. The active heterodimeric enzyme can be separated into its subunits by SDS-PAGE. Immunoblotting with antibodies raised against specific caspases will therefore reveal the processed caspase subunits if activation has occurred (Protocol 9.1). Many such antibodies are now commercially available (see *Table 9.1*). Caspase activation is accompanied by loss of the pro-form of the enzyme, which should correlate with the appearance of the processed subunits. However, the processed forms of caspases can be difficult to detect depending on the antibody being used and the concentration of the protein samples.

As caspase activation is widely recognised as a pivotal event in apoptosis, demonstration of caspase activation by western blotting is often used as a readout for apoptosis. When using this technique to detect caspase activity in cell cultures, at least 50% of the cells should be undergoing apoptosis by morphological criteria, since the processed forms of caspases tend to be unstable and so are difficult to detect.

Protocol 9.1: Direct detection of caspase activation status by SDS-PAGE/immunoblot analysis

EQUIPMENT, MATERIALS AND REAGENTS

Cells to be analysed.

Ice.

SDS lysis buffer (50 mM Tris pH 6.8, 2% SDS, 10% glycerol).

PROTEAN cell vertical electrophoresis system (western blotting) plus Power Pac.

Dry heating block unit.

0.2 μm nitrocellulose membranes.

12% or 15% SDS polyacrylamide gels.

5% NFDM (non-fat dried milk) in TBST (Tris-buffered saline, 0.1% Tween 20).

Primary caspase antibody (see *Table 9.1*).

Horseradish peroxidase-coupled species-specific secondary antibody.

Chemiluminescent-based detection reagent.

ECL film.

METHOD

1. Following treatment, cells are typically lysed at a concentration of $10^7\,\text{ml}^{-1}$ in 1X SDS lysis buffer, heated at 95°C for 7 minutes and loaded on 12% or 15% SDS polyacrylamide gels.

2. Electrophoresis is carried out at 55–70 V, followed by transfer of proteins onto 0.2 μm nitrocellulose membranes (Schleicher & Schuell) overnight at 30 mA. Faster transfers can be achieved by blotting for 3–4 hours at 150 mA however this can lead to loss of small proteins through the membrane.

3. Following transfer of proteins, the membranes are blocked by incubating with 5% NFDM in TBST for 30 minutes at room temperature on a rocking platform.

4. Incubation with the relevant primary caspase antibody (see *Table 9.1*), usually at a 1:1000 dilution in 5% NFDM, is carried out for 2 hours at room temperature.

5. Following 3 ×10 minute washes in TBST, the membrane is incubated with a horseradish peroxidase-coupled species-specific secondary antibody for 1 hour at room temperature.

6. Three further 10 minute washes in TBST are carried out before incubating the membrane with hydrogen peroxide substrate and chemiluminescent-based detection. Blotting membranes are then covered with Saran wrap followed by exposure to radiographic film.

Note: Detection of the processed form of caspases can sometimes be difficult. If loading larger sample volumes or more concentrated samples does not improve detection, fixation of proteins to the nitrocellulose membrane with glutaraldehyde can sometimes enhance detection [10].

Table 9.1   Recommended caspase antibodies for western blotting

Antibody	Species/clonality	Commercial source	Detection
anti-Caspase-1	Rabbit polyclonal	Santa Cruz Biotechnology	44 kDa pro-form and ~29 kDa processed form
anti-Caspase-2/ICH-1$_L$	Mouse monoclonal	BD Transduction Laboratories	48 kDa pro-form and ~30 kDa processed form
anti-Caspase-3/CPP32/ Yama/Apopain	Mouse monoclonal	BD Transduction Laboratories	32 kDa pro-form and 20/12 kDa processed forms
anti-Caspase-4/TX/ ICH-2/ICE$_{rel}$II	Mouse monoclonal	MBL (Medical and Biological Laboratories Co.)	43 kDa pro-form and ~20 kDa processed form
anti-Caspase-5/ TY/ICE$_{rel}$III	Mouse monoclonal	MBL	47 kDa pro-form
anti-Caspase-6/Mch-2	Rabbit polyclonal	Upstate Biotechnology	32 kDa pro-form-difficult to detect 19 kDa processed form
anti-Caspase-7/Mch-3/ ICH-LAP3	Mouse monoclonal	BD Transduction Laboratories	33 kDa pro-form and 17 kDa processed form
anti-Caspase-8/FLICE/ Mch-5/MACH	Mouse monoclonal	Upstate Biotechnology	53 kDa pro-form and ~35 kDa processed form
anti-Caspase-9/Mch-6/ ICH-LAP6	Mouse monoclonal	Upstate Biotechnology	46 kDa pro-form and 18/15 kDa processed forms

Table 9.2   Source of commercially available antibodies specific to the active form of caspases or caspase-cleaved substrates that may be used for immunohistochemical analysis

Antibody specific to	Supplier
Active caspase-3	New England BioLabs (NEB), Promega, Cambridge Bioscience, Biovision
Active caspase-7	NEB, Biovision
Active caspase-9	NEB, Biovision
Cleaved Human PARP	NEB
Cleaved DFF45	NEB

Antibodies are also available that are specific for active caspases, reacting very poorly with the pro-forms (see *Table 9.2*). Such antibodies are particularly useful for the *in situ* detection of caspase activation in individual cells or tissue sections.

9.3   Indirect detection of caspase activity by immunoblotting for caspase substrate proteins

Active caspases promote cellular demolition primarily by targeting key structural and regulatory proteins within the cell for degradation and also by activating other enzymes such as DNases [11]. Caspases can also promote cytochrome c release from mitochondria via Bid cleavage [12], and can facilitate the release of other mitochondrial pro-apoptotic proteins such as SMAC/DIABLO [13].

Numerous caspase substrates have been identified to date, such as inhibitor of caspase-activated deoxyribonuclease (ICAD), PARP, X-linked inhibitor of apoptosis (XIAP) and Lamin A [14]. However, the impact of many of these substrate cleavage events on the apoptotic phenotype has yet to be determined. Demonstration of caspase substrate proteolysis is often used as an indirect readout for caspase activity during apoptosis (Protocol 9.2; *Figure 9.2*). There are a number of commercially available

Protocol 9.2: Indirect detection of caspase activity by immunoblotting for caspase substrate proteins

EQUIPMENT, MATERIALS AND REAGENTS

As for Protocol 9.1.

METHOD

As for Protocol 9.1, except membranes are probed with substrate antibodies (*Table 9.3*).

Table 9.3   Recommended caspase substrate antibodies for western blotting

Antibody	Species/clonality	Commercial source	Detection
XIAP/hILP-1 (clone 48)	Mouse monoclonal	BD Transduction Laboratories	57 kDa protein and 45/29 kDa breakdown products
Gelsolin	Mouse monoclonal	BD Transduction Laboratories	93 kDa protein and 50 kDa breakdown product
Vimentin	Mouse monoclonal	Boehringer Mannheim	60 kDa protein only
Stat-1	Mouse monoclonal	BD Transduction Laboratories	91/84 kDa protein only
DFF 45/ICAD	Rabbit polyclonal	Upstate Biotechnology	45 kDa protein only
PARP	Mouse monoclonal	BD Pharmingen	116 kDa protein, weakly detects 89 kDa breakdown product

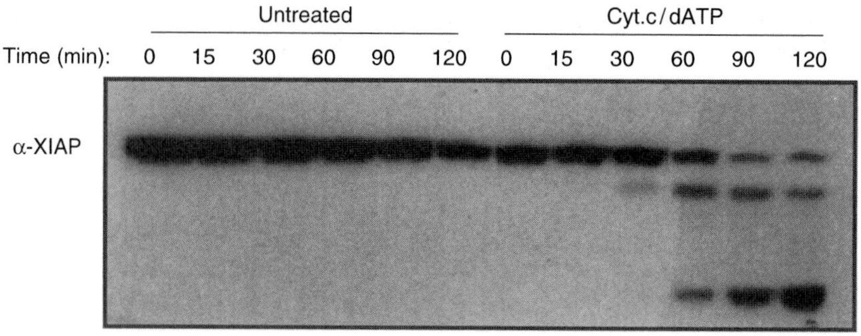

Figure 9.2

Kinetics of cytochrome c/dATP-inducible XIAP processing in Jurkat extracts. Jurkat cell-free extracts were incubated at 37°C for the indicated times with or without cytochrome c (50 μg ml⁻¹) and dATP (1 mM). Reactions were analysed by SDS-PAGE and immunoblotted for XIAP (*Table 9.2*).

caspase substrate antibodies, summarised in *Table 9.3*. In general, substrate processing is an acceptable readout for caspase activity, however this technique is not recommended if it is intended to demonstrate the activity of a particular caspase, as the specific array of substrates that each caspase is capable of targeting is still largely undefined. However, caspase-3 does appear to be the primary executioner caspase, as relatively few substrates for caspases-6 and -7 have been identified thus far [14].

9.4 Detection of active caspases or cleaved caspase substrates in individual cells by immunostaining

The recent commercialisation of antibodies that are specific for the active (i.e. processed) forms of certain caspases, or to the cleaved forms of caspase substrates, has made it possible to detect endogenous activated caspases or caspase-mediated proteolytic events within individual cells. When combined with standard immunohistochemical techniques, these antibodies offer an insight into caspase activation events within an individual cell that may otherwise be undetectable at the level of bulk cell populations by western blot analysis.

In Protocol 9.3 we describe a conventional immunostaining protocol and detail sources of commercially available antibodies for the immunohistochemical detection of active caspases and cleaved caspase substrates in individual cells. Typically, cells are plated on sterile coverslips 12–24 hours prior to treatment with a pro-apoptotic stimulus. Following induction of apoptosis, cell monolayers are washed in PBS to remove serum proteins, followed by fixation and permeabilisation to allow entry of antibodies into the cells and then immunostained with the appropriate primary antibody. After probing with fluorochrome-coupled secondary antibodies, cell monolayers are fixed with a glass cover (using a commercially available anti-fade reagent) and examined by fluorescence microscopy.

Protocol 9.3: Detection of active caspases or cleaved caspase substrates in individual cells by immunostaining

EQUIPMENT, MATERIALS AND REAGENTS

Cells to be analysed.

Coverslips (22 × 22 mm).

6-well plates.

PBS pH 7.2 (150 mM NaCl, 2.7 mM KCl, 8 mM Na_2HPO_4, 1.5 mM KH_2PO_4).

PFA.

Triton X-100.

Primary antibody (see *Table 9.3*).

BSA.

Fluorochrome-coupled secondary antibody.

Antifade reagent mounting medium.

METHOD

Note: The volumes indicated are sufficient to treat cells plated on a 22 × 22 mm coverslip in 6-well tissue culture plates and should be adjusted according to the scale of the experiment.

1. Plate cells on sterile coverslips ~12–24 hours prior to transfection or treatment with a pro-apoptotic stimulus. As a guideline, for 6-well plates, seed cells at a density of $5 \times 10^4 - 1 \times 10^5$ per well. Note that it may take longer for cells to adhere properly onto glass coverslips than conventional tissue culture plastic. The length of time required to facilitate proper cell attachment/spreading should be determined in advance.

2. After inducing apoptosis, wash the coverslips three times (2 ml per well of a 6-well plate) in chilled PBS pH 7.2 to remove traces of serum proteins present in the culture media. It is convenient to process the coverslips within the 6-well plate itself.

3. Fix cells for 10 minutes at room temperature in 3% PFA in PBS pH 7.2 (2 ml per well).

4. Rehydrate/wash cells in three changes of PBS, pH 7.2 (15 minutes/2 ml per wash) at room temperature.

5. Permeabilise cells for 15 minutes (in 2 ml per well of 0.15% Triton X-100, in PBS pH 7.2) at room temperature to facilitate access of the antibodies to intracellular compartments.

6. Probe cells with primary antibody (see *Table 9.2*) diluted in 2% BSA/PBS pH 7.2. As a guideline, try using a 1:100 dilution of your primary antibody to probe cells for 1 hour at room temperature. 100 μl of diluted antibody applied drop-wise is sufficient for staining an individual coverslip.

7. Perform three washes (2 ml per wash for 10 minutes each) in 2% BSA/PBS pH 7.2 to remove non-specifically bound antibodies. This is best performed at low speed on a rocking platform.

8. Probe cells with an appropriate fluorochrome-coupled secondary antibody, diluted in 2% BSA in PBS pH 7.2. As a guideline, try probing for 45 minutes at room temperature with a 1:250 dilution of secondary antibody. As fluorescent probes can be bleached by light, the coverslips should be shielded from light during this and subsequent steps.

9. Perform three wash steps (2 ml per wash for 10 minutes each) using PBS pH 7.2.

10. Carefully blot off excess PBS and mount cells onto microscope slides in a commercially available anti-fade reagent. A narrow gauge syringe needle can be used to lever coverslips out of the wells for mounting.

11. Allow the surface of the coverslips to dry for 5 minutes. To prevent dehydration and/or damage to the samples caused by movement during microscopy, it is advisable to seal the edges of the coverslip to the microscope slide. A clear nail varnish is suitable for this purpose. After sealing, slides should be protected from light and stored at 4°C. Samples should retain their fluorescence for several weeks under these conditions. Note that prolonged UV illumination during microscopy will bleach the slides in the locality being visualised, thus attenuating the fluorescent signal.

9.5   Use of cell-free extracts as a source of cytochrome c-inducible active caspases and caspase substrates

Many stimuli that cause cell stress, such as exposure to cytotoxic drugs or UV irradiation, trigger cytochrome c release from mitochondria, thereby activating a cascade of caspase activation events instigated by a caspase-9/Apaf-1 complex, referred to as the apoptosome. As detailed in Protocol 9.4, cell-free systems comprising cytoplasmic extracts prepared from viable cells are valuable tools for the study of events that occur downstream of caspase activation [15–18]. Thus, addition of cytochrome c and dATP (or ATP), or granzyme B, to cytosolic extracts can initiate caspase activation in extracts derived from a variety of cell lines (e.g. Jurkat T lymphoblasts, HEK 293T, U937, HeLa and MCF-7 cells) and primary cells such as neutrophils. Typically, extracts are incubated at 37°C to initiate apoptosis and samples are taken at specific time points thereafter (usually from 0–120 min).

The advantages of this system are that the cell extracts can be manipulated before the addition of cytochrome c/dATP and the effects on subsequent events, such as caspase activation, can be readily assessed. Usually, a cytochrome c concentration of $50\,\mu g\,ml^{-1}$ is sufficient to trigger caspase activation in the extracts, but threshold concentrations (5–$10\,\mu g\,ml^{-1}$) can be used to reveal kinetic differences or inhibitory effects more readily. The system can be manipulated to study caspases by the addition of defined recombinant proteins or caspases, or by the immunodepletion of specific caspases to identify those responsible for processing other caspases and substrates.

Protocol 9.4: Use of cell-free extracts as a source of cytochrome c-inducible active caspases and caspase substrates

EQUIPMENT, MATERIALS AND REAGENTS

Cells to be analysed.

PBS pH 7.2.

Dounce-type homogeniser.

Cell extract buffer (CEB): 20 mM Hepes-KOH pH 7.5, 10 mM KCl, 1.5 mM MgCl$_2$, 1 mM EDTA, 1 mM EGTA, 1 mM DTT, 100 μM PMSF, 10 μg ml^{-1} leupeptin, 2 μg ml^{-1} aprotinin.

Light microscope.

Micro-centrifuge.

Bovine heart cytochrome c.

dATP.

2X SDS loading dye: 100 mM Tris pH 6.8, 4% SDS, 0.2% bromophenol blue, 20% glycerol and 5% β-mercaptoethanol (from a 14.4 M stock).

15% SDS-PAGE gels.

Caspase antibody (see *Table 9.1*).

METHOD

(a) Preparation of cell-free extracts

1. Cells ($\sim$5 $\times$ 10^8) are pelleted at 400 g and washed twice with PBS pH 7.2. When preparing the extracts, note that it is important to minimise PBS carry-over from cell washes prior to homogenisation as cytochrome c-dependent caspase activation in the extracts is antagonised by increasing ionic strength.

2. The cell pellet is resuspended in 2 ml PBS and transferred to a 2 ml Dounce-type homogeniser and pelleted at 800 g to ensure formation of a tight cell pellet.

3. Two volumes of CEB to the volume of the packed pellet is used to resuspend the pellet.

4. Cells are allowed to swell under the hypotonic conditions for 30 minutes on ice.

5. Cells are disrupted with $\sim$40 strokes of a B-type pestle. Cell lysis can be confirmed by examination of a small aliquot of the suspension under a light microscope.

6. The crude lysate is centrifuged at 15 000 g for 15 minutes at 4°C to generate post-nuclear extracts. The supernatant is taken and stored at −70°C until ready to carry out cell-free reactions (extracts should be stored in aliquots to avoid repeated freeze-thaw).

(b) Assembly of cell-free reactions

1. Reactions are typically set up in 100 μl volumes, using 50 μl cell extract, bovine heart cytochrome c (50 μg ml^{-1} final concentration) and dATP (1 mM final concentration) and bringing to 100 μl with CEB.

2. Extracts are incubated at 37°C to allow cell-free apoptosis to proceed. At desired time points, aliquots are taken (~20 μl) and added to an equal volume of 2X SDS loading dye.

3. Samples are then loaded on 15% SDS-PAGE gels and immunoblotted with the relevant caspase antibody (see *Table 9.1*).

9.6 Use of radiolabelled caspases or caspase substrates as a readout of caspase activity

The addition of ^{35}S-labelled caspases can be employed to observe processing where antibodies are not available (*Figure 9.3*). Many commercially available bacterial, yeast and mammalian expression vectors possess an RNA polymerase binding site (such as T3, T7 or SP6) upstream of the multiple cloning site. This enables their use in conjunction with commercial *in vitro* transcription and translation (ITT) kits to produce ^{35}S-labelled proteins for *in vitro* experiments.

In the absence of a suitable source of antibody, ^{35}S-labelled translated caspases or their substrates can be used as an alternative readout of caspase activity *in vitro*. Thus, when used in conjunction with cell-free extracts (see above) the cleavage of ^{35}S-labelled caspases into their processed form, or the cleavage of caspase substrates (such as [^{35}S]PARP) can be used as an indicator of caspase activation.

Typically, a source of amino-acid stripped cell-free lysate (such as rabbit reticulocyte lysate from Promega) is programmed to produce radiolabelled proteins by the addition of RNA-free plasmid template, [^{35}S]methionine, a minimal amino acid mixture and the appropriate RNA polymerase. Reactions are typically set up on a 50–100 µl scale and should employ nuclease-free microcentrifuge tubes and pipettes in order to minimise degradation of template message during ITT. Following translation, radiolabelled proteins are resolved by standard SDS-PAGE and visualised either by direct radiography, or following a fluorogenic amplification step (Protocol 9.5).

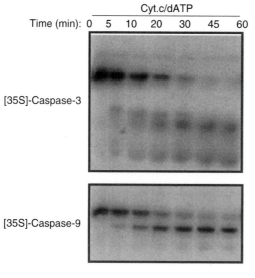

Figure 9.3

Cytochrome c/dATP-initiated processing of caspases-3 and -9. [^{35}S]methionine-labelled caspases were incubated with Jurkat cell-free extracts in the presence or absence of cytochrome c/dATP. Following incubation at 37°C for the indicated times, reactions were stopped and caspase processing was analysed by SDS-PAGE/fluorography.

Protocol 9.5: Use of radiolabelled caspases or caspase substrates to detect caspase activity

EQUIPMENT, MATERIALS AND REAGENTS

ITT kit containing a source of cell-free lysate (i.e. rabbit reticulocyte lysate) plus a methionine-free amino acid stock.

Source of high-quality, RNA-free plasmid DNA.

RNA polymerase (such as T3/T7/SP6) appropriate to the RNA polymerase binding site on your plasmid of interest.

Nuclease-free water.

Translation grade [^{35}S]methionine ($\sim$10 mCi ml^{-1}).

Fixative solution (45% methanol, 10% glacial acetic acid).

Fluorographic reagent (i.e. Amplify from Amersham Pharmacia Biotech).

Gel vacuum dryer.

3 MM chromatography paper from Whatman.

Kodak X-film.

METHOD

Note: The following protocol is based on the ITT kit supplied by Promega.

1. Assemble the ITT reaction shown in *Table 9.4* and allow translation to proceed for 1–2 hours at 37°C. After translation, freeze ITT products at −70°C.

2. Following translation, resolve 0.5–2 μl of translation products by standard SDS-PAGE.

3. Incubate gels in fixative solution for 1–2 hours, or until the bromophenol blue dye front has turned green/yellow in colour. Wash gels briefly in distilled water.

4. If required, an optional signal amplification step can be included by incubating gels in a commercially available fluorographic reagent. Otherwise, proceed directly to step 5.

5. Vacuum dry the gels onto 3 MM chromatography blotting paper and expose to film for 12–36 hours. Exposure at −70°C can help to sharpen the quality of the image and the use of intensifying screens within the exposure cassette will enhance autoradiographic signals.

Note: as a guideline, 2 μl of translation product should be sufficient to visualise a single band on an SDS-PAGE gel following a 12-hour fluorographic exposure, although the yield of your radiolabelled protein should be determined empirically. It is worth noting that proteins larger than 60 kDa may translate poorly, although increased incubation times during ITT may help to remedy this problem.

Table 9.4   ITT reaction mixture for use with radiolabelled caspases or caspase substrates in the detection of caspase activity

Component	Volume
2X Reticulocyte lysate	25 µl
25X reaction buffer	2 µl
RNA polymerase	1 µl
RNAsin	1 µl
Amino acid mix (without methionine)	1 µl
RNA-free plasmid template	1 µg
[^{35}S]methionine (translation grade 10 mCi ml^{-1})	1–2 µl (10–20 mCi)
Nuclease-free water	To 50 µl total

9.7   Measurement of caspase activity using fluorogenic/colorimetric substrate peptides

The ability of caspases to cleave tetrapeptide fragments containing 'optimal' substrate recognition motifs, was originally identified using combinatorial peptide library screening approaches [19]. Caspase substrate peptides coupled either to the chromogenic probe *p*-nitroaniline (pNA) or the fluorometric compounds 7-amino-4-fluoromethyl coumarin (AFC) or 7-amino-4-methyl coumarin (AMC), can be incubated in the presence of apoptotic cell lysates or cytochrome c-stimulated cell-free reactions. Upon caspase-mediated peptide substrate cleavage, the increased absorbance of the free chromophore or emission of the free fluorophore can be detected by spectrophotometry or fluorimetry, respectively.

One advantage that this assay offers is that quantitative comparisons can be made about the relative levels of caspase activity under defined experimental conditions. Additionally, the more sensitive fluorogenic probes allow real time kinetic comparisons to be made between individual treatments. This approach may be particularly useful in studies examining the ability of a given molecule to modulate apoptosis by either repressing or enhancing caspase activity.

It is important to note several of the limitations of this assay. While it is possible to compare caspase activity in terms of the *relative* rate or extent of substrate proteolysis between different treatments, it is important to combine this analysis with an indicator of the *actual* extent of apoptosis in the culture (such as simple morphological assessment). Another important caveat is that caspases share an overlap in peptide substrate preference (see *Table 9.5*). So in spite of what some manufacturers may claim, there is no such thing as a peptide substrate specific for an individual caspase activity and so the assay should not be used to attribute a specific proteolytic activity to an individual caspase. That notwithstanding, when used with caution the *in vitro* peptide substrate assay (Protocol 9.6) provides a reliable readout of caspase-like activity.

Table 9.5   The substrate preference of some of the human caspases, as determined by combinatorial peptide library screening*

Caspases	Peptide substrate preference
−2, −3, −7	DExD
−6, −8, −9, −10	I/L/VExD

*Determined from [19]

Protocol 9.6: Measurement of caspase activity using fluorogenic/colorimetric substrate peptides

EQUIPMENT, MATERIALS AND REAGENTS

Cells to be analysed.

Micro-centrifuge.

Lysis buffer: 150 mM NaCl, 20 mM Tris pH 7.5, 1% Triton X-100, 10 μg ml^{-1} leupeptin, 5 μg ml^{-1} aprotinin, 100 μM PMSF.

2X protease reaction buffer: 100 mM Hepes pH 7.4, 150 mM NaCl, 0.2% CHAPS, 4 mM DTT.

Substrate appropriate to the caspase activity being assayed (see *Table 9.6* for guidelines).

METHOD

Note: The following protocol is optimised for use with chromogenic substrates.

1. Stimulate cells to undergo apoptosis as required, remembering to include an untreated sample to control for background caspase activity in healthy cells.

2. Harvest 2×10^6 cells per treatment by centrifugation at 400 g for 10 minutes. Resuspend the cell pellets in 50 μl of ice-cold lysis buffer. Allow 10 minutes for complete lysis to occur, then pellet insoluble material for 20 minutes at top speed at 4°C in a microfuge.

3. Add an equal volume (50 μl) of 2X protease reaction buffer. If a time course analysis is being conducted, samples can be frozen at −70°C, or kept on ice at this stage until all of the time points have been collected.

4. Add 10 μl (from a 10X stock) of a substrate appropriate to the caspase activity being assayed (see *Table 9.6* for guidelines) to a final concentration of 50 μM. Peptides should be dissolved in DMSO to a final concentration of 500 μM and stored in aliquots at −70°C. Incubate the samples for 30 minutes at 37°C. At this stage, include a control reaction composed of lysis/protease reaction buffer plus the appropriate peptide, to control for spontaneous peptide hydrolysis during incubation.

5. Dilute the samples with 1 ml of an ice-cold 'stop' solution (i.e. dilute HCl) and determine the absorbency/emission of the free probe as indicated in *Table 9.7*.

Table 9.6 The peptide sequences of commercially synthesised caspase substrates, available either as conjugates of the chromogenic probe pNA, or the fluorogenic probes, AFC and AMC

Caspase	Chromogenic/fluorogenic substrate
Caspase-2	Ac-VDVAD
Caspase-3	Ac-DEVD
Caspase-6	Ac-VEID
Caspase-7	Ac-DEVD
Caspase-8	Ac-IETD
Caspase-9	Ac-LEHD

Table 9.7 The absorption or excitation/emission wavelengths for the measurement of caspase-cleaved free pNA, AMC and AFC

Free probe	Absorbance (nm)	Excitation (nm)	Emission (nm)
pNA	405		
AMC		360–380	460
AFC		400	505

9.8 References

1. Yuan, J., Shaham, S., Ledoux, S., Ellis, H.M. and Horvitz, H.R. (1993) The *C. elegans* cell death gene ced-3 encodes a protein similar to mammalian interleukin-1 beta-converting enzyme. *Cell* **75**: 641–652.

2. Ekert, P.G., Silke, J. and Vaux, D.L. (1999) Caspase inhibitors. *Cell Death Differ.* **6**: 1081–1086.

3. Zheng, T.S., Hunot, S., Kuida, K. and Flavell, R.A. (1999) Caspase knockouts: matters of life and death. *Cell Death Differ.* **6**: 1043–1053.

4. Creagh, E.M. and Martin, S.J. (2001) Caspases: cellular demolition experts. *Biochem. Soc. Trans.* **29**: 676–701.

5. Adrain, C. and Martin, S.J. (2001) The mitochondrial apoptosome: a killer unleashed by the cytochrome seas. *Trends Biochem. Sci.* **26**: 390–397.

6. Zou, H., Li, Y., Liu, X. and Wang, X. (1999) An Apaf-1 cytochrome c multimeric complex is a functional apoptosome that activates procaspase-9. *J. Biol. Chem.* **274**: 11549–11556.

7. Boldin, M.P., Gincharov, T.M., Goldsev, Y.V. and Wallach, D. (1996) Involvement of MACH, a novel MORT1/FADD-interacting protease, in Fas/APO-1 and TNF receptor-induced cell death. *Cell* **85**: 803–815.

8. Ashkenazi, A. and Dixit, V.M. (1998) Death receptors: signaling and modulation. *Science* **281**: 1305–1308.

9. Slee, E.A., Harte, M.T., Kluck, R.M., Wolf, B.B., Casiano, C.A., Newmeyer, D.D., Wang, H.G., Reed, J.C., Nicholson, D.W., Alnemri, E.S., Green, D.R. and Martin, S.J. (1999) Ordering the cytochrome c-initiated caspase cascade: hierarchical activation of caspases-2, -3, -6, -7, -8, and -10 in a caspase-9-dependent manner. *J. Cell. Biol.* **144**: 281–292.

10. Connern, C.P. and Halestrap, A.P. (1996) Chaotrophic agents and increased matrix volume enhance binding of mitochondrial cyclophilin to the inner mitochondrial membrane and sensitise the mitochondrial permeability transition to [Ca^{2+}]. *Biochemistry* **35**: 8172–8180.

11. Liu, X., Zou, H., Slaughter, C. and Wang, X. (1997) DFF, a heterodimeric protein that functions downstream of caspase-3 to trigger DNA fragmentation during apoptosis. *Cell* **90**: 405–413.

12. Li, H., Zhu, H., Xu, C.J. and Yuan, J. (1998) Cleavage of BID by caspase-8 mediates the mitochondrial damage in the Fas pathway of apoptosis. *Cell* **94**: 491–501.

13. Adrain, C., Creagh, E.M. and Martin, S.J. (2001) Apoptosis-associated release of Smac/DIABLO from mitochondria requires active caspases and is blocked by Bcl-2. *EMBO J.* **20**: 6627–6636.

14. Slee, E.A., Adrain, C. and Martin, S.J. (2001) Executioner caspase-3, -6, and -7 perform distinct, non-redundant roles during the demolition phase of apoptosis. *J. Biol. Chem.* **276**: 7320–7326.

15. Martin, S.J., Newmeyer, D.D., Mathias, S., Farschon, D., Wang, H-G., Reed, J.C., Kolesnick, R.N. and Green, D.R. (1995) Cell-free reconstitution of Fas-, UV radiation- and ceramide-induced apoptosis. *EMBO J.* **14**: 5191–5200.

16. Martin, S.J., Amarante-Mendes, G.P., Shi, L., Chuang, T-H., Casiano, C.A., O'Brien, G.A., Fitzgerald, P., Tan, E.M., Bokoch, G.M., Greenberg, A.H. and Green, D.R. (1996) The cytotoxic cell protease granzyme B initiates apoptosis in a cell-free system by proteolytic processing and activation of the ICE/CED-3 family protease, CPP32, via a novel two-step machanism. *EMBO J.* **15**: 2407–2416.

17. Kluck, R.M., Martin, S.J., Hoffman, B.M., Zhou, J.S., Green, D.R. and Newmeyer, D.D. (1997) Cytochrome c activation of CPP32-like proteolysis plays a critical role in a Xenopus cell-free apoptosis system. *EMBO J.* **16**: 4639–4649.

18. Ellerby, H.M., Martin, S.J., Ellerby, L.M., Naiem, S.S., Rabizadeh, S., Salvesen, G.S., Casiano, C.A., Cashman, N.R., Green, D.R. and Bredesen, D.E. (1997) Establishment of a cell-free system of neuronal apoptosis: comparison of premitochondrial, mitochondrial, and postmitochondrial phases. *J. Neuroscience* **17**: 6165–6178.

19. Thornberry, N.A., Rano, T.A., Peterson, E.P., Rasper, D.M., Timkey, T., Garcia-Calvo, M., Houtzager, V.M., Nordstrom, P.A., Roy, S., Vaillancourt, J.P., Chapman, K.T. and Nicholson, D.W. (1997) A combinatorial approach defines specificities of members of the caspase family and granzyme B. Functional relationships established for key mediators of apoptosis. *J. Biol. Chem.* **272**: 17907–17911.

Mitochondrial Outer-Membrane Permeabilization in Apoptosis

10

Nigel J. Waterhouse, Jean-Ehrland Ricci,
Helen M. Beere, Joseph A. Trapani and
Douglas R. Green

Contents

Cell Proliferation and Apoptosis, David Hughes and Huseyin Mehmet (Eds)
© 2003 BIOS Scientific Publishers Ltd, Oxford

10.1　Introduction

Mitochondria are complex energy-producing organelles enclosed by an inner and an outer membrane. During apoptosis, the outer membrane becomes permeable and proteins contained within the mitochondrial intermembrane space are released into the cytosol. These include cytochrome c, AIF, SMAC (also known as DIABLO) and htrA2, all of which have been shown to exert pro-apoptotic activities. Oncogenes such as Bcl-2, which prevents permeabilisation of the outer membrane and the release of intermembrane space proteins, block apoptotic cell death and maintain the clonogenic potential of the cell. In contrast, inhibition of apoptosis downstream of mitochondrial outer membrane permeabilisation only delays eventual cell death. Mitochondrial outer membrane permeabilisation has therefore received much interest as a critical point of regulation in the apoptotic process. In this chapter we will discuss some of the studies that have established a role for mitochondria in apoptosis and the methods that have been used to dissect some of the complex issues surrounding mitochondrial outer membrane permeabilisation.

10.2　Pro-apoptotic proteins in the mitochondria are released into the cytoplasm during apoptosis

Apoptotic cell death can be induced by a variety of stimuli including death receptor ligation (e.g. Fas), proteases contained within cytotoxic granules (e.g. granzyme B) and treatments that damage or stress the cell. All these inducers trigger activation of caspases, a family of proteases that mediate the organised dismantling of a cell by cleavage of target substrates (see Chapter 9). Death receptor ligation induces the formation of a DISC, which results in the activation of caspase-8, while granzyme B directly cleaves and activates caspases. Although caspase inhibitors (e.g. zVAD-fmk) can block receptor-induced apoptosis [1, 2], they are only able to delay the apoptosis induced by granzyme B [3, 4] or cytotoxic drugs [1]. This observation demonstrates that it is likely not the activation of caspases *per se* that commits a cell to die but instead, an earlier upstream event that is key in ensuring its inevitable demise. This it seems may be regulated by mitochondrial involvement. All of these pro-apoptotic stimuli induce the activation and translocation of various pro-apoptotic members of the Bcl-2 family to the mitochondria where they induce mitochondrial outer membrane permeabilisation and the release of pro-apoptotic proteins, some of which are involved in caspase activation (*Figure 10.1*). For reviews see [5] and [6].

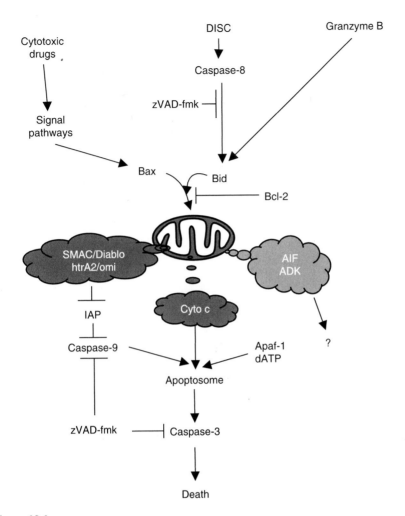

Figure 10.1

Mitochondrial pathways to caspase activation during apoptosis. This figure serves as a reference cartoon for some of the molecules that will be discussed in this chapter. Cytotoxic drugs utilise various signalling pathways to activate pro-apoptotic Bcl-2 family members (e.g. Bax and Bak) that target the mitochondria to release various proteins from the mitochondrial intermembrane space. One of these proteins, cytochrome c, initiates formation of the apoptosome, a proteinaceous complex composed of apaf-1, cytochrome c and caspase-9. This results in the activation of caspase-9, however the activation of caspase-9 is kept in check by inhibitor of apoptosis (IAP) proteins. SMAC/DIABLO and htrA2/Omi, which are also released from the mitochondria, deregulate the IAPs ensuring efficient activation of caspase-9. While ligation of death receptors (e.g. Fas) and granzyme B may activate caspases directly, both caspase-8 (activated by Fas ligation) and granzyme B cleave and activate the pro-apoptotic Bcl-2 family member Bid to induce mitochondrial outer membrane permeabilisation and the activation of caspases via apoptosome assembly. The involvement that other molecules released from the mitochondrial intermembrane space have in caspase activation is not known. The points of inhibition of the caspase inhibitor zVAD-fmk and the anti-apoptotic Bcl-2 family members (e.g. Bcl-2) are indicated.

The role mitochondria play in apoptosis and their link with caspase activation were initially identified in separate laboratories using quite different approaches. Spontaneous caspase activation was initially observed in protein extracts isolated from Xenopus eggs. Importantly, however, this caspase activity was not present if the heavy membrane fraction, containing mitochondria, was removed [7]. Cytochrome c was later identified as the mitochondrial component required for caspase activation in this system [8]. In a similar manner, caspase activation could be induced by adding dATP to total extracts isolated from HeLa cells [9]. Biochemical fractionation of these extracts identified cytochrome c as a key component for efficient caspase activation, and that addition of both cytochrome c and dATP to cytosolic extracts (which contained no endogenous cytochrome c) resulted in an increase in caspase activity. This simple *in vitro* reconstruction of cellular events represents the widely used 'cell-free system' for the study of caspase activation (see Protocol 10.1).

Biochemical fractionation of the cytosolic fraction revealed that another protein, APAF-1 [10, 11], was also required for cytochrome c-dependent caspase activation. It is now known that when cytochrome c is released into the cytoplasm, it binds to APAF-1 to initiate its oligomerisation. This in turn mediates the recruitment of pro-caspase-9 together with a number of other molecules to form the 'apoptosome' [12, 13]. As a consequence, pro-caspase-9 is activated which in turn recruits and activates pro-caspase-3. The cell-free system of caspase activation has been invaluable in investigating the formation and regulation of the apoptosome (see Chapter 9).

Cytosolic extracts prepared in buffers containing detergents were found to have greater ability to activate caspases than extracts prepared in the absence of detergents [14]. Direct addition of detergents to isolated extracts did not increase the activity of the extracts; however, addition of soluble membrane extracts (containing mitochondrial proteins) increased the activity of extracts prepared in the absence of detergents (see Protocol 10.2). One protein isolated from the soluble membrane extracts was SMAC or DIABLO. This protein de-regulates inhibitor of apoptosis proteins (IAPs), which bind to and block caspases directly [15]. Database searching has revealed a second homologue of SMAC (SMACβ) that contains pro-apoptotic activity, however it does not have an IAP-binding domain [16] and the IAP-binding domain of SMAC has been shown to be dispensable for its pro-apoptotic activity, suggesting that the reversal of IAP-mediated caspase inhibition may not be its only mode of activity. Recently several groups have also identified an additional mitochondrial IAP-binding protein called htrA2/Omi [17–19] that is also released upon mitochondrial outer membrane permeabilisation during apoptosis.

Pro-apoptotic proteins liberated from mitochondria during apoptosis have also been identified by treating isolated mitochondria in various ways to induce outer membrane permeabilisation (see Protocols 10.3 and 10.4). One method by which this is achieved is by treating isolated mitochondria with calcium or chemicals (e.g. atractyloside or lonidamide) to induce a phenomenon called permeability transition (PT). In this case, a pore opens in the mitochondria resulting in changes in osmotic potential, the mitochondria swell and the mitochondrial outer membrane ruptures releasing the contents of the mitochondrial space (see Protocol 10.5). Mitochondria treated in this way were shown to release an activity that induced morphological changes in isolated nuclei similar to those observed during apoptosis [20, 21]. The protein responsible for this activity has been characterised as AIF [22]. Treatment of isolated mitochondria with PT inducers has been shown to liberate over 70 proteins [23], some of which do not reside in the mitochondrial intermembrane space. It is not clear whether any of these additional proteins are actually released from mitochondria during apoptosis or whether any of these proteins have pro-apoptotic activity. While the induction of PT is one way to induce the release of pro-apoptotic proteins from mitochondria (and it has

Protocol 10.1: Isolation of cytosolic extracts

EQUIPMENT, MATERIALS AND REAGENTS

Cells to be analysed.

Ice.

Homogenisation buffer: 20 mM Hepes–KOH pH 7.5, 10 mM KCl, 1.5 mM MgCl$_2$, 1 mM sodium EDTA, 1 mM sodium EGTA, 1 mM DTT, and protease inhibitors (0.1 mM PMSF, 5 µg ml^{-1} pepstatin A, 10 µg ml^{-1} leupeptin, 2 µg ml^{-1} aprotinin, 25 µg ml^{-1} N-acetyl-leu-leu-norleucinal (ALLN)). (Note: include 68 mM sucrose and 200 mM mannitol in the buffer if intact mitochondria are required as an end point.)

Kontes dounce homogeniser with a B pestle or a 22G needle.

Micro-centrifuge.

Ultra centrifuge.

METHOD

1. Incubate 5×10^8 cells (15 minutes on ice) in 2 volumes of homogenisation buffer. (Include 68 mM sucrose and 200 mM mannitol in the buffer to maintain intact mitochondria.)

2. Homogenise the cells (15 strokes) in a Kontes dounce with a B pestle or by 15 passages through a 22G needle. The number of strokes must be optimised for each cell type under each condition by evaluating lysis using Trypan blue exclusion.

3. Centrifuge the cells (10 000 g for 5 minutes at 4°C).

4. Centrifuge the supernatant (100 000 g for 1 hour at 4°C). Alternatively, steps 3 and 4 can be omitted and crude lysates prepared by centrifugation at 10 000 g for 15 minutes at 4°C.

5. Store the supernatant in small aliquots at −80°C.

Protocol 10.2: Isolation of soluble membrane extracts

EQUIPMENT, MATERIALS AND REAGENTS

Cells to be analysed.

Ice.

Cell homogenisation buffer: 20 mM Hepes–KOH pH 7.5, 10 mM KCl, 1.5 mM MgCl$_2$, 1 mM sodium EDTA, 1 mM sodium EGTA, 1 mM DTT, 68 mM sucrose, 200 mM mannitol and 0.1 mM PMSF, 5 μg ml^{-1} pepstatin A, 10 μg ml^{-1} leupeptin, 2 μg ml^{-1} aprotinin and 25 μg ml^{-1} ALLN. Membrane extraction buffer: 0.5% (w/v) CHAPS, 20 mM Hepes–KOH pH 7.5, 10 mM KCl, 1.5 mM MgCl$_2$, 1 mM sodium EDTA, 1 mM sodium EGTA, 1 mM DTT, 0.1 mM PMSF, 5 μg ml^{-1} pepstatin A, 10 μg ml^{-1} leupeptin, 2 μg ml^{-1} aprotinin and 25 μg ml^{-1} ALLN.

Kontes dounce homogeniser with a B pestle or a 22G needle.

Micro-centrifuge.

Ultra-centrifuge.

METHOD

1. Incubate 5×10^8 cells (15 minutes on ice) in 2 volumes of homogenisation buffer.

2. Homogenise the cells (15 strokes) in a Kontes dounce with a B pestle or by 15 passages through a 22G needle.

3. Centrifuge the cells (1000 g for 5 minutes at 4°C).

4. Centrifuge the supernatant (10 000 g for 30 minutes at 4°C).

5. Resuspend the pellet in membrane extraction buffer.

6. Centrifuge the lysate (100 000 g for 1 hour at 4°C).

7. Store the supernatant at −80°C.

Protocol 10.3: Isolation of mouse liver mitochondria

EQUIPMENT, MATERIALS AND REAGENTS

Mouse.

Ice.

PBS pH 7.4.

15 ml dounce homogeniser with a tight-fitting teflon pestle.

Mitochondrial isolation buffer (MIB): 220 mM mannitol, 68 mM sucrose, 10 mM Hepes-KOH pH 7.4, 10 mM KCl, 1 mM EGTA, 1 mM EDTA, 0.1% BSA and protease inhibitors (0.1 mM PMSF, 5 μg ml^{-1} pepstatin A, 10 μg ml^{-1} leupeptin, 2 μg ml^{-1} aprotinin and 25 μg ml^{-1} ALLN).

Mitochondrial storage buffer: 200 mM mannitol, 50 mM sucrose, 10 mM succinate, 10 mM Hepes-KOH pH 7.4, 0.1% BSA and 5 mM potassium phosphate pH 7.4.

Benchtop centrifuge.

Micro-centrifuge.

METHOD

1. Rapidly excise the liver from a freshly sacrificed adult (2–6 months old) mouse, and place it in ice-cold PBS pH 7.4.

2. Cut the liver into small pieces using sterile scissors and separate into two parts.

3. Homogenise each half (20–40 strokes of a 15 ml dounce with a tight-fitting teflon pestle) in 10 ml of ice-cold MIB.

4. Centrifuge the lysates (600 g for 10 minutes at 4°C) in a swinging bucket rotor using round bottom 15 ml polypropylene tubes.

5. Centrifuge the supernatant (3500 g for 15 minutes at 4°C).

6. Resuspend the pellet in 15 ml of fresh MIB and centrifuge (1500 g for 5 minutes at 4°C).

7. Centrifuge the supernatant (5500 g for 10 minutes at 4°C).

8. Repeat steps 6 and 7 twice.

9. Resuspend the pellet (mitochondria) in 500 μl of ice-cold mitochondrial storage buffer. Note: the mitochondria can be stored on ice for a maximum of 4 hours.

Protocol 10.4: Isolation of mitochondria from cultured cells

EQUIPMENT, MATERIALS AND REAGENTS

Cells to be analysed.

Ice.

PBS pH 7.4.

2 ml dounce homogeniser with a loose-fitting B-type pestle.

Mitochondrial isolation buffer (MIB): 220 mM mannitol, 68 mM sucrose, 10 mM Hepes-KOH pH 7.4, 10 mM KCl, 1 mM EGTA, 1 mM EDTA, 0.1% BSA and protease inhibitors (0.1 mM PMSF, 5 μg ml^{-1} pepstatin A, 10 μg ml^{-1} leupeptin, 2 μg ml^{-1} aprotinin and 25 μg ml^{-1} ALLN).

Micro-centrifuge.

Mitochondrial storage buffer: 200 mM mannitol, 50 mM sucrose, 10 mM succinate, 10 mM Hepes-KOH pH 7.4, 0.1% BSA and 5 mM potassium phosphate pH 7.4.

METHOD

1. Harvest at least 5 $\times 10^8$ cells by centrifugation (800 g for 10 minutes) in ice-cold PBS (pH 7.4).

2. Resuspend the cells in three volumes of ice-cold MIB.

3. Homogenise the cells (60–80 strokes in a 2 ml dounce using a loose-fitting B-type pestle). Note; number of strokes needs to be titrated for each cell line.

4. Centrifuge the lysate (600 g for 10 minutes at 4°C).

5. Centrifuge the supernatant (3500 g for 15 minutes at 4°C).

6. Resuspend the pellet in 10 volumes of MIB and centrifuge (1500 g for 5 minutes at 4°C).

7. Centrifuge the supernatant (5500 g for 10 minutes at 4°C).

8. Repeat steps 6 and 7 twice.

9. Resuspend the pellet (mitochondria) in 100 μl of ice-cold mitochondrial storage buffer. Note: the mitochondria can be stored on ice for a maximum of 4 hours.

Protocol 10.5: Induction of permeability transition (PT) in isolated mitochondria

EQUIPMENT, MATERIALS AND REAGENTS

Freshly isolated mitochondria.

PT buffer: 125 mM KCl, 2.5 mM potassium phosphate pH 7.4, 2.5 mM succinate and 20 mM Hepes-KOH.

PT inducer e.g. atractyloside (5 mM), $CaCl_2$ (100 μM) or lonidamine (50 μM).

METHOD

1. Resuspend freshly isolated mitochondria (100 μg mitochondrial protein ml^{-1}) in PT buffer containing the PT inducer e.g. atractyloside (5 mM), $CaCl_2$ (100–500 μM), lonidamine (50 μM). Note: adding cyclosporin A (40 μM) to the assay buffer can transiently inhibit PT as a negative control.

Protocol 10.6: Induction of mitochondrial outer membrane permeabilisation

EQUIPMENT, MATERIALS AND REAGENTS

Freshly isolated mitochondria.

Pure truncated Bid (tBid) ($10\,\mu g\,ml^{-1}$) or pure Bax (30–$40\,\mu g\,ml^{-1}$).

Mitochondria storage buffer with $80\,mM$ KCl.

METHOD

1. Incubate freshly isolated mitochondria in mitochondria storage buffer with $80\,mM$ KCl (equivalent of $100\,\mu g$ mitochondrial protein ml^{-1}) with tBid ($10\,\mu g\,ml^{-1}$) or Bax (30–$40\,\mu g\,ml^{-1}$) for 5–60 minutes at 37°C.

been suggested that this is the mechanism of outer membrane permeabilisation in cells), the release of proteins from the mitochondria during the early stages of apoptosis does not have to involve PT [24–26]. To isolate molecules with pro-apoptotic activity from isolated mitochondria, it may therefore be more appropriate to induce outer membrane permeabilisation in a more physiological manner.

It is clear that whatever the mechanism of mitochondrial outer membrane permeabilisation, this event is regulated by Bcl-2 family proteins and treating isolated mitochondria with pro-apoptotic Bcl-2 family proteins (e.g. Bax, Bak or Bid) can release proteins from the mitochondrial intermembrane space [27–29] (see Protocol 10.6). Full-length Bax, however, is relatively insoluble and so we routinely use a recombinant truncated form lacking the C-terminal 19 amino acids (encoded in pGEX-KG-BaxC19 from S. Korsmeyer). Full length Bid is also less active (10-fold) than caspase-cleaved Bid and so we generally use a truncated form of Bid (tBid) that is missing 7 kDa from the N-terminus. This is either produced by cleavage of full length Bid by caspase-8 or by expression and purification of the recombinant protein that lacks the N-terminal 7 kDa.

10.3   Detecting the translocation of proteins to and from mitochondria in cells

One common characteristic of the mitochondrial intermembrane space proteins mentioned above is that they must be released into the cytoplasm to exhibit their pro-apoptotic activity. While the methods described above can be used to follow and determine the effects of these proteins *in vitro*, it is often necessary to follow the translocation of these proteins in cells undergoing apoptosis. Cellular fractionation in combination with western blotting is the most common way by which this is achieved (see Protocol 10.7). This technique requires good antibodies that recognise the protein of interest. While antibodies for immunodetection of cytochrome c (clone 7H8.2C12) and Bax by immunoblot analysis can be obtained from BD/Pharmingen (San Diego) or other sources, antibodies to other mitochondria-associated proteins may not be readily available and it may be necessary to contact the original investigators to obtain these. The methods reported here have been used successfully to follow Bax translocation to and cytochrome c release from the mitochondria. However, minor adaptations may be required to follow the translocation of other proteins.

For immunoblotting, the cytosolic and mitochondrial fractions from whole cell lysates are isolated using a method designed to minimise mitochondrial disruption. The relative distribution of the protein of interest between these two fractions is then compared in untreated versus dying cells. Simple homogenisation is the most commonly used method to fractionate cells (Protocol 10.7a). Parameters such as number of dounce strokes and force used must be optimised for each cell line and each pestle, such that only minimal amounts of mitochondrial proteins can be detected in the cytosol of control cells. To achieve this, many cells may remain intact. The homogenisation method can yield inconsistent results since each sample must be generated individually (the time between the processing of the first and last sample can be large), everyone has their own dounce technique and bubbles generated in the lysate can cause damage to the mitochondrial outer membranes. Nitrogen cavitation may be used as an alternative method to overcome some of the inconsistencies associated with homogenisation (Protocol 10.7b). While it has been reported that this technique is less damaging to mitochondria [30–32], some reports show up to 30% of total cellular cytochrome c in the cytoplasm isolated from control cells [30] and as with homogenisation large numbers of cells are required.

Protocol 10.7: Cellular fractionation to detect cytochrome c release

EQUIPMENT, MATERIALS AND REAGENTS

Cells for analysis.

Ice-cold PBS pH 7.2.

Homogenisation buffer: 220 mM mannitol, 68 mM sucrose, 50 mM Pipes-KOH pH 7.4, 80 mM KCl, 5 mM EGTA, 2 mM MgCl$_2$, 1 mM EDTA, 1 mM DTT and protease inhibitors.

Dounce homogeniser and a B-type pestle.

Micro-centrifuge.

Cavitation medium: tissue culture medium containing 150 mM NaCl, 5 mM KCl and 10 mM Tris-HCl pH 7.4.

Nitrogen cavitation buffer: 210 mM mannitol, 70 mM sucrose, 1 mM EGTA, 0.5% BSA, 5 mM Hepes-KOH buffer pH 7.2.

Nitrogen chamber.

Plasma membrane lysis buffer: 80 mM KCl, 250 mM sucrose, and 200 µg ml^{-1} digitonin in PBS *or* 20 mM Hepes-KOH pH 7.5, 250 mM sucrose, 80 mM KCl, 1.5 mM MgCl$_2$, 1 mM Na-EDTA, 1 mM EGTA, 1 mM DTT, proteinase inhibitors and 60 U of streptolysin O.

METHOD

(a) Homogenisation

1. Wash 2.5×10^7 cells in ice-cold PBS pH 7.2.

2. Incubate the cells (30 minutes on ice, with intermittent shaking) in 500 µl of homogenisation buffer.

3. Homogenise the cells (60 strokes) with a glass dounce and a B-type pestle.

4. Centrifuge the lysate (800 g for 10 minutes).

5. Centrifuge the supernatant (14 000 g for 10 minutes).

6. Store the supernatant (cytosol) and pellet (mitochondria) at −70°C for subsequent analysis.

(b) Nitrogen cavitation

1. Wash 6×10^9 cells three times in cavitation medium.

2. Wash the cells once in nitrogen cavitation buffer.

3. Make a slurry of the washed cells and place in the nitrogen chamber.

4. Apply pressure (N$_2$ 20 kg cm^{-2} for 30 minutes).

5. Centrifuge the lysate (800 g for 10 minutes).

6. Centrifuge the supernatant (14 000 g for 10 minutes).

7. Store the supernatant (cytosol) and pellet (mitochondria) at $-70°C$ for subsequent analysis.

(c) Selective lysis of the plasma membrane

1. Wash 1 $\times 10^6$ cells in ice-cold PBS pH 7.2.

2. Incubate the cells (5 minutes on ice) in 100 μl of 80 mM KCl, 250 mM sucrose and 200 μg ml^{-1} digitonin in PBS.

or

2. Incubate the cells (20 minutes at 37°C) in 100 μl of 20 mM Hepes-KOH pH 7.5, 250 mM sucrose, 80 mM KCl, 1.5 mM MgCl$_2$, 1 mM Na-EDTA, 1 mM EGTA, 1 mM DTT, proteinase inhibitors and 60 U of streptolysin O.

3. Centrifuge the cells (10 000 g for 5 minutes).

4. Store the recovered pellet (cells and mitochondria) and supernatant (cytosol) at $-70°C$ for subsequent analysis.

Recently, more studies have reported using pore-forming proteins to permeabilise the plasma membrane while leaving the mitochondrial membranes intact [3, 26, 33, 34] (Protocol 10.7c). Proteins that translocate to the mitochondria are collected in the pellet and proteins that translocate from the mitochondria are collected in the permeabilisation buffer. Digitonin from *Digitalis purpurea* and streptolysin O have both been used. This technique has been used for a variety of cell lines and can be scaled up for isolation of cytosol from large numbers of cells if required. It is wise to be aware that commercial sources of these lytic agents may contain large amounts of contaminants, so the efficiency of lysis should be monitored by incubating small aliquots of the ongoing lysing reaction with Trypan blue and counting the percentage of blue (i.e. permeabilised) cells. The method should be standardised in control cells such that ~95% of cells are Trypan blue positive with all cytochrome c remaining in mitochondria. This technique is inexpensive, rapid (~15 min), efficient (all samples are treated at the same time) and the majority of cells are lysed (a single centrifugation step can be used to separate cytosol from mitochondria). Further, only a small number of cells are required for analysis. This is the method of choice in our laboratories for detecting cytochrome c release during apoptosis. It may also be used to evaluate the translocation of proteins from the cytosol to the mitochondria (e.g. Bax).

While these methods provide simple fractionation procedures to allow for western blot analysis of proteins that translocate to and from the mitochondria during apoptosis, if antibodies are not available for the protein of interest, functional assays may be required. AIF was one of the first pro-apoptotic proteins seen to translocate from mitochondria during apoptosis. The recombinant protein induces chromatin condensation in isolated nuclei and, until recently, chromatin condensation was used as a functional assay for this protein [20, 21] (see Protocol 10.8). Recently, antibodies against amino acids 150–170, 166–185 and 181–200 of AIF have been produced and used successfully for western blot analysis [22]. Adenylate kinase (ADK) can also be detected by functional assay [35] (see Protocol 10.9). In Jurkat cells, the majority of ADK activity can be attributed to the mitochondrial form of ADK, ADK2. Therefore measuring total ADK activity has been used as a functional assay to monitor ADK2 release from the mitochondria during apoptosis.

While cellular fractionation followed by either western blotting or functional assay is sufficient for detecting gross changes in the location of proteins with respect to mitochondria, they are not quantitative and can be misleading. If only the cytoplasmic fraction is assayed, without the mitochondrial fractions for comparison, an increase in apoptosis from 5% to 10%, which would result in a two-fold increase in cytochrome c in the cytoplasmic fraction, may appear significant. Single cell analysis is a more useful way to obtain an accurate estimation of the number of cells in which a specific protein has translocated to or from the mitochondria. This may be achieved using immunocytochemistry, although limitations are imposed by the effects of cell fixation on antibody specificity and sensitivity to protein of interest (see Protocol 10.10). This protocol has been used successfully to follow the translocation of Bax to and cytochrome c from mitochondria in HeLa cells. Using a combination of antibodies to different proteins (each conjugated to different fluorophores or isolated from different species, to allow selectivity with a secondary antibody) this technique can be used to compare the localisation of cytochrome c with the localisation of other proteins or organelles in the same cell. The distribution of mitochondrial proteins can also be compared to other apoptotic events such as phosphatidylserine exposure or chromatin condensation by using probes attached to different fluorophores. Since this technique requires fixation prior to antigen visualisation, the samples can be stored for subsequent analysis and do not have to

Protocol 10.8: Detection of apoptosis inducing factor

EQUIPMENT, MATERIALS AND REAGENTS

Freshly isolated mitochondria.

Isolated HeLa cell nuclei.

PT buffer: 125 mM KCl, 2.5 mM potassium phosphate pH 7.4, 2.5 mM succinate and 20 mM Hepes-KOH.

PT inducer e.g. atractyloside (5 mM), $CaCl_2$ (100–500 μM), lonidamine (50 μM).

AIF buffer: 220 mM mannitol, 68 mM sucrose, 2 mM NaCl, 2.5 mM PO_4H_2K, 0.5 mM EGTA, 2 mM $MgCl_2$, 5 mM Na pyruvate, 1 mM PMSF, 1 mM DTT, 10 mM Hepes-KOH pH 7.4.

10 μM DAPI.

Fluorescence microscope.

METHOD

1. Incubate mitochondria (200 μg ml^{-1}) in PT buffer with PT inducer (Protocol 10.5) for 90 minutes at 37°C.

2. Remove mitochondria by centrifugation (5500 g for 10 minutes at 4°C).

3. Add mitochondria supernatant to isolated HeLa nuclei (10^3 nuclei μl^{-1}) in AIF buffer.

4. Stain the nuclei with 10 μM DAPI or similar DNA stain.

5. Examine the nuclei by fluorescence microscopy for small, condensed nuclei.

Protocol 10.9: Detecting adenylate kinase

EQUIPMENT, MATERIALS AND REAGENTS

Freshly isolated mitochondria.

Mitochondrial lysis buffer: 1% Triton-X-100, 210 mM mannitol, 60 mM sucrose, 10 mM Hepes-KOH pH 7.4, 10 mM KCl, 5 mM EGTA, 10 mM succinate, 0.5 mM DTT.

100 mM ATP.

100 mM phosphoenol pyruvate.

1 mM rotenone.

1.5 mM oligomycin.

A 1:1 mix of pyruvate kinase and lactate dehydrogenase (80 U ml^{-1} each).

100 mM NADH.

Adenylate kinase assay buffer: 130 mM KCl, 6 mM MgSO$_4$, 100 mM Tris-HCl pH 7.5.

20 μM diadenosine pentaphosphate.

Chicken muscle myokinase.

96-well microtitre plate.

Spectrophotometric microtitre plate reader.

METHOD

1. Lyse isolated mitochondria in small volumes of mitochondrial lysis buffer.

2. Prepare adenylate kinase assay buffer by adding 5 μl each of 100 mM ATP, 100 mM phosphoenol pyruvate, 1 mM rotenone, 1.5 mM oligomycin, a mix of pyruvate kinase and lactate dehydrogenase (80 U ml^{-1} each) and 15 μl of 100 mM NADH to 1 ml of 130 mM KCl, 6 mM MgSO$_4$, 100 mM Tris-HCl pH 7.5.

3. Add 200 μl of this assay buffer to an equal volume of mitochondrial lysate or cytosol in a 96-well microtitre plate.

4. Measure the consumption of NADH by following the decrease in absorbance at 366 nm over a 10 minute period.

5. Subtract the background values obtained in the presence of the adenylate kinase inhibitor diadenosine pentaphosphate (20 μM).

6. Calibrate adenylate kinase activity against chicken muscle myokinase with defined enzymatic activity.

Protocol 10.10: Measurement of protein translocation by immunocytochemistry

EQUIPMENT, MATERIALS AND REAGENTS

Cells to be analysed.

Either LabTek chamber slides or sterile coverslips in 24-well plates.

Apoptotic inducer of choice.

4% (w/v) PFA.

Cytochrome c antibody (clone 6H2.B4) or anti-Bax antibody diluted 1:200 in blocking buffer.

Blocking buffer (0.05% Saponin, 3% BSA in PBS pH 7.2).

Fluorescently labelled secondary antibody.

Fluorescence microscope.

METHOD

1. Grow cells to ~70% confluence, either in LabTek chamber slides or on sterile coverslips and treat them with the apoptotic inducer of choice.

2. Incubate the cells (15 minutes at room temperature with gentle rocking) in 4% PFA.

3. Wash the cells 6 times in blocking buffer (0.05% saponin, 3% BSA in PBS pH 7.2).

4. Incubate the cells (overnight at 4°C with gentle rocking) with anti-cytochrome c (clone 6H2.B4) or anti-Bax (Pharmingen, San Diego CA) diluted 1:200 in blocking buffer.

5. Wash the cells 6 times in blocking buffer.

6. Incubate the cells (1 hour at room temperature with gentle rocking) with the appropriate fluorescently labelled secondary antibody (e.g. Ig Texas Red or FITC) diluted 1:200 in blocking buffer.

7. The fluorescence of the conjugated antibody can be detected by confocal or fluorescence microscopy. Texas Red can be detected using 595 nm excitation and 615 nm emission, while FITC can be detected using 494 nm excitation and 519 nm emission.

Protocol 10.11: Quantitative analysis of cytochrome c-GFP redistribution in cells by flow cytometry

EQUIPMENT, MATERIALS AND REAGENTS

Cells expressing cytochrome c-GFP.

CLAMI Buffer (250 mM sucrose, 80 mM KCl, 50 μg ml^{-1} digitonin in PBS pH 7.2).

FACS.

METHOD

1. Incubate 2×10^4 cells for 5 minutes in 100 μl of ice-cold cell *lysis* and *mitochondria intact* (CLAMI) buffer (250 mM sucrose, 80 mM KCl, 50 μg ml^{-1} digitonin) in PBS.

2. Dilute each sample with 200 μl ice-cold digitonin-free CLAMI buffer and immediately analyse the content of GFP in the cells by flow cytometry (488 nm, FL-1).

be processed immediately. Immunocytochemistry may also be combined with microscopic evaluation to quantitate the percentage of cells in a population that have released their cytochrome c into the cytosol. This is indicated by a clearly visible change from a punctate (mitochondrial) distribution to a diffuse (cytoplasmic) staining pattern for cytochrome c.

Immunocytochemistry, however, is labour intensive and relatively expensive. Recently GFP-tagged cytochrome c stably expressed in cells has been shown to mimic the redistribution of endogenous cytochrome c during apoptosis [33]. Using cytochrome c-GFP, the percentage of cells that have released cytochrome c can be estimated without laborious fixing and staining. The release of cytochrome c can also be assessed over time by time-lapse microscopy and can be directly compared with other apoptotic events including any change in mitochondrial transmembrane potential, phosphatidylserine exposure and loss of plasma membrane integrity [26, 36–38]. Mitochondrial dyes such as tetramethylrhodamine ethyl ester (TMRE, discussed below) can be used to evaluate mitochondrial localisation, and immunocytochemistry and western blot can be used to monitor the concomitant release of endogenous and expressed cytochrome c. Cells can be treated with an apoptotic stimulus and examined for the release of cytochrome c-GFP by confocal microscopy. The percentage of cells that have released cytochrome c-GFP from mitochondria can be quantitated by counting cells with punctate versus diffuse green fluorescence, or by flow cytometry by permeabilising the outer membrane and allowing the non-mitochondrial cytochrome c to exit the cell (see Protocol 10.11). In this case, all cytosolic protein is released from the cells and those cells in which cytochrome c has been released from the mitochondria will have lower overall fluorescence compared to control cells. Other proteins that translocate to and from the mitochondria have also been tagged with GFP (e.g. Bax-GFP) and can also be used for this type of analysis.

10.4   The impact of mitochondrial outer membrane permeabilisation on mitochondrial function

Mitochondria serve several functions in the cell including the production of ATP. Although they have their own DNA and can synthesise proteins, they primarily rely on the import of nuclear-encoded proteins for bioenergetics and metabolism. Some of these functions, including the production of ATP by oxidative phosphorylation and protein import, rely upon a proton gradient across the mitochondrial inner membrane. This gradient is called the mitochondrial membrane potential ($\Delta\Psi_m$) and its generation is a complex process that involves transport of electrons along the electron transport chain (although if required, $\Delta\Psi_m$ can also be generated by reversing this process). Cytochrome c is an essential component of this process (*Figure 10.2*), and therefore permeabilisation of the outer membrane is likely to impact on specific mitochondrial functions. In addition, $\Delta\Psi_m$ has been suggested to transiently increase [39] and to decrease [40, 41] during apoptosis. Indeed some hypotheses suggest that these changes in $\Delta\Psi_m$ or alterations in mitochondrial bioenergetics are directly or indirectly responsible for mitochondrial outer membrane permeabilisation during apoptosis.

Measurement of $\Delta\Psi_m$ can be performed using a variety of commercially available cationic dyes that accumulate in the mitochondria in a $\Delta\Psi_m$-dependent manner (*Table 10.1*). Many researchers are adamant about the dye they use to monitor $\Delta\Psi_m$ and this has sparked many debates. In our hands, most of the dyes yield

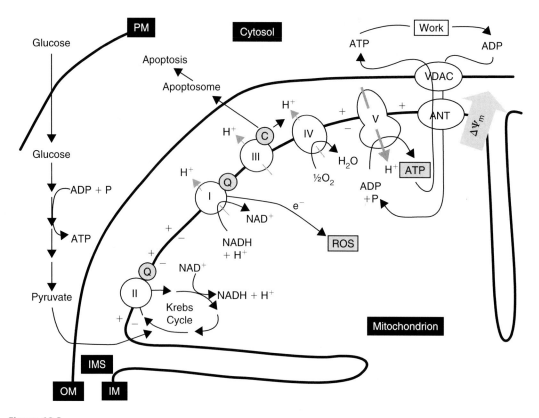

Figure 10.2

Parameters of a functional mitochondrion, ATP production, generation of $\Delta\Psi_m$ and production of reactive oxygen species. Glucose that is taken up by the cell is converted to pyruvate. This is taken up by the mitochondria and utilised in the Krebs/citric acid cycle. Complex II catalyses a step of the Krebs cycle where succinate is converted to fumarate and an electron is passed to coenzyme Q (Q). Complex I, which is involved in NADH metabolism also passes an electron to Q. Coenzyme Q transfers the electron to complex III where it is picked up by cytochrome c (C) and transferred to complex IV. At the end of this pathway, complex IV converts oxygen to water. In the reactions catalysed by complexes I, III and IV, protons are transferred from the matrix to the intermembrane space creating a negative potential called the mitochondrial transmembrane potential ($\Delta\Psi_m$). Complex V, also known as ATP synthase, utilises this potential (transfers protons in the opposite direction) to convert ADP and free phosphate to ATP. The flow of protons is indicated by the thin grey arrows, the net effect is a negative potential of approximately $-180\,mV$ as indicated by the large grey arrow. The ATP produced is then transferred to the cytoplasm via the adenine nucleotide transporter (ANT) and the voltage-dependent anion channel (VDAC) where it can be utilised in ATP-dependent processes. The resulting ADP is transferred back into the mitochondria, again by the ANT and VDAC, completing the cycle. Reactive oxygen species are generated as a by-product of this process. (Abbreviations: IM, inner membrane; OM, outer membrane; IMS, intermembrane space; PM, plasma membrane.)

similar results when quantifying the percentage of cells that have lost $\Delta\Psi_m$ during apoptosis. Problems arise, however, when these dyes are used to assess mitochondrial function since some of the dyes inhibit electron transport, undergo self-quenching or do not reflect subtle changes in $\Delta\Psi_m$. Some of these problems are listed in *Table 10.1*. In our experience, TMRE is the most robust and sensitive dye for measuring $\Delta\Psi_m$ in both cells and isolated mitochondria (see Protocol 10.12). Cells can survive for extended periods (days) in the presence of TMRE, and the dye responds

Table 10.1   Fluorescent dyes used to measure mitochondrial membrane potential

Dye	EX/EM	Staining conditions	Potential problem
JC-1 (green)*	510/527	5–10 μg ml^{-1}	Does not respond to
(red)*	485–585/590	(20 min at 37°C)	subtle changes in $\Delta\Psi_m$
DioC$_6$(3)	484/500	40 nM (20 min at 37°C)	Large response to plasma membrane potential
			Inhibits complex I
MitoTracker Red	578/599	150 nM (20 min at 37°C)	Photosensitiser
MitoTracker Orange	551/576	150 nM (20 min at 37°C)	May induce PT
Rhodamine 123	507/529	2–5 μM (20 min at 37°C)	Autofluorescence
			Inhibits ATP synthase
TMRE	550/573	50–100 nM	Responds to plasma membrane
		(20 min at 37°C)	potential

All of the dyes listed above can be obtained from Molecular Probes (Eugene, OR).
*In untreated cells with high $\Delta\Psi_m$, JC-1 becomes concentrated and forms J-aggregates that fluoresce orange/red, while in depolarised cells, JC-1 is less concentrated and fluoresces green. These two states are easily distinguishable by flow cytometry or fluorescence microscopy. (Abbreviations: EX/EM, excitation/emission wavelength, nm.)

to loss and recovery of $\Delta\Psi_m$. Although we have seen no major problems when measuring $\Delta\Psi_m$ in cells undergoing apoptosis, some problems have been reported when using this dye, including its binding to polystyrene [42]. A control sample that includes carbonyl cyanide p-(trifluoromethoxy)phenylhydrazone (FCCP 5 μM) or carbamoyl cyanide n-chlorophenylhydrazone (CCCP 10 μM) in the buffer should be included. These chemicals transfer protons across the membranes, dissipating the potential, thereby giving an indication of the level of fluorescence in a cell with depolarised mitochondria. If dyes other than TMRE are used, a control population should be depolarised with FCCP or CCCP before staining the cells. To obtain good separation of polarised and depolarised cells, it may be necessary to titrate the dye or the number of cells per ml to obtain the greatest difference between cells in the presence or absence of FCCP.

Some of the dyes used to evaluate $\Delta\Psi_m$ also accumulate across non-mitochondrial membranes that possess a potential. The inclusion of KCl (137 mM) in staining media will largely neutralise the plasma membrane potential ensuring that $\Delta\Psi_m$ is the predominant potential being measured. Some reports have used dyes diluted in PBS to stain cells; however, since mitochondria in cells react quickly to environmental changes, such as removal of glucose and the addition of mitochondrial toxins to the medium, the removal of cells from growth medium may alter the $\Delta\Psi_m$ in some instances. The growth medium does not affect flow cytometry. TMRE, however, is not suitable for use on fixed cells and MitoTracker dyes (red or orange from Molecular Probes Inc.) are more suitable for this particular application (see Protocol 10.13).

$\Delta\Psi_m$ can also be measured using a tetraphenylphosphonium (TPP) electrode. TPP$^+$ is a lipophilic cation, which distributes across the inner mitochondrial membrane in a Nernstian fashion and provides qualitative information about $\Delta\Psi_m$. So far, no TPP$^+$ electrodes are commercially available, however the Ca^{2+}-selective microelectrode (ETH 1001) has been successfully used to measure $\Delta\Psi_m$ using TPP$^+$ [43] and can also be coupled to measurement of oxygen consumption (see below).

Mitochondrial membrane potential is utilised for a number of functions in the mitochondria. However, it is mainly used by ATP synthase to produce ATP from ADP and free phosphate. Apoptosis is an energy-dependent process that can occur only if ATP is available, such that if intracellular ATP levels are depleted, cells are forced to a necrotic death. A simple kit from Roche (HSII) can be used to measure ATP in cells (see Protocol 10.14).

Protocol 10.12: Measurement of $\Delta\Psi_m$ in live cells

EQUIPMENT, MATERIALS AND REAGENTS

Cells.

50 nM tetramethyl rhodamine ethyl ester (TMRE) in media.

5 μM FCCP or 10 μM CCCP in 50 nM TMRE media (for negative control).

FACS or confocal microscope.

METHOD

1. Incubate 2.5×10^5 cells (20 minutes at 37°C in the dark) in 500 μl of media containing 50–100 nM TMRE.

2. Analyse cells for fluorescence directly by FACS (488 nm excitation is fine and collect emission in Fl-2), by fluorescence microscopy or confocal microscopy (excitation 550 nm, emission 573 nm).

Protocol 10.13: Measurement of $\Delta\Psi_m$ in cells that are to be fixed

EQUIPMENT, MATERIALS AND REAGENTS

Cells.

150 nM MitoTracker orange/red in media.

5 μM FCCP or 10 μM CCCP in media.

4% PFA.

PBS pH 7.2.

FACS or confocal microscope.

METHODS

1. Incubate a negative control sample (20 minutes at 37°C) in media containing FCCP (5 μM) or CCCP (10 μM).

2. Incubate 2.5×10^5 cells (20 minutes at 37°C in the dark) in 500 μl of media containing 150 nM Mitotracker Red. (FCCP or CCCP should also be in the negative control at this time.)

3. Fix the cells in 4% PFA (20 minutes at room temperature with gentle rocking).

4. Wash the cells in PBS pH 7.2.

5. Analyse cells for fluorescence directly by FACS (488 nm excitation is fine and collect emission in Fl-2), by fluorescence microscopy or confocal microscopy (excitation 578 nm, emission 599 nm).

Protocol 10.14: ATP measurements in cells

EQUIPMENT, MATERIALS AND REAGENTS

Cells.

ATP measuring kit (HSII, Roche).

Luminescence reader.

METHOD

1. Lyse 1×10^5 cells in 100 μl of lysis buffer. (The lysate can be stored at −70°C.)

2. Add 100 μl of luciferase agent.

3. After a delay of 1 second, integrate the luminescence over 10 seconds on a luminescence reader.

4. Concentrations can be determined from a log-log plot of an ATP standard curve performed at the same time.

5. If the lysates at step 1 are too concentrated, they can be diluted further in lysis buffer.

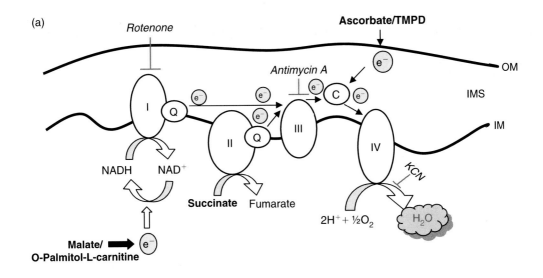

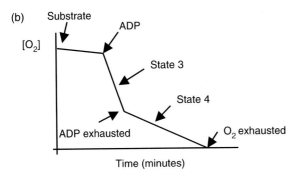

Figure 10.3

(a) Electrons passed from complex I or II to complex III are transferred via cytochrome c to complex IV. At this stage O_2 is consumed and therefore O_2 measurement provides a good readout of the contribution of the various complexes to the electron transport chain. The electron transport chain can be initiated by adding substrates (bold), while addition of inhibitors (italic) can prevent the contribution of the individual complexes. (b) Taking the contribution of complex II, for example, inhibition of complex I with rotenone while adding succinate (substrate) ensures the measurement of complex II without the contribution of complex I. When ADP is added, the electron transport chain is initiated and respiration increases (O_2 is consumed quickly; state 3). This continues until all of the ADP is converted to ATP and state 4 (slow consumption of O_2) respiration occurs. Addition of antimycin removes the contribution of complex II by blocking complex III and basal respiration is measured. To measure the contribution of complex I, malate/O-palmitol-L-carnitine is used as a substrate, and to measure the contribution of cytochrome c, ascorbate/TMPD is used as a substrate and antimycin is used as an inhibitor to prevent the contribution of complex I, II or III.

The efficient production of ATP by respiration requires the conversion of oxygen and hydrogen to water by complex IV (*Figures 10.2* and *10.3*). The consumption of oxygen is therefore a useful method for determining the functional capabilities of mitochondria. This is achieved using an oxygen electrode that can be coupled to a TPP electrode to simultaneously measure $\Delta\Psi_m$ (see Protocol 10.15). For this kind of measurement an Oxigraph apparatus composed of an oxygen electrode coupled to a temperature-regulated sample chamber is required. The oxygen concentration in the

Protocol 10.15: Measurement of oxygen consumption by isolated mitochondria

EQUIPMENT, MATERIALS AND REAGENTS

Freshly isolated mitochondria.

Respiration buffer (140 mM KCl, 10 mM $MgCl_2$, 10 mM MOPS pH 7.4, 5 mM KH_2PO_4, 1 mM EGTA and 0.2% fatty acid free BSA).

Oxygraph apparatus.

2.5 mM Malate/40 μM O-palmityl-L-carnitine.

2 μM rotenone.

5 mM succinate.

150 μM ADP.

1 μM Antimycin A.

1 mM KCN.

Experimental buffer (137 mM KCl, 10 mM Hepes pH 7.2, 2.5 mM $MgCl_2$).

30 μM TPP+.

ETH 1001 electrode.

METHOD

(a) Contribution of complex I

1. Equilibrate 600 μl of respiration buffer (140 mM KCl, 10 mM $MgCl_2$, 10 mM 3-(N-morpholino) propane sulphonic acid (MOPS) pH 7.4, 5 mM KH_2PO_4, 1 mM EGTA and 0.2% fatty acid-free BSA) to 30°C in the chamber of an oxygraph apparatus.

2. Set the recorder to 200 (representative of 100% O_2).

3. Add 300–600 μg of mitochondria in a volume up to 50 μl.

4. Record O_2 for 1 minute.

5. Add complex I substrate (malate (2.5 mM)/O-Palmitol-L-carnitine (40 μM)).

6. Record O_2 for 2 minutes.

7. Add ADP (150 μM) and record O_2 until a good state 3 (actively respiring state:fast)/state 4 (slower rate after all the ADP has been phosphorylated to form ATP) measurement can be made (between 2 to 4 minutes).

8. Add rotenone (2 μM) to inhibit complex I and determine the basal respiration.

9. Record O_2 for a further 2 minutes.

An example trace (cartoon format) of respiring mitochondria is presented in *Figure 10.3*.

(b) Contribution of complex II

1. Equilibrate the chamber as in part (a) step 1.

2. Add 300 μg of mitochondria up to 50 μl.

3. Add rotenone (2 μM) to block complex I, and succinate (5 mM) as a substrate for complex II.

4. Measure O_2 for 3 minutes.

5. Add ADP (150 μM) and record until a good state 3/state 4 measurement can be made (2–4 minutes).

6. Add antimycin A (1 μM) to inhibit complex III and determine the basal respiration.

7. Record O_2 for 2 minutes.

(c) Contribution of cytochrome c

1. Once the chamber is equilibrated (part (a) step 1), add 150 μg of mitochondria up to 50 μl.

2. Add 1 μM Antimycin A to inhibit complex III and then 1 mM ascorbate/0.4 mM N,N,N',N'-tetramethyl-p-phenylenediamine dihydrochloride (TMPD) as a substrate for cytochrome c.

3. Record O_2 for 2 minutes.

4. Add 150 μM of ADP and record until a good state 3/state 4 measurement can be made (2–4 minutes).

5. Add 1 mM of KCN to inhibit complex IV and determine the basal respiration.

6. Record O_2 for 2 minutes.

(d) Coupling O_2 measurement with TPP to measure $\Delta\Psi_m$

1. Equilibrate the experimental buffer (137 mM KCl, 10 mM Hepes pH 7.2, 2.5 mM $MgCl_2$) to 30°C.

2. Add 1 mg of mitochondria to the chamber.

3. Add mitochondrial substrate (e.g. 2.5 mM malate/40 μM O-Palmitol-L-carnitine).

4. Add TPP^+ to a final concentration of 30 μM.

5. Record mitochondrial transmembrane potential using an ETH 1001 electrode.

sealed incubation chamber (containing mitochondria) is continuously monitored, and the effects of adding specific mitochondrial substrates and inhibitors to the chamber can be observed (*Figure 10.3*). The oxygen consumption by complex IV in response to substrates given to complex I, II or cytochrome c can be monitored. This can be used to evaluate the function of these specific complexes. With minor adaptations, oxygen consumption by mitochondria in permeabilised cells can also be measured (see Protocol 10.16).

Reactive oxygen species (ROS) are generated as a by-product of mitochondrial oxidative phosphorylation. While various compounds and dyes can be used to follow the generation of free radical species, one fluorescent probe that we have successfully used for ROS is dihydroethidium (see Protocol 10.17). In its reduced form, it is a colourless dye that competes with the cell's natural ROS scavengers for ROS. Upon oxidation, it turns into ethidium and intercalates with DNA, thereby staining the nuclei red. The degree of staining correlates with the amount of ROS present in the cell. Cells pretreated with ROS scavengers such as N-acetyl cysteine, serve as a control. At high concentrations, these compounds may be toxic and should be titrated for each experimental system.

Protocol 10.16: Measurement of oxygen consumption in permeabilised cells

EQUIPMENT, MATERIALS AND REAGENTS

Cells.

Digitonin buffer (250 mM sucrose, 80 mM KCl 50 µg/ml digitonin) in PBS pH 7.2.

Respiration buffer (250 mM Sucrose, 2 mM EDTA, 30 mM KH_2PO_4, 5 mM $MgCl_2$ and 50 mM Tris pH 7.4).

Equipment from Protocol 10.15.

METHOD

1. Permeabilise cells with digitonin (Protocol 10.7, part (c)).

2. Follow the method described for isolated mitochondria (Protocol 10.15), using 250 mM sucrose, 2 mM EDTA, 30 mM KH_2PO_4, 5 mM $MgCl_2$ and 50 mM Tris pH 7.4 as respiration buffer.

Protocol 10.17: Measurement of reactive oxygen species in intact cells

EQUIPMENT, MATERIALS AND REAGENTS

Cells.

2 µM dihydroethidium in media.

7.5 mM N-acetyl cysteine (as a negative control).

FACS or confocal microscope.

METHOD

1. Incubate 2.5×10^5 cells (20 minutes at 37°C in the dark) with 2 µM dihydroethidium in medium.

2. Analyse cells for fluorescence directly by FACS (488 nm excitation is fine and collect emission in Fl-2), by fluorescence microscopy or confocal microscopy (excitation 518 nm, emission 605 nm).

Cells can be pretreated with ROS scavengers such as N-acetyl cysteine, for controls. At high concentrations, these compounds may be toxic and should be titrated for each experimental system.

10.5 Physical changes in mitochondria

During apoptosis, the outer membrane of mitochondria becomes permeable to small molecules. This permeabilisation can be measured in isolated mitochondria by taking advantage of the fact that, after it is released, cytochrome c can pass back into the mitochondria. Since cytochrome c functions to transfer electrons from complex III to complex IV, both of which are located on the mitochondrial inner membrane, accessibility of purified cytochrome c to the inner membrane can be measured by the rate at which cytochrome c is either reduced by complex III or oxidised by complex IV (see Protocols 10.18 and 10.19). Intact mitochondria will exhibit minimal oxidation or reduction of exogenously added cytochrome c.

The permeability transition (see above) is one way by which mitochondrial permeabilisation is believed to occur during apoptosis and methods to evaluate this event have been reported in many studies. This generally involves detecting a drop in $\Delta\Psi_m$ since PT involves the opening of a pore in the mitochondrial membranes that allows passage of molecules (including protons) less than 1.5 kDa across both the inner and outer mitochondrial membranes. This results in a loss of $\Delta\Psi_m$, gross swelling and outer membrane rupture. Loss of $\Delta\Psi_m$, however, can occur for reasons other than PT, for example in the presence of protonophores or loss of oxidative phosphorylation. Mitochondrial swelling, seen as reduction in absorbance at a wavelength of 540 nm, is therefore often used as a measurement of PT in isolated mitochondria (see Protocol 10.20). This swelling is best observed in hypotonic buffers. Thus, changes in mitochondrial mass can also be monitored with the use of dyes such as MitoTracker Green which accumulates in mitochondria in a $\Delta\Psi_m$-independent manner and Nonyl-acridine orange which binds cardiolipin, which have been used to measure mitochondrial mass (see Protocol 10.21). Alternatively, PT can be detected by observing the change in green fluorescence of mitochondria stained with calcein-acetoxymethyl ester (calcein-AM) (see Protocol 10.22). Upon induction of PT, calcein-AM will exit the mitochondria resulting in a detectable loss of fluorescence. With slight adaptations, calcein-AM fluorescence can also be used to detect PT in cells (see Protocol 10.23).

Protocol 10.18: Measurement of complex III accessibility

EQUIPMENT, MATERIALS AND REAGENTS

Freshly isolated mitochondria.

Assay buffer (125 mM sucrose, 60 mM KCl, 20 mM Tris-HCl, pH 7.4 and 2 mM KCN).

2.5 mM ferricytochrome c.

3.5 mM decyl benzoquinol.

Spectrophotometer.

METHOD

1. Add 25 μl mitochondrial preparation (1 mg ml^{-1}) to 300 μl of 125 mM sucrose, 60 mM KCl, 20 mM Tris-HCl pH 7.4 and 2 mM potassium cyanide (complex IV inhibitor).

2. Add 10 μl of 2.5 mM ferricytochrome c (horse heart).

3. Add 5 μl of 3.5 mM decyl benzoquinol (DBH$_2$) and mix to start the reaction. DBH$_2$ is a complex III substrate.

4. Measure the rate of cytochrome c reduction (increase in A$_{550}$ over 30 seconds) to determine the level of outer membrane permeability.

Protocol 10.19: Measurement of complex IV accessibility

EQUIPMENT, MATERIALS AND REAGENTS

Freshly isolated mitochondria.

Sodium dithionate crystals.

Assay buffer (125 mM sucrose, 60 mM KCl, 20 mM Tris-HCl, pH 7.4, 100 uM cytochrome c [horse heart], 1 μM CCCP, 1 μM rotenone, 1 μM antimycin).

Spectrophotometer.

METHOD

1. Add sodium dithionate crystals to the assay buffer (125 mM sucrose, 60 mM KCl, 20 mM Tris-HCl pH 7.4, 100 μM cytochrome c [horse heart], 1 μM CCCP, 1 μM rotenone (inhibitor of complex I), 1 μM antimycin (inhibitor of complex III)) until the A_{550}/A_{565} is between 6 and 10 (so that the majority of the cytochrome c is reduced).

2. Add 2 μl of mitochondrial preparation (1 mg ml^{-1}) to 450 μl of assay buffer.

3. Measure the rate of cytochrome c oxidation (decrease in A_{550} over 30 seconds) as a measure of outer membrane permeability.

Rates of complex III and IV accessibility can be compared to 100% outer membrane permeability produced by pretreatment of mitochondria with 1% Triton-X-100 (final assay concentration <0.1%).

Protocol 10.20: Measurement of mitochondrial swelling

EQUIPMENT, MATERIALS AND REAGENTS

PT Buffer (125 mM KCl, 2.5 mM potassium phosphate pH 7.4, 2.5 mM succinate and 20 mM Hepes-KOH).

PT inducer, e.g. atractyloside (5 mM) $CaCl_2$ (100 µM) or Ionidamine (50 µM).

40 µM Cyclosporin A.

Spectrophotometer.

METHOD

1. Incubate 300 µl of PT buffer (125 mM KCL, 2.5 mM potassium phosphate pH 7.4, 2.5 mM succinate and 20 mM Hepes-KOH containing the PT-inducer, toxins, inhibitors, Bcl-2 family member etc. being assayed) in individual wells of a microtitre plate until the temperature is stable at 37°C.

2. Add mitochondria (10–50 µl, equivalent of 50–100 µg of isolated mitochondrial protein from the same preparation) to the wells and mix.

3. Measure the absorbance at 540 nm. Swelling is observed as a reduction in A_{540}.

4. Cyclosporin A (40 µM) can be added as a control to block PT, however the inhibition of PT is only transient.

PT-induced swelling can also be measured at room temperature using spectrophotometers without temperature control. In instruments with horizontal light beams, mitochondria may fall out of the light path over long periods of time due to gravity. Addition of cyclosporin A to the assay buffer will act as a control to indicate non-PT-induced swelling (due to buffer composition) or settling of mitochondria during the assay.

Protocol 10.21: Detection of PT in isolated mitochondria using calcein-AM fluorescence

EQUIPMENT, MATERIALS AND REAGENTS

Freshly isolated mitochondria.

Calcein buffer (200 mM mannitol, 50 mM sucrose, 10 mM succinate, 10 mM Hepes-KOH, pH 7.4, 0.1% BSA, 5 mM potassium phosphate pH 7.4, containing 5 μM calcein-AM).

Spectrophotometer.

METHOD

1. Incubate isolated mitochondria (15 minutes at room temperature) in 200 mM mannitol, 50 mM sucrose, 10 mM succinate, 10 mM Hepes-KOH pH 7.4, 0.1% BSA, 5 mM potassium phosphate pH 7.4, containing 5 μM calcein-AM.

2. Wash the mitochondria in calcein-free buffer and treat the cells with a PT inducer.

3. Wash the mitochondria at the end of the assay period and determine the calcein fluorescence in mitochondria using a fluorescence spectrophotometer (excitation 488 nm/emission 520 nm). PT is seen as a loss of fluorescence.

Protocol 10.22: Detection of PT in cells using calcein-AM fluorescence

EQUIPMENT, MATERIALS AND REAGENTS

1 μM Calcein AM in media.

1 mM CoCl$_2$ in media.

Fluorescence microscope.

METHOD

1. Incubate cells (10 minutes at 37°C) in media containing calcein-AM (1 μM).

2. Wash the cells to remove extracellular stain.

3. Incubate the cells (1 h at 37°C) with CoCl$_2$ (1 mM). Co^{2+} chelates the calcein localised in the nucleus and cytosol resulting in a disappearance of fluorescence in these areas. Mitochondria exclude Co^{2+} and the green fluorescence remains.

4. Co^{2+} enters the mitochondria upon induction of PT resulting in a fall in fluorescence. This change in distribution can be followed by fluorescence microscopy (excitation 488 nm/emission 520 nm).

Protocol 10.23: Determination of mitochondrial mass

EQUIPMENT, MATERIALS AND REAGENTS

100 mM nonyl acridine orange

or

150 nM MitoTracker Green.

Fluorescence microscope.

METHOD

1. Incubate 2.5×10^5 cells (20 minutes at 37°C in the dark) in 500 µl of media containing 100 nM nonyl-acridine orange (NAO) or 150 nM MitoTracker green.

2. Analyse cells for fluorescence directly by FACS (488 nm excitation is fine and collect emission in Fl-1). For confocal or fluorescence microscopy, use excitation 490 nm, emission 516 nm for MitoTracker green or excitation 495 nm, emission 519 nm for NAO).

10.6   References

1. Amarante-Mendes, G.P., Finucane, D.M., Martin, S.J., Cotter, T.G., Salvesen, G.S. and Green, D.R. (1998) Anti-apoptotic oncogenes prevent caspase-dependent and independent commitment for cell death. *Cell Death Differ.* **5**: 298–306.

2. Sun, X.M., MacFarlane, M., Zhuang, J., Wolf, B.B., Green, D.R. and Cohen, G.M. (1999) Distinct caspase cascades are initiated in receptor-mediated and chemical-induced apoptosis. *J. Biol. Chem.* **274**: 5053–5060.

3. Heibein, J.A., Barry, M., Motyka, B. and Bleackley, R.C. (1999) Granzyme B-induced loss of mitochondrial inner membrane potential (Delta Psi m) and cytochrome c release are caspase independent. *J. Immunol.* **163**: 4683–4693.

4. Trapani, J.A., Jans, D.A., Jans, P.J., Smyth, M.J., Browne, K.A. and Sutton, V.R. (1998) Efficient nuclear targeting of granzyme B and the nuclear consequences of apoptosis induced by granzyme B and perforin are caspase-dependent, but cell death is caspase-independent. *J. Biol. Chem.* **273**: 27934–27938.

5. Von Ahsen, O., Waterhouse, N.J., Kuwana, T., Newmeyer, D.D. and Green, D.R. (2000) The 'harmless' release of cytochrome c. *Cell Death Differ.* **7**: 1192–1199.

6. Waterhouse, N.J. and Green, D.R. (1999) Mitochondria and apoptosis: HQ or high-security prison? *J. Clin. Immunol.* **19**: 378–387.

7. Newmeyer, D.D., Farschon, D.M. and Reed, J.C. (1994) Cell-free apoptosis in Xenopus egg extracts: inhibition by Bcl-2 and requirement for an organelle fraction enriched in mitochondria. *Cell* **79**: 353–364.

8. Kluck, R.M., Bossy-Wetzel, E., Green, D.R. and Newmeyer, D.D. (1997) The release of cytochrome c from mitochondria: a primary site for Bcl-2 regulation of apoptosis. *Science* **275**: 1132–1136.

9. Yang, J., Liu, X., Bhalla, K., Kim, C.N., Ibrado, A.M., Cai, J., Peng, T.I., Jones, D.P. and Wang, X. (1997) Prevention of apoptosis by Bcl-2: release of cytochrome c from mitochondria blocked. *Science* **275**: 1129–1132.

10. Li, P., Nijhawan, D., Budihardjo, I., Srinivasula, S.M., Ahmad, M., Alnemri, E.S. and Wang, X. (1997) Cytochrome c and dATP-dependent formation of Apaf-1/caspase-9 complex initiates an apoptotic protease cascade. *Cell* **91**: 479–489.

11. Zou, H., Henzel, W.J., Liu, X., Lutschg, A. and Wang, X. (1997) Apaf-1, a human protein homologous to *C. elegans* CED-4, participates in cytochrome c-dependent activation of caspase-3. *Cell* **90**: 405–413.

12. Tsujimoto, Y. (1998) Role of Bcl-2 family proteins in apoptosis: apoptosomes or mitochondria? *Genes Cells* **3**: 697–707.

13. Zou, H., Li, Y., Liu, X., and Wang, X. (1999) An APAF-1 cytochrome c multimeric complex is a functional apoptosome that activates procaspase-9. *J. Biol. Chem.* **274**: 11549–11556.

14. Du, C., Fang, M., Li, Y., Li, L. and Wang, X. (2000) Smac, a mitochondrial protein that promotes cytochrome c-dependent caspase activation by eliminating IAP inhibition. *Cell* **102**: 33–42.

15. Verhagen, A.M., Ekert, P.G., Pakusch, M., Silke, J., Connolly, L.M., Reid, G.E., Moritz, R.L., Simpson, R.J. and Vaux, D.L. (2000) Identification of DIABLO, a mammalian protein that promotes apoptosis by binding to and antagonizing IAP proteins. *Cell* **102**: 43–53.

16. Roberts, D.L., Merrison, W., MacFarlane, M. and Cohen, G.M. (2001) The inhibitor of apoptosis protein-binding domain of Smac is not essential for its proapoptotic activity. *J. Cell Biol.* **153**: 221–228.

17. Hegde, R., Srinivasula, S.M., Zhang, Z., Wassell, R., Mukattash, R., Cilenti, L., DuBois, G., Lazebnik, Y., Zervos, A.S., Fernandes-Alnemri, T. and Alnemri, E.S. (2001) Identification of Omi/HtrA2 as a mitochondrial apoptotic serine protease that disrupts IAP-caspase interaction. *J. Biol. Chem.* **277**: 432–438.

18. Suzuki, Y., Imai, Y., Nakayama, H., Takahashi, K., Takio, K. and Takahashi, R. (2001) A serine protease, HtrA2, is released from the mitochondria and interacts with XIAP, inducing cell death. *Mol. Cell* **8**: 613–621.

19. Verhagen, A.M., Silke, J., Ekert, P.G., Pakusch, M., Kaufmann, H., Connolly, L.M., Day, C.L., Tikoo, A., Burke, R., Wrobel, C., Moritz, R.L., Simpson, R.J. and Vaux, D.L. (2001) HtrA2 promotes cell death through its serine protease activity and its ability to antagonise inhibitor of apoptosis proteins. *J. Biol. Chem.* **277**: 445–454.

20. Susin, S.A., Zamzami, N., Castedo, M., Hirsch, T., Marchetti, P., Macho, A., Daugas, E., Geuskens, M. and Kroemer, G. (1996) Bcl-2 inhibits the mitochondrial release of an apoptogenic protease. *J. Exp. Med.* **184**: 1331–1341.

21. Zamzami, N., Susin, S.A., Marchetti, P., Hirsch, T., Gomez-Monterrey, I., Castedo, M. and Kroemer, G. (1996) Mitochondrial control of nuclear apoptosis. *J. Exp. Med.* **183**: 1533–1544.

22. Susin, S.A., Lorenzo, H.K., Zamzami, N., Marzo, I., Snow, B.E., Brothers, G.M., Mangion, J., Jacotot, E., Costantini, P., Loeffler, M., Larochette, N., Goodlett, D.R., Aebersold, R., Siderovski, D.P., Penninger, J.M. and Kroemer, G. (1999) Molecular characterization of mitochondrial apoptosis-inducing factor. *Nature* **397**: 441–446.

23. Patterson, S.D., Spahr, C.S., Daugas, E., Susin, S.A., Irinopoulou, T., Koehler, C., and Kroemer, G. (2000) Mass spectrometric identification of proteins released from mitochondria undergoing permeability transition. *Cell Death Differ.* **7**: 137–144.

24. Finucane, D.M., Bossy-Wetzel, E., Waterhouse, N.J., Cotter, T.G. and Green, D.R. (1999a) Bax-induced caspase activation and apoptosis via cytochrome c release from mitochondria is inhibitable by Bcl-xL. *J. Biol. Chem.* **274**: 2225–2233.

25. Finucane, D.M., Waterhouse, N.J., Amarante-Mendes, G.P., Cotter, T.G. and Green, D.R. (1999b) Collapse of the inner mitochondrial transmembrane potential is not required for apoptosis of HL60 cells. *Exp. Cell Res.* **251**: 166–174.

26. Waterhouse, N.J., Goldstein, J.C., von Ahsen, O., Schuler, M., Newmeyer, D.D. and Green, D.R. (2001) Cytochrome c maintains mitochondrial transmembrane potential and ATP generation after outer mitochondrial membrane permeabilisation during the apoptotic process. *J. Cell Biol.* **153**: 319–328.

27. Kluck, R.M., Esposti, M.D., Perkins, G., Renken, C., Kuwana, T., Bossy-Wetzel, E., Goldberg, M., Allen, T., Barber, M.J., Green, D.R. and Newmeyer, D.D. (1999) The pro-apoptotic proteins, Bid and Bax, cause a limited permeabilisation of the mitochondrial outer membrane that is enhanced by cytosol. *J. Cell Biol.* **147**: 809–822.

28. Kuwana, T., Smith, J.J., Muzio, M., Dixit, V., Newmeyer, D.D. and Kornbluth, S. (1998) Apoptosis induction by caspase-8 is amplified through the mitochondrial release of cytochrome c. *J. Biol. Chem.* **273**: 16589–16594.

29. von Ahsen, O., Renken, C., Perkins, G., Kluck, R.M., Bossy-Wetzel, E. and Newmeyer, D.D. (2000) Preservation of mitochondrial structure and function after Bid- or Bax-mediated cytochrome c release. *J. Cell Biol.* **150**: 1027–1036.

30. Adachi, S., Gottlieb, R.A. and Babior, B.M. (1998) Lack of release of cytochrome C from mitochondria into cytosol early in the course of Fas-mediated apoptosis of Jurkat cells. *J. Biol. Chem.* **273**: 19892–19894.

31. Gottlieb, R.A. and Adachi, S. (2000) Nitrogen cavitation for cell disruption to obtain mitochondria from cultured cells. *Meth. Enzymol.* **322**: 213–221.

32. Kjeldsen, L., Sengelov, H. and Borregaard, N. (1999). Subcellular fractionation of human neutrophils on Percoll density gradients. *J. Immunol. Methods* **232**: 131–143.

33. Goldstein, J.C., Waterhouse, N.J., Juin, P., Evan, G.I. and Green, D.R (2000b) The coordinate release of cytochrome c during apoptosis is rapid, complete and kinetically invariant. *Nat. Cell Biol.* **2**: 156–162.

34. Vantieghem, A., Assefa, Z., Vandenabeele, P., Declercq, W., Courtois, S., Vandenheede, J.R., Merlevede, W., de Witte, P. and Agostinis, P. (1998) Hypericin-induced photosensitization of HeLa cells leads to apoptosis or necrosis. Involvement of cytochrome c and procaspase-3 activation in the mechanism of apoptosis. *FEBS Lett.* **440**: 19–24.

35. Single, B., Leist, M. and Nicotera, P. (1998) Simultaneous release of adenylate kinase and cytochrome c in cell death. *Cell Death Differ.* **5**: 1001–1003.

36. Goldstein, J.C., Kluck, R.M. and Green, D.R. (2000a) A single cell analysis of apoptosis. Ordering the apoptotic phenotype. *Ann. N. Y. Acad. Sci.* **926**: 132–141.

37. Pinkoski, M.J., Waterhouse, N.J., Heibein, J.A., Wolf, B.B., Kuwana, T., Goldstein, J.C., Newmeyer, D.D., Bleackley, R.C. and Green, D.R. (2001) Granzyme B-mediated apoptosis proceeds predominantly through a Bcl-2-inhibitable mitochondrial pathway. *J. Biol. Chem.* **276**: 12060–12067.

38. Zimmermann, K.C., Waterhouse, N.J., Goldstein, J.C., Schuler, M. and Green, D.R. (2000) Aspirin induces apoptosis through release of cytochrome c from mitochondria. *Neoplasia* **2**: 505–513.

39. Vander Heiden, M.G., Chandel, N.S., Williamson, E.K., Schumacker, P.T. and Thompson, C.B. (1997) Bcl-xL regulates the membrane potential and volume homeostasis of mitochondria. *Cell* **91**: 627–637.

40. Zamzami, N., Marchetti, P., Castedo, M., Decaudin, D., Macho, A., Hirsch, T., Susin, S.A., Petit, P.X., Mignotte, B. and Kroemer, G. (1995a) Sequential reduction of mitochondrial transmembrane potential and generation of reactive oxygen species in early programmed cell death. *J. Exp. Med.* **182**: 367–377.

41. Zamzami, N., Marchetti, P., Castedo, M., Zanin, C., Vayssiere, J.L., Petit, P.X. and Kroemer, G. (1995b) Reduction in mitochondrial potential constitutes an early irreversible step of programmed lymphocyte death in vivo. *J. Exp. Med.* **181**: 1661–1672.

42. Scaduto, R.C., Jr., and Grotyohann, L.W. (1999) Measurement of mitochondrial membrane potential using fluorescent rhodamine derivatives. *Biophys. J.* **76**: 469–477.

43. Mootha, V.K., French, S. and Balaban, R.S. (1996) Neutral carrier-based 'Ca(2+)-selective' microelectrodes for the measurement of tetraphenylphosphonium. *Anal. Biochem.* **236**: 327–330.

Nuclear Changes in Dying Cells

11

Eric H. Baehrecke and Huseyin Mehmet

Contents

11.1 Introduction

The role of apoptosis in the programmed cell death that is essential for normal embryonic development is now well established [1]. Moreover, an expanding number of clinical disorders have been identified in which apoptosis forms a major component [2]. These include neurodegenerative disorders such as Alzheimer's disease

Cell Proliferation and Apoptosis, David Hughes and Huseyin Mehmet (Eds)
© 2003 BIOS Scientific Publishers Ltd, Oxford

and ischaemic stroke, autoimmune diseases and cancer. Apoptosis involves a regulated series of events to efficiently terminate the cellular function and to dismantle the cell while maintaining homeostasis. To accomplish this complicated task, critical organelles and structures are rapidly modified.

The nucleus is logically central to the activation of cell death, as this is the site where the genome is stored, and one of the primary locations where genetic information is regulated. Some of the earliest morphological changes in programmed cell death are seen in the nucleus, where chromatin condenses and moves to the margin such that it appears to associate with the nuclear envelope [3]. Along with chromatin condensation and nuclear shrinkage, specific proteases and nucleases are activated giving rise to a characteristic pattern of DNA cleavage, and protein disassembly, causing an ordered nuclear fragmentation.

Since the biochemical defects in these diseases can also result in death by necrosis, it has become increasingly important to distinguish between apoptotic and necrotic cell death. A number of methods have been developed to specifically detect apoptosis or necrosis. At the level of the cell nucleus, such techniques have employed light or electron microscopy to monitor morphological changes, electrophoretic analysis of nuclear DNA to determine patterns of fragmentation characteristic for each mode of cell death and FACS to detect changes in DNA content and nuclear density.

11.2   Distinguishing between apoptosis and necrosis

Apoptosis is an active process usually requiring the transcription and translation of specific genes as well as the utilisation of intracellular energy sources. In contrast, necrosis is the degeneration of cells that have been killed by insupportable insult from an external source. Differences in the underlying mechanisms of these two types of cell death are reflected in the morphology of dead cells. Thus morphological appearance is the most reliable method of distinguishing between apoptosis and necrosis. Apoptotic cells have reduced cytoplasmic volume, nuclei are shrunken (pyknotic) and cell membranes and organelles remain intact. In contrast, necrotic cells are swollen, their nuclei undergo lysis, the plasma and nuclear membranes are destroyed and organelles disintegrate.

Although methods for distinguishing apoptosis from necrosis are available, none of these have been shown to be completely reliable except for histological analysis. For example, DNA ladders resulting from the activation of an endogenous apoptotic endonuclease are often associated with apoptosis. However, an increasing number of models of apoptosis do not display detectable fragmentation of DNA to small (200 bp) fragments [4] and, furthermore, nucleosome ladders have been reported in cell populations undergoing necrosis [5]. FACS analysis based on nuclear density and DNA content has been used successfully (see Chapter 12) but, in addition to its complexity, FACS may not be able to differentiate accurately the mode of cell death in heterogeneous populations, since the DNA content of healthy cells can vary according to cell type and their position in the cell cycle [6].

While it is not entirely clear how or why the specific morphological and biochemical changes occur in a cell that is programmed to die, it is evident that nuclear characteristics can be used to distinguish programmed cell death from non-physiological cell death that is commonly known as necrosis. Here we describe several methods that have been used to define nuclear changes during programmed cell death.

11.3   Morphological changes in nuclear structure

Apoptosis was first defined as a distinct programme of cell death 30 years ago in a seminal paper by Kerr, Wyllie and Currie [7]. Although cells with the characteristics of apoptosis had been described many years before, Kerr and his colleagues were the first to distinguish apoptotic cell death from necrosis exclusively on the basis of morphological criteria. At the nuclear level, one of the major apoptotic changes is the condensation of chromatin to form distinctive crescents or caps abutted against the nuclear membrane. This is closely associated with the progressive shrinkage of the entire nucleus into a single ball (pyknosis) or, in some cases, nuclear fragmentation (karyorrhexis). These fragments are surrounded by the nuclear envelope that remains largely intact even though it may display an increased permeability to cytosolic molecules [8]. These apoptotic changes in the nucleus are probably due to both DNA fragmentation and proteolysis of key nuclear polypeptides (see below) and can be readily distinguished from necrotic events by simple histological staining methods (Protocols 11.1, 11.2). Moreover, the changes in nuclear structure and permeability also enable the use of specific dyes as rapid assays of the onset of apoptotic cell death.

11.3.1 Nuclear membrane integrity, pyknosis, and fragmentation

The integrity of the nuclear membrane appears to be compromised during programmed cell death, although the exact changes and order of changes during this process are not well understood. Few studies have attempted to take advantage of a possible change in nuclear membrane integrity even though this may provide an important point in the degradation of the nucleus. McCall and Steller [9] utilised the 'leaking' of a nuclear reporter gene to monitor the effect of caspase mutations on nuclear structure. In their experiments, caspase mutants retained β-galactosidase activity in the nucleus of *Drosophila* nurse cells, while controls that possessed caspase activity lost nuclear activity as reflected by β-galactosidase activity in the cytoplasm. While this technique has not been tested in other cells, the principle of this technique could be applied to other systems and is particularly attractive for *in vivo* analyses of programmed cell death. Simpler techniques based on the increased cell and nuclear membrane permeability of apoptotic cells utilise DNA-intercalating dyes such as AO (*Figure 11.1*) or PI that are excluded from healthy cells, in combination with cell-permeable nuclear dyes such as DAPI or Hoechst 33342 (Protocols 11.3, 11.4, 11.5). The mechanism of cell death can be determined by measuring the loss of membrane integrity and microscopic examination of changes in nuclear morphology using combinations of these DNA-binding dyes [10, 12].

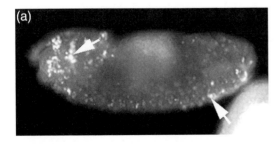
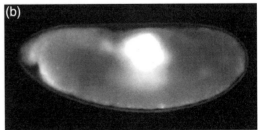

Figure 11.1

Detection of programmed cell death in whole embryo mounts by nuclear staining with AO. (a) Wild-type *Drosophila* embryos possess nuclear AO staining in the head and ventral nerve cord (arrows).
(b) Homozygous mutant embryos that are deficient in programmed cell death exhibit no nuclear AO staining.

Protocol 11.1: Haematoxylin and eosin staining of cell monolayers

EQUIPMENT, MATERIALS AND REAGENTS

Cells grown on 24-well plates.

8% (w/v) paraformaldeyde (PFA) (pH 7.4) made up fresh.

PBS.

Mayer's haematoxylin.

1% eosin.

Circular cover glasses.

Non-aqueous mountant.

Light microscope.

METHOD

1. Add an equal volume of 8% PFA to cell monolayers in culture medium to give a final concentration of 4%.

2. Incubate for 20 minutes at 22°C (room temperature).

3. Wash twice in PBS.

4. Stain in Mayer's haematoxylin for 10 minutes at 22°C (room temperature).

5. Wash 3 times in PBS to remove stain (be sure to remove precipitate).

6. Differentiate in running tap water for approximately 1 minute.

7. Rinse twice in PBS.

8. Stain in 1% eosin for 5 to 30 minutes at 22°C (room temperature) depending on cell type.

9. Rinse briefly (a few seconds) in PBS. Note: If cells are overstained in eosin, the excess can be washed out with more PBS washes.

10. Allow monolayers to air dry and mount with circular cover glasses using non-aqueous mountant.

11. View with light microscope (cell nuclei appear blue and the cytoplasm appears pink).

Protocol 11.2: Histological staining of tissue sections with haematoxylin and eosin

EQUIPMENT, MATERIALS AND REAGENTS

Paraffin or frozen tissue sections.

Wax solvent (e.g. Histo-Clear® or xylene).

100% ethanol.

Deionised water.

Cole's haematoxylin.

Eosin (1% w/v in water).

Non-aqueous mountant.

Light microscope.

METHOD

1. De-wax paraffin tissue sections (5 to 15 μm) by incubating slides in wax solvent for approximately 5–10 minutes at 22°C (room temperature) until the wax has dissolved; this can be checked by eye.

2. Rehydrate by immersing slides in 100% ethanol, then into a second preparation of 100% ethanol, then into a preparation of 70% ethanol and finally in running tap water. Note: Frozen sections can be fixed in acetic acid/methanol (3%/95%) for 2 minutes at 22°C (room temperature) then rinsed in running tap water.

3. Stain using Cole's haematoxylin for 5 to 15 minutes (depending on thickness of sections) at 22°C (room temperature).

4. Wash slides in running tap water.

5. Differentiate colour in acetic acid/ethanol (1%/95%) for a few seconds.

6. Put slides in running tap water until dye is 'blue'.

7. Stain with eosin for 5 to 15 minutes (depending on thickness of sections) at 22°C (room temperature).

8. Wash slides in running tap water.

9. Air dry completely and mount with a cover glass containing a drop of non-aqueous mountant.

10. View with a light microscope.

Notes:
(i) Apoptotic cells typically exhibit intense, uniform nuclear basophilia accompanied by chromatin condensation and nuclear pyknosis or fragmentation.
(ii) Necrotic cells are identified by intense cytoplasmic eosinophilia with reduced plasma and nuclear membrane integrity or by the dispersion of chromatin as a fine web into the cytoplasm (usually associated with increased cell volume).

Protocol 11.3: PI staining of nuclei in cell monolayers

EQUIPMENT, MATERIALS AND REAGENTS

Cells grown on 24-well plate.

Methanol:acetic acid (3:1 ratio).

PBS.

Propidium iodide (PI) stock (2 mg ml^{-1} in PBS).

RNAse stock (10 mg ml^{-1} in PBS).

Aqueous fluorescence mountant (e.g. 'Citifluor').

Circular glass cover slips.

Fluorescence microscope.

METHOD

1. Fix cells by adding one volume (e.g. 500 μl) of methanol:acetic acid (3:1).

2. Incubate 3 minutes at room temperature (19–22°C).

3. Aspirate and replace with one volume of methanol:acetic acid (3:1).

4. Incubate a further 3 minutes at room temperature.

5. Aspirate and wash twice with PBS, leaving in second wash.

6. Prepare PI by diluting 1:500 in PBS.

7. Prepare RNAse by diluting 1:100 in PI/PBS.

8. Add PI/RNAse/PBS solution to cover monolayer.

9. Incubate at 37°C for 30 minutes.

10. Aspirate and wash twice with PBS.

11. Gently place coverslip into each well, using a drop of mountant.

12. View with a fluorescence microscope (under a rhodamine/Texas red filter).

Note: Some mountants come with nuclear dyes added to facilitate visualisation of nuclear morphology (e.g. VECTASHIELD® Mounting Medium with DAPI from Vector Laboratories).

Protocol 11.4: Hoechst 33342/PI staining of cells in culture

EQUIPMENT, MATERIALS AND REAGENTS

Hoechst/PI mixture (Hoechst 33342 (0.4 mg ml^{-1}) and PI (0.4 mg ml^{-1}) in PBS.

PBS.

Aqueous mountant.

Fluorescence microscope.

METHOD

1. Add Hoechst/PI mixture to culture wells (10% of culture medium volume).

2. Incubate 15 minutes at 37°C.

3. Wash cells gently with PBS.

4. Mount with circular cover glasses and non-aqueous mountant.

5. View under a fluorescence microscope.

Note: Hoechst 33342 is cell permeable while PI only enters the nuclei if the plasma membrane is breached. This stain can therefore differentiate between necrotic and (early) apoptotic cells. This assay works particularly well in conjunction with flow cytometry.

Protocol 11.5: AO staining to assay apoptosis in *Drosophila* embryos

EQUIPMENT, MATERIALS AND REAGENTS

Drosophila embryos or larval tissues.

50% (v/v) bleach.

Heptane.

Acridine orange (AO) (5 μg ml^{-1} in PBS).

Series 700 Halocarbon oil (Halocarbon Products Corp., Hackensack, NJ).

Fluorescence microscope.

METHOD

1. Collect embryos or larval tissues following standard protocol.

2. Remove embryo chorions with 50% bleach.

3. Rinse with water.

4. Place in an equal volume of heptane and AO.

5. Incubate for 5 minutes at room temperature while shaking.

6. Carefully remove embryos from interface.

7. Transfer embryos to series 700 Halocarbon oil.

8. Visualise cell death by fluorescence microscopy (see *Figure 11.1*).

11.4   Proteolysis of nuclear proteins

Several of the morphological and biochemical changes that occur as cells initiate programmed cell death have been linked to the activity of caspases that cleave protein substrates associated with the nucleus. Thus, defined molecular weight changes in nuclear proteins can serve as a valuable approach to detect and understand the regulation of programmed cell death.

11.4.1   Nuclear matrix proteins

In addition to cleavage of cytoskeletal proteins such as vimentin and fodrin, a large number of nuclear matrix proteins (NMPs) have been shown to be caspase substrates [11, 15]. Specifically, the nuclear lamins A/C and B, NuMA, hnRNP proteins C1 and C2, the 70 kDa component of U1 small ribonucleoprotein, the scaffold-attachment factor A and the special AT-rich sequence-binding protein 1 have all been identified as caspase targets. Indeed, proteomic analysis has revealed that changes in more than 65 different NMPs could be detected following induction of apoptosis by distinct triggers [14]. The localised expression of NMPs and loss of their expression due to caspase cleavage provides a rapid method to assess programmed cell death and caspase activity in cells, tissues or intact animals [15]. Antibodies against NMPs exist for a variety of model systems and can be used to detect changes in expression both *in situ* and in extracts of cells or tissues. For *in situ* detection of changes in NMP expression, cells, tissues or animals must be fixed and permeabilised such that the antibodies have appropriate access to antigenic epitopes. Ideally, caspase cleavage of such substrates should also be monitored by western blotting; however, the synchrony of a dying cell population may limit this type of analysis to cultured cells or highly synchronised cell deaths. While these approaches have not been widely used, they offer great potential since both cell death and the biochemical activity of caspases can be monitored.

For detection of specific substrate cleavage in cell extracts standard methods of protein extraction, SDS-PAGE, western blotting, and detection can be used. Antibodies against nuclear intermediate filament proteins (e.g. lamins) have been produced by several investigators and commercial companies. The technique selected for visualisation of expression depends on levels of expression in the desired cell type. Secondary antibodies that are conjugated to either a specific fluorochrome or horseradish peroxidase (HRP) are commonly used to visualise protein expression. HRP has the advantage of enabling the use of light microscopy to observe cell morphology, while fluorescent antibodies combined with fluorescent DNA dyes also provide an appealing way to visualise changes in both protein cleavage and changes in nuclear morphology.

11.4.2   Poly (ADP ribose) polymerase cleavage

The nuclear repair enzyme, poly (ADP ribose) polymerase (PARP) catalyses the attachment of ADP ribose units from NAD to nuclear proteins. However, excessive activation of PARP can deplete NAD and ATP (which is consumed in NAD regeneration) leading to an increase in the NADH/NAD ratio and eventually cell death by energy depletion [16]. Among the most common (although not ubiquitous) of apoptotic events is the caspase-3-mediated cleavage of PARP [17] which inactivates the enzyme, presumably conserving cellular energy. Moreover, genetic disruption of the PARP gene can provide protection against apoptosis [18]. Western blotting analysis of PARP cleavage can therefore give an indirect measure of both caspase activation and nuclear disassembly in model cell systems [19] and in tissues undergoing apoptosis (Protocol 11.6).

Protocol 11.6: Western blotting analysis of PARP cleavage

EQUIPMENT, MATERIALS AND REAGENTS

1% SDS (in deionised water).

Bicinchoninic acid (BCA) protein quantitation kit (Pierce Biotechnology, USA).

21G needle.

8% polyacrylamide gel (prepared using a standard protocol).

Electrophoresis loading buffer: 62.5 mM Tris-Hcl pH 6.8, 1% sodium dodecyl sulphate, 20% glycerol, 0.005% bromophenol blue and 10% 2-mercaptoethanol.

Electrophoresis running buffer: 25 mM Tris, 192 mM glycine, 0.1% SDS, pH 8.3.

Electrophoresis tank.

Electrical power pack.

Electro-transfer apparatus.

Blocking buffer (10% (w/v) non-fat milk powder in PBS/Tween (0.1%)).

Transfer buffer: 25 mM Tris, 192 mM glycine, 20% methanol, pH 8.3.

PVDF membrane.

Ponceau S (a 0.1% solution made up in 5% acetic acid).

Mouse monoclonal anti-PARP antibody C-2-10 (Transduction Laboratories, USA).

Secondary anti-mouse antibody linked to HRP.

Enhanced chemiluminescence reagent.

Autoradiographic film.

METHOD

1. Lyse cells in 1% SDS heated to 95°C for 5 minutes.

2. For fresh tissue, snap freeze in liquid nitrogen, homogenise in 1% SDS by trituration through a 21G needle, then heat to 90°C for 5 minutes.

3. Clarify lysates by centrifugation (10 000 g for 5 minutes).

4. Determine protein content by the BCA method (this is relatively insensitive to SDS).

5. Load equal amounts of protein in each lane on an 8% polyacrylamide gel by electrophoresis.

6. Electro-transfer gel onto PVDF membrane.

7. Check transfer with Ponceau S.

8. Incubate membrane with blocking buffer for 2 hours at 22°C (room temperature).

9. Wash three times in PBS/Tween.

10. Incubate membrane with primary antibody diluted 1:3000 in blocking buffer for 60 minutes at 22°C.

11. Wash three times in PBS/Tween.

12. Incubate membrane with secondary antibody diluted 1:1000 in blocking buffer for 60 minutes at 22°C.

13. Visualise HRP with enhanced chemiluminescence according to manufacturer's instructions.

11.5   DNA fragmentation in apoptosis

The discovery that programmed cell death is accompanied by cleavage of nuclear DNA at nucleosome linkers provided a biochemical assay to measure apoptotic cell death. One approach is to follow the translocation of active nucleases such as AIF and endonuclease G (EndoG) from mitochondria to the nucleus, either by western blotting or immunocytochemical analysis (Protocol 11.7). Another is to analyse DNA fragmentation in apoptotic nuclei and for this purpose two general approaches have been used: electrophoresis (Protocols 11.8, 11.9, 11.10) and *in situ* end-labelling (Protocols 11.11, 11.12, 11.13). Electrophoresis of fragmented DNA provides clear evidence of 'DNA ladders' [4, 20]. However, this approach is limited to studies of cultured cells or cases where large numbers of cells are dying in an organism, such as when an entire tissue is destroyed.

The comet assay [21] overcomes some of the liabilities associated with detecting fragmented DNA by gel electrophoresis (Protocol 11.14), as it can be used to detect altered DNA structure in single cells and can distinguish apoptotic from necrotic nuclei (*Figure 11.2*). End-labelling of fragmented DNA *in situ* has the distinct advantage that it can be used to study programmed cell death in an intact organism in conjunction with analyses of cell morphology. This approach can be used to study isolated cell deaths that occur during normal development, or when homeostasis is perturbed by infection or trauma. Two enzymes have been used to label the fragmented ends of DNA.

Both terminal deoxynucleotidyl transferase (TdT) and DNA polymerase I have been used to attach labelled nucleotides to the ends of DNA (*Figure 11.3*). TdT specifically binds to 3′ protruding ends of double-stranded DNA, and then attaches nucleotides which contain the specific modification desired for visualisation [22]. In contrast, DNA polymerase I recognises 3′ recessed ends of DNA, and then attaches nucleotides which contain the specific modification desired for visualisation [23]. The disadvantage of DNA polymerase I is that 3′ recessed ends commonly occur as nicks in DNA of viable cells, and may lead to non-specific labelling of viable cells. It should be noted that *in situ* end-labelling of DNA does not enable one to distinguish high molecular weight DNA from nucleosome ladders, and it has also been suggested that *in situ* end-labelling cannot reliably distinguish DNA fragments arising from apoptosis and the random DNA degradation that accompanies necrosis [5, 24].

11.5.1   Relocalisation of endonucleases

Apoptotic DNA fragmentation was discovered before the identification of specific endonucleases, although it was only a matter of time before the link between caspases and DNA laddering was established as a specialised enzyme called caspase-activated DNAse (CAD) [25]. This protein is typically bound to the 45 kDa subunit of DNA fragmentation factor (DFF45), also known as inhibitor of caspase-activated DNAse (ICAD), and activation of CAD may occur as a result of caspase-3-mediated cleavage of ICAD into 30 kDa and 11 kDa fragments. Cleaved ICAD is no longer able to bind CAD which is free to translocate from the cytosol to the nucleus [25, 26]. However, other studies suggest that that the ICAD/CAD complex is already localised in the nucleus and it is caspase-3 that translocates upon activation. Once in the nucleus, active caspase-3 can cleave ICAD, thus releasing active CAD for DNA disassembly [27, 28]. Since it is not entirely clear which of these possibilities is correct, it is probably imprudent to rely on translocation of CAD as a measure of nuclear apoptosis.

Protocol 11.7: Localisation of EndoG or AIF in cell nuclei

EQUIPMENT, MATERIALS AND REAGENTS

Cell monolayers grown on 8-well chamber slides.

PBS.

PFA (4% w/v) + picric acid (0.19% v/v of a saturated solution) in PBS.

0.1% (v/v) Triton X-100 in PBS.

Blocking buffer: 2% BSA in PBS.

Rabbit polyclonal antiserum against EndoG.

Fluorescein-conjugated goat anti-rabbit antibody.

Cover glass.

Fluorescence mountant.

Fluorescence microscope.

METHOD

1. Plate cells at approximately 10^4 cells per chamber slide.

2. At the end of the (apoptosis) treatment, wash cell monolayers three times in PBS.

3. Fix in freshly prepared PFA for 10 to 30 minutes at room temperature (19–22°C). Note: The time required will depend on the cell type.

4. Wash monolayers three times in PBS for 15 minutes each wash.

5. Permeabilise by incubating in 0.1% Triton X-100 for 5 to 15 minutes at room temperature.
   Notes:
    (i) The time required will depend on the cell type.
   (ii) In some cell types, better results may be achieved by permeabilising with SDS (0.1% v/v in PBS) for 5 to 10 minutes.

6. Wash monolayers three times in PBS for 15 minutes each wash.

7. Block for 60 minutes at room temperature (19–22°C) in blocking buffer.

8. Incubate for 4 hours at room temperature with antiserum against EndoG (diluted 1:200).

9. Wash three times for 10 minutes each in blocking buffer.

10. Incubate for 1 hour with fluorescein-conjugated goat anti-rabbit antibody (diluted 1:500).

11. Wash monolayers three times in PBS for 10 minutes each wash.

12. Gently place coverslip into each well, using a drop of mountant.

13. Examine using a fluorescence microscope.

Note: The same procedure can be followed for AIF, using appropriate primary and secondary antibodies.

Protocol 11.8: Detection of nucleosome ladders by agarose gel electrophoresis

EQUIPMENT, MATERIALS AND REAGENTS

Cells or tissues to be analysed.

PBS.

Mortar and pestle (for tissues only).

Lysis buffer (20 mM Tris (pH 8.0), 20 mM EDTA (pH 8.0), 200 mM NaCl, 1% SDS).

Pancreatic RNase (500 μg ml^{-1}).

Proteinase K (5 mg ml^{-1}; pretreat by incubating at 37°C for 30 minutes prior to use).

Tris borate EDTA (TBE) buffer (1X = 89 mM Tris borate, pH 8.3, 25 mM EDTA).

Gel loading buffer (10X = 30% Ficoll, 0.25% bromophenol blue, 0.25% xylene cyanole, in 10X TBE).

Low melting temperature agarose.

Ethidium bromide solution (1 μg ml^{-1} in 1X TBE).

Horizontal submarine electrophoresis tank.

Electrical power pack.

UV transilluminator.

METHOD

1. Rinse cells/tissues in PBS. For the isolation of DNA from tissues (or entire organisms, such as *Drosophila* embryos), they should first be frozen in liquid nitrogen and homogenised with a mortar and pestle. Note: DNA extraction is much more efficient if small numbers of cells ($<1 \times 10^6$) or small amounts of tissue are used.

2. Add 50 μl of lysis buffer to the cells or homogenised tissue in a 1.5 ml polypropylene tube, and mix with a pipette tip. Note: do not vortex or this will shear DNA.

3. Add 2 μl of pancreatic RNase, and incubate at 37°C for 1 hour.

4. Meanwhile, prepare a 2% (w/v) agarose gel for electrophoresis in 1X TBE.

5. Add 2.5 μl of pretreated Proteinase K to cells/tissue and incubate at 50°C for 1 hour.

6. Add 10 μl loading buffer (10X) to sample and load half of sample volume into agarose gel for electrophoresis.

7. Electrophorese agarose gel at 50 V until the bromophenol blue has run approximately 75% into the gel.

8. Incubate gel for 30 minutes at 22°C in ethidium bromide solution.

9. Rinse in TBE. Visualise DNA with UV transilluminator (*Figure 11.5*).

Protocol 11.9: Detection of high molecular weight DNA fragments by gel electrophoresis

EQUIPMENT, MATERIALS AND REAGENTS

Cells or tissues to be analysed.

Ice-cold PBS.

Mortar and pestle (for tissues only).

Lysis buffer (10 mM Tris pH 7.6, 100 mM EDTA pH 8.0, 20 mM NaCl).

Clean glass microscope slides.

Scalpel blade.

Proteinase K (5 mg ml^{-1} stock).

Sarkosyl.

Tris-EDTA (TE) buffer (1X = 10 mM Tris-HCl (pH 7.6), 1 mM EDTA).

TBE buffer (1X = 89 mM Tris-borate, pH 8.3, 25 mM EDTA).

Low melting point agarose.

Ethidium bromide solution (1 μg ml^{-1} in 1X TBE).

Horizontal submarine electrophoresis tank.

Electrical power pack.

UV transilluminator.

METHOD

1. Rinse cells or homogenised tissues (frozen in liquid nitrogen and homogenised with a mortar and pestle) in ice-cold PBS.

2. Resuspend cells (5 × 10^7) or tissues in 1 ml ice-cold lysis buffer.

3. Prepare an equal volume of 1% low melting temperature agarose in lysis buffer, and cool it to 42°C.

4. Warm the cell suspension to 42°C and mix well with agarose such that the cells are evenly distributed and pour the mixture onto microscope slides.

5. Allow agarose to solidify and cut the agarose gel containing cells/tissues into 5–10 mm square plugs using a sterile scalpel blade (each plug should contain 5–20 μg of DNA).

6. Place the agarose plugs in lysis buffer containing 1 mg ml^{-1} proteinase K and 1% Sarkosyl and incubate at 50°C for 24 hours. Note: to isolate DNA

for detection of high molecular weight fragments (50 kilobases or greater) caution must be taken not to shear the DNA. For this reason, cells/tissues are lysed *in situ* in an agarose plug.

7. Replace the lysis buffer containing 1 mg ml^{-1} proteinase K and 1% Sarkosyl and incubate at 50°C for an additional 24 hours.

8. Incubate each plug in 50–100 volumes of TE (pH 7.6) containing 40 μg ml^{-1} PMSF at 50°C for 1 hour.

9. Replace the TE buffer, and incubate at 50°C for an additional hour. Note: the plugs can be stored at 4°C for days in TE buffer, or for longer periods of time in 0.5 M EDTA (pH 8.0).

10. Perform pulse field gel electrophoresis or variants of this method (e.g. field inversion gel electrophoresis) to separate high molecular weight DNA fragments in accordance with manufacturer's instructions. The method below is only given as an example.

    (a) Place each agarose-cell plug carefully onto a tooth of the gel comb to be used when pouring the agarose gel (reserve one tooth for DNA size markers).

    (b) Place the comb in the levelled tray used for the gel, and carefully surround the plugs with 1% agarose in 0.5X TBE to hold the plugs in place.

    (c) Once the agar holding the plugs is solidified, pour a 1% agarose gel in 0.5X TBE (being careful not to disturb the plugs).

    (d) Cool the gel, carefully remove the comb, add 1X TBE running buffer, and run a pulse field gel as suggested by the manufacturer for the resolution of DNA fragments of 50–300 kilobases.

    (e) Incubate gel for 30 minutes at 22°C in ethidium bromide solution.

    (f) Rinse gel in TBE and visualise DNA with UV transilluminator (*Figure 11.4*).

Protocol 11.10: Detection of low molecular weight DNA laddering in cultured cells

EQUIPMENT, MATERIALS AND REAGENTS

Cells.

Tris-HCl (100 mM, pH 8.5).

Lysis buffer (100 mM Tris-HCl pH 8.5, 5 mM EDTA, 0.2% SDS and 200 mM NaCl) containing 100 μg ml^{-1} proteinase K.

One volume isopropranol.

Centrifuge.

TE buffer: 1X = 10 mM Tris-HCl (pH 7.6), 1 mM EDTA.

Spectrophotometer.

Electrophoresis apparatus.

1% agarose gel.

TBE: 100 mM Tris borate (1X = 89 mM Tris-borate, pH 8.3, 25 mM EDTA).

Ethidium bromide (20 μg ml^{-1} in TBE).

METHOD

1. Wash cells in Tris-HCl and resuspend in minimal volume of lysis buffer.

2. Incubate 16 hours at 55°C and precipitate DNA by adding one volume of isopropranol with agitation for 15 minutes at 22°C (room temperature).

3. Following centrifugation at 10 000 g for 5 minutes, air dry pellet and resuspend in TE buffer.

4. Determine DNA content by spectrophotometry.

5. Electrophorese DNA (5 μg per lane) in a 1% agarose gel in TBE (70 V for 4 hours).

6. Visualise by ethidium bromide staining.

Protocol 11.11: TUNEL analysis of fragmented DNA in whole mount *Drosophila* embryos

EQUIPMENT, MATERIALS AND REAGENTS

Embryonic tissue to be analysed.

4% PFA in PBS (pH 7.4).

Proteinase K (10 μg ml^{-1} in PBS).

Citrate buffer (100 mM sodium citrate, 0.1% Triton X-100).

TdT.

TdT buffer (30 mM Tris (pH 7.2), 140 mM sodium cacodylate (pH 7.2), 1 mM cobalt chloride, 0.1 mM DTT).

Digoxigenin-labelled nucleotide (e.g. dUTP).

1% (w/v) BSA in PBS.

FITC-labelled secondary antibody.

Cover glasses.

Mounting medium.

Fluorescence microscope.

METHOD

1. Fix freshly isolated and dechorinated embryos with 4% PFA in PBS (pH 7.4) for 30 minutes at 22°C.

2. Rinse and incubate specimen in proteinase K (10 μg ml^{-1} in PBS) at 50°C for as long as possible without disrupting cell morphology. Note: the conditions for fixation and permeabilisation are critical, as they dictate penetration of the TdT enzyme. *Drosophila* embryos are typically incubated for 5 minutes in Proteinase K, although this must be determined empirically for each tissue/batch of enzyme.

3. Rinse specimen in PBS and post-fix in 4% PFA for a further 20 minutes at 22°C.

4. After several rinses in PBS, incubate specimen in citrate buffer at 65°C for 30 minutes.

5. Rinse with at least 3 changes of PBS, and incubate spcimen in TdT buffer for 1 hour.

6. Add the labelled nucleotide of choice (at the concentration recommended by the manufacturer) and TdT at 0.3 units μl^{-1} of the final reaction solution volume.

7. Incubate the reaction at 37°C for 3–16 hours then wash the specimen 3 times in PBS.

8. Following the incorporation of digoxigenin-labelled dUTP, block with 1% BSA for 1 hour at 22°C.

9. Incubate the specimen with appropriately diluted digoxigenin antibody for either 2 hours at room temperature or at 4°C overnight.

10. Wash the specimens 4 times in PBS (20 minutes each wash) and incubate with FITC-labelled secondary antibody.

11. Wash the specimens 4 times in PBS (20 minutes each wash) and mount slide using a cover glass containing a drop of mounting medium.

12. Visualise using a fluorescence microscope.

Protocol 11.12: *In situ* end-labelling of fragmented DNA in cell monolayers

EQUIPMENT, MATERIALS AND REAGENTS

8% (v/v) PFA in PBS.

Cell monolayers.

Proteinase K (100 μg ml^{-1} in PBS).

ISEL buffer (50 mM Tris-HCl buffer (pH 7.5), 5 mM MgCl$_2$, 10 mM 2-mercaptoethanol, 0.005% (w/v) BSA).

End-labelling mix: 10 mM each of dGTP, dTTP and dATP and 4 mM biotin-14-dATP in ISEL buffer, containing Klenow fragment of DNA polymerase I (5 units ml^{-1}).

Deionised water.

PBS.

Hydrogen peroxide (0.3% v/v in PBS).

Horseradish peroxidase (HRP) detection kit (e.g. from Vector Laboratories).

Non-aqueous mountant (e.g. DPX).

Light microscope.

METHOD

1. Add one volume of 8% PFA to cell monolayers.

2. Wash twice in PBS.

3. Digest with proteinase K at 22°C (room temperature) for at least 15 minutes (needs to be empirically determined for each cell line) but for as long as possible, without disrupting cell morphology.

4. Wash twice in PBS and add enough end-labelling mix to cover monolayer of cells.

5. Incubate for 2 hours at 37°C and then terminate the reaction by sequential washes in deionised water ($\times$1) and PBS ($\times$2).

6. Add enough hydrogen peroxide to cover monolayer and incubate for 15 minutes at 22°C.

7. After four washes in PBS detect incorporated biotinylated nucleotide using a commercial kit based on streptavidin-HRP (see above).

8. Wash four times in PBS after developing colour and mount slide using a cover glass containing a drop of mounting medium.

9. Visualise using a light microscope.

Note: for negative controls, Klenow fragment of DNA polymerase I can be omitted from the end-labelling mixture.

Protocol 11.13: *In situ* end-labelling of tissue sections

EQUIPMENT, MATERIALS AND REAGENTS

Wax solvent (e.g. Histo-Clear® or xylene).

Proteinase K (20 μg ml^{-1} in PBS).

Deionised water.

Ethanol: 100% and 70% (v/v) in deionised water.

ISEL Buffer (50 mM Tris-HCl buffer (pH 7.5), 5 mM MgCl$_2$, 10 mM 2-mercaptoethanol, 0.005% (w/v) BSA).

End-labelling mix: 10 mM each of dGTP, dTTP and dCTP and 4 mM biotin-14-dATP in ISEL buffer, containing Klenow fragment of DNA polymerase I (5 units ml^{-1}).

PBS.

Hydrogen peroxide (0.3% v/v in PBS).

HRP detection kit (e.g. from Vector Laboratories).

Non-aqueous mountant (e.g. DPX).

Light microscope.

METHODS

1. De-wax tissue sections by incubating slides in wax solvent for approximately 5–10 minutes at 22°C (room temperature) until the wax has dissolved; this can be checked by eye.

2. Drain off excess wax solvent and immerse slides in 100% ethanol, then into a second preparation of 100% ethanol, then into a preparation of 70% ethanol and finally in running tap water to remove all traces of alcohol.

3. Digest with proteinase K (20 μg ml^{-1}) at 22°C (room temperature) for 5–15 minutes (needs to be empirically determined for each tissue) but for as long as possible, without disrupting cell morphology.

4. After washing, incubate slides in 0.3% (v/v) hydrogen peroxide for 15 minutes at room temperature. Wash by immersing slides in PBS (four times) then add end-labelling mix.

5. Incubate for 60 minutes at 37°C and wash by immersing slides in PBS (four times).

6. After further washing, detect incorporated biotinylated nucleotide using a commercial kit based on streptavidin-HRP (see above).

7. Wash by immersing slides in PBS (four times) after developing colour.

8. Counterstain sections with haematoxylin and eosin (Protocol 11.2).

9. Mount slide using a cover glass containing a drop of mounting medium and visualise using a light microscope.

Note: for negative controls, Klenow Fragment of DNA polymerase I can be omitted from the end-labelling mixture.

Protocol 11.14: Comet assay for detection of apoptotic and necrotic nuclei

EQUIPMENT, MATERIALS AND REAGENTS

Single cell suspensions or tissue samples to be analysed.

Low melting temperature agarose.

Electrophoresis film.

Comet lysing solution (2.5 M NaCl, 100 mM EDTA, 10 mM Tris, 1% SLS, 10% DMSO, 1% TritonX-100).

Alkaline buffer (300 mM NaOH, 1 mM EDTA, pH 13).

Horizontal gel electrophoresis apparatus.

Tris-HCl buffer (0.4 M pH 7.4).

Ethidium bromide (20 μg ml^{-1}).

Fluorescence microscope.

Analysis software (e.g. 'Comet 2.2' from Kinetic Imaging Ltd., Liverpool).

METHOD

1. Prepare single cell suspensions using appropriate methods (e.g. enzyme digestion).

2. Mix cells with low melting temperature agarose and load onto a strip electrophoresis film.

3. Add a second layer of agarose and immerse the gel in comet lysing solution and incubate for 1 hour at 4°C.

4. Transfer the gels to alkaline buffer and incubate for 20 minutes before electrophoresing in this buffer for a further 20 minutes.

5. Wash gels three times in Tris-HCl buffer.

6. Stain gels with 20 μg ml^{-1} ethidium bromide and view with a fluorescence microscope (*Figure 11.2*).

7. Quantitate apoptosis and necrosis using appropriate analysis software.

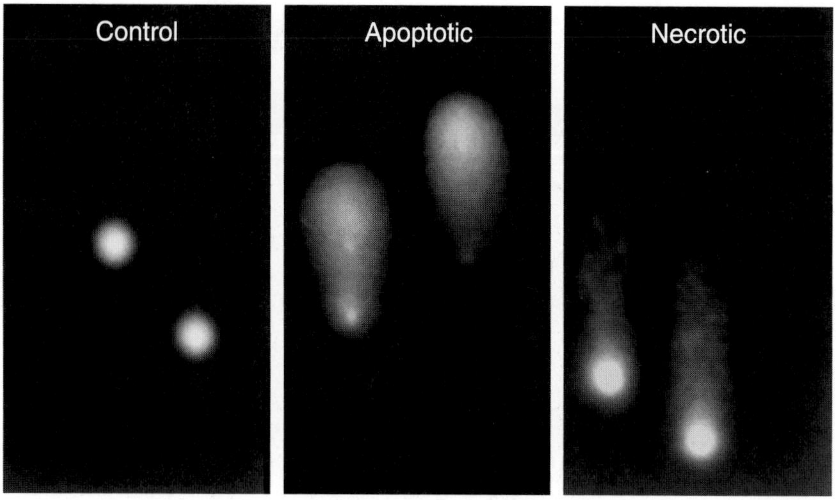

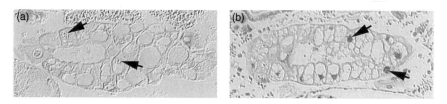

Figure 11.2

Microgel electrophoresis of healthy, apoptotic and necrotic Swiss 3T3 fibroblast nuclei. Quiescent cells were treated for 15 minutes with 10% conditioned medium at 37°C (healthy), or at 50°C (necrotic) or for 16 hours with 10% conditioned medium (at 37°C) containing 100 nM staurosporine (apoptotic). At the end of the experiment, nuclei were prepared and analysed by the comet assay.

Figure 11.3

Detection of fragmented DNA during programmed cell death by the TUNEL procedure. (a) Paraffin section of a wild-type *Drosophila* salivary gland 6 hours after puparium formation and before the onset of cell death (arrows mark unstained nuclei). (b) Paraffin section of a wild-type *Drosophila* salivary gland 12 hours after puparium formation and following the onset of cell death (arrows mark TUNEL-positive nuclei).

The subcellular relocalisation of two other nucleases, AIF and EndoG, has clearly been shown to change following apoptotic commitment and is well established as an indicator of imminent chromatin cleavage. In the case of the endonuclease AIF, this 50 kDa mitochondrial protein is released upon disruption of the membrane permeability transition and translocates to the nucleus. Although it was first mistakenly identified as a protease [29] it is now clear that AIF is an endonuclease, capable of giving rise to high molecular weight DNA fragments, an event that is independent of caspase activity [30]. Rather intriguingly, AIF has also been reported to exhibit NADH oxidase activity, which can be dissociated from its apoptosis-inducing function [31] and it remains to be determined which of these is the most prominent function of AIF.

The existence of a low molecular weight laddering endonuclease was hinted at in transgenic mice lacking a functional CAD gene; although these animals were phenotypically normal, they still exhibited residual DNA fragmentation. This was

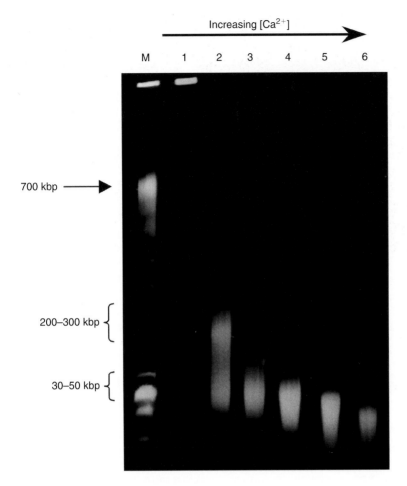

Figure 11.4

Field inversion gel electrophoresis (FIGE) of rat liver nuclei after activation of endogenous endonucleases using increasing concentrations of calcium (10 to 200 μM) in the presence of 4 mM magnesium. Note that the molecular weight markers (M) refer to kilobase pairs (kbp). We are grateful to Dr K. Cain and Prof G.M. Cohen of the MRC Toxicology Unit, Leicester, UK for supplying this image.

due to the activity of EndoG, isolated as a mitochondrion-specific nuclease that is released during apoptosis. Once released from mitochondria, EndoG translocates to the nucleus where it cleaves DNA into nucleosomal fragments through a caspase-independent apoptotic pathway [32]. Protocols to measure the translocation of AIF and EndoG are presented below.

11.5.2 Detection of DNA fragmentation by agarose gel electrophoresis

Whether they reside in the nucleus or are translocated there following an apoptotic stimulus, activated nucleases degrade DNA into fragments of 50 to 300 kilobases (*Figure 11.4*), and subsequently into smaller pieces that are approximately 200 base pairs in length, giving rise to nucleosome ladders (*Figure 11.5*). DNA can be extracted from dying cells, separated by agarose gel electrophoresis, and visualised to detect

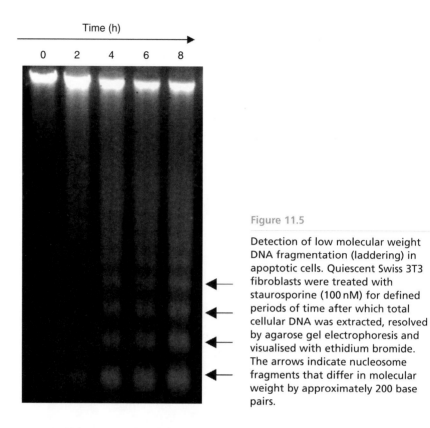

Figure 11.5

Detection of low molecular weight DNA fragmentation (laddering) in apoptotic cells. Quiescent Swiss 3T3 fibroblasts were treated with staurosporine (100 nM) for defined periods of time after which total cellular DNA was extracted, resolved by agarose gel electrophoresis and visualised with ethidium bromide. The arrows indicate nucleosome fragments that differ in molecular weight by approximately 200 base pairs.

'DNA ladders' that are diagnostic of apoptotic cell death. To separate larger DNA fragments, either pulse field or field inversion gel electrophoresis is required. However, smaller DNA fragments can be readily separated by standard agarose gel electrophoresis. DNA fragmentation serves as a hallmark that can be used to distinguish apoptotic cell death from necrosis. In apoptosis, clear nucleosomal ladders can often be observed, whereas in necrotic cells the DNA is generally degraded and forms a smear when analysed by agarose gel electrophoresis. It should be noted that not all apoptotic cells proceed to low molecular weight nucleosomal ladders and this method should not be relied upon to definitively exclude apoptosis. Nevertheless, the ability to combine classical analyses of cell morphology with the detection of degraded DNA provides a powerful approach to visualise cell death in the context of intact organisms.

Apoptotic DNA fragmentation is distinct from the random degradation observed in necrosis. Applying electrophoresis to intact nuclei, microgel electrophoresis (also known as the comet assay) is a method of detecting and measuring DNA strand breaks in individual cells and can discriminate between apoptotic and necrotic nuclei [33]. The technique allows the identification of cells with differing degrees of DNA damage and has successfully been applied to quantitative studies of drug- or radiation-induced DNA damage and carcinogenesis. The assay is simple to perform with minimal handling. Briefly, cell populations are subjected to electrophoresis in a thin layer of agarose and the nuclei visualised by fluorescence microscopy following ethidium bromide labelling (*Figure 11.2*). The DNA migrates away from the nucleus, depending on the degree of DNA damage, so that cells resemble a comet, with the head made up of the nuclear remnants and the tail of damaged DNA (reviewed in [21]). The application of computer-based image analysis to the comet can then quantitate DNA damage, since the intensity of the dye is proportional to the amount of

DNA present. Not only the number of DNA strand breaks within a single cell can be measured, but also any heterogeneity of DNA damage within a cell population can be identified. This aspect makes the comet assay particularly powerful, since it is possible to distinguish apoptotic nuclei from those undergoing necrosis [34].

11.6 Concluding remarks

The nucleus exhibits some of the earliest signs of programmed cell death, and several techniques exist to monitor changes that distinguish this highly regulated form of active death from passive necrosis. However, caution should be utilised in the selection of techniques for detection of cell death and more than one method should be used if possible. While new methods continue to be developed to analyse apoptosis, the morphology of the cell is ultimately the best indicator that this process has been activated. For example, techniques now exist to monitor whole genome transcription, and increased transcription of one or more cell death genes is an indication that cell death has been activated. However, in the absence of morphological markers of programmed cell death, this evidence is like finding a bloody knife at the scene of a murder without a corpse. One of the greatest challenges that faces cell death researchers is applying what has been learned in cultured cell lines to specific cell types that die *in vivo*. While we have presented several techniques that can be used to detect cell death in fixed specimens, perhaps the next advance in the detection of dying cells will be to visualise apoptosis *in vivo*; indeed, to monitor programmed cell death in intact live animals in real time under normal physiological conditions.

11.7 References

1. Jacobson, M.D., Weil, M. and Raff, M.C. (1997) Programmed cell death in animal development. *Cell* **88**: 347–354.
2. Thompson, C.B. (1995) Apoptosis in the pathogenesis and treatment of disease. *Science* **267**: 1456–1462.
3. Martelli, A.M., Zweyer, M., Ochs, R.L., Tazzar, P.L., Tabellini, G., Narducci, P. and Bortul, R. (2001) Nuclear apoptotic changes: an overview. *J. Cell. Biochem.* **82**: 634–646.
4. Wyllie, A.H. (1980) Glucocorticoid-induced thymocyte apoptosis is associated with endogenous endonuclease activation. *Nature* **284**: 555–556.
5. Collins, R.J., Harmon, B.V., Gobe, G.C. and Kerr, J.F. (1992) Internucleosomal DNA cleavage should not be the sole criterion for identifying apoptosis. *Int. J. Radiat. Biol.* **61**: 451–453.
6. Ormorod, M.G., Sun, X-M., Brown, D., Snowden, R.T. and Cohen, G.M. (1994) Quantification of apoptosis and necrosis by flow cytometry. *Acta Oncologica* **3**: 417–424.
7. Kerr, J.F., Wyllie, A.H. and Currie, A.R. (1972) Apoptosis: a basic biological phenomenon with wide-ranging implications in tissue kinetics. *Br. J. Cancer* **26**: 239–257.
8. Earnshaw, W.C. (1995) Nuclear changes in apoptosis. *Curr. Biol.* **7**: 337–343.
9. McCall, K. and Steller, H. (1998) Requirement for DCP-1 caspase during *Drosophila* oogenesis. *Science* **279**: 230–234.
10. Tinnemans, M.M., Lenders, M.H., ten Velde, G.P., Ramaekers, F.C. and Schutte, B. (1995) Alterations in cytoskeletal and nuclear matrix-associated proteins during apoptosis. *Eur. J. Cell Biol.* **68**: 35–46.
11. Slee, E.A., Adrain, C. and Martin, S.J. (2001) Executioner caspase-3, -6, and -7 perform distinct, non-redundant roles during the demolition phase of apoptosis. *J. Biol. Chem.* **276**: 7320–7326.
12. Abrams, J.M., White, K., Fessler, L.I. and Steller, H. (1993) Programmed cell death during *Drosophila* embryogenesis. *Development* **117**: 29–43.

13. Sun, X.M., Snowden, R.T., Skilleter, D.N., Dinsdale, D., Ormerod, M.G. and Cohen, G.M. (1992) A flow-cytometric method for the separation and quantitation of normal and apoptotic thymocytes. *Anal. Biochem.* **204**: 351–356.

14. Gerner, C., Gotzmann, J., Frohwein, U., Schamberger, C., Ellinger, A. and Sauermann, G. (2002) Proteome analysis of nuclear matrix proteins during apoptotic chromatin condensation. *Cell Death Differ.* **9**: 671–681.

15. Rao, L., Perez, D. and White, E. (1996) Lamin proteolysis facilitates nuclear events during apoptosis. *J. Cell Biol.* **135**: 1441–1455.

16. Eliasson, M.J., Sampei, K., Mandir, A.S., Hurn, P.D., Traystman, R.J., Bao, J., Pieper, A., Wang, Z.Q., Dawson, T.M., Snyder, S.H. and Dawson, V.L. (1997) Poly(ADP-ribose) polymerase gene disruption renders mice resistant to cerebral ischemia. *Nat. Med.* **3**: 1089–1095.

17. Lazebnik, Y.A., Kaufmann, S.H., Desnoyers, S., Poirier, G.G. and Earnshaw, W.C. (1994) Cleavage of poly(ADP-ribose) polymerase by a proteinase with properties like ICE. *Nature* **371**: 346–347.

18. Bolanas, J.P., Almeida, A., Stewart, V., Peuchen, S.L., Clark, J.B. and Heales, S.J.R. (1997) Nitric oxide-mediated mitochondrial damage in the brain: mechanisms and implication for neurodegenerative diseases. *J. Neurochem.* **68**: 2227–2240.

19. Kaufmann, S.H., Desnoyers, S., Ottaviano, Y., Davidson, N.E. and Poirier, G.G. (1993) Specific proteolytic cleavage of poly(ADP-ribose) polymerase: an early marker of chemotherapy-induced apoptosis. *Cancer Res.* **53**: 3976–3985.

20. Barry, M.A. and Eastman, A. (1993) Identification of deoxyribonuclease II as an endonuclease involved in apoptosis. *Arch. Biochem. Biophys.* **300**: 400–450.

21. Olive, P.L., Johnston, P.J., Banath, J.P. and Durand, R.E. (1998) The comet assay: a new method to examine heterogeneity associated with solid tumors. *Nat. Med.* **4**: 103–105.

22. Gavrieli, Y., Sherman, Y. and Ben-Sasson, S.A. (1992) Identification of programmed cell death in situ via specific labelling of nuclear DNA fragmentation. *J. Cell Biol.* **119**: 493–501.

23. Wijsman, J.H., Jonker, R.R., Keijzer, R., van de Velde, C.J., Cornelisse, C.J. and van Dierendonck, J.H. (1993) A new method to detect apoptosis in paraffin sections: in situ end-labeling of fragmented DNA. *J. Histochem. Cytochem.* **41**: 7–12.

24. Edwards, A.D., Yue, X., Cox, P., Hope, P.L., Azzopardi, D.V., Squier, M.V. and Mehmet, H. (1997) Apoptosis in the brains of infants suffering intrauterine cerebral injury. *Pediatr. Res.* **42**: 684–689.

25. Enari, M., Sakahira. H., Yokoyama, H., Okawa, K., Iwamatsu, A. and Nagata, S. (1998) A caspase-activated DNase that degrades DNA during apoptosis, and its inhibitor ICAD. *Nature* **391**: 43–50.

26. Zhang, C., Raghupathi, R., Saatman, K.E., LaPlaca, M.C. and McIntosh, T.K. (1999) Regional and temporal alterations in DNA fragmentation factor (DFF)-like proteins following experimental brain trauma in the rat. *J. Neurochem.* **73**: 1650–1659.

27. Chen, D., Stetler, R.A., Cao, G., Pei, W., O'Horo, C., Yin, X.M. and Chen, J. (2000) Characterization of the rat DNA fragmentation factor 35/Inhibitor of caspase-activated DNase (Short form). The endogenous inhibitor of caspase-dependent DNA fragmentation in neuronal apoptosis. *J. Biol. Chem.* **275**: 38508–38517.

28. Lechardeur, D., Drzymala, L., Sharma, M., Zylka, D., Kinach, R., Pacia, J., Hicks, C., Usmani, N., Rommens, J.M. and Lukacs, G.L. (2000) Determinants of the nuclear localization of the heterodimeric DNA fragmentation factor (ICAD/CAD). *J. Cell Biol.* **150**: 321–334.

29. Susin, S.A., Zamzami, N., Castedo, M., Hirsch, T., Marchetti, P., Macho, A., Daugas, E., Geuskens, M. and Kroemer, G. (1996) Bcl-2 inhibits the mitochondrial release of an apoptogenic protease. *J. Exp. Med.* **184**: 1331–1341.

30. Cande, C., Cohen, I., Daugas, E., Ravagnan, L., Larochette, N., Zamzami, N. and Kroemer, G. (2002) Apoptosis-inducing factor (AIF): a novel caspase-independent death effector released from mitochondria. *Biochimie* **84**: 215–222.

31. Miramar, M.D., Costantini, P., Ravagnan, L., Saraiva, L.M., Haouzi, D., Brothers, G., Penninger, J.M., Peleato, M.L., Kroemer, G. and Susin, S.A. (2001) NADH

oxidase activity of mitochondrial apoptosis-inducing factor. *J. Biol. Chem.* **276**: 16391–16398.

32. Li, L.Y., Luo, X. and Wang, X. (2001) Endonuclease G is an apoptotic DNase when released from mitochondria. *Nature* **412**: 95–99.

33. Singh, N.P., McCoy, M.T., Tice, R.R. and Schneider, E.L. (1988) A simple technique for quantification of low levels of DNA damage in individual cells. *Exp. Cell Res.* **175**: 184–191.

34. Olive, P.L. and Banath, J.P. (1995) Sizing highly fragmented DNA in individual apoptotic cells using the comet assay and a DNA cross-linking agent. *Exp. Cell Res.* **221**: 19–26.

Flow Cytometric Studies of Cell Death

12

Marco Ranalli, Andrew Oberst, Marco Corazzari
and Vincenzo De Laurenzi

Contents

12.1 Introduction

The microscope has long been the symbol of the biologist, perhaps not surprisingly since, for much of the last century, the observation of cellular phenomena has relied primarily on microscopic analysis. Despite the venerable history and continuing relevance of the microscope in biology, microscopic studies carry an inherent limitation; while unparalleled for analysing individual cells, the limited number of cells observable in such studies precludes the quantitative, cell-by-cell analysis of entire populations. Flow cytometry overcomes this limitation by allowing the analysis of a large number of cells in a very short time, greatly simplifying the quantitative, real-time observation

Cell Proliferation and Apoptosis, David Hughes and Huseyin Mehmet (Eds)
© 2003 BIOS Scientific Publishers Ltd, Oxford

of population-wide phenomena such as apoptosis. Flow cytometry has been applied to the study of apoptosis in a variety of ways. This chapter seeks to describe some of the most relevant techniques developed to date.

12.1.1 The flow cytometer

Fundamentally, the flow cytometer is a machine that accepts an input of liquid-suspended cells, passes them one by one through a light source, and measures a variety of light scattering and fluorescence properties for each cell. Input cells are passed through a capillary at a rate of thousands per second, and individually subjected to light, generally in the form of a laser beam or equivalent monochromatic light source. The light scattering or fluorescence (when fluorescent dyes are employed) characteristics of each cell are collected by light sensors, and these data can subsequently be analysed by various software packages. A variety of information can be gleaned from such analysis. A sensor positioned at 180° from the light source measures *forward scatter* for each cell, which reflects the cell's size, while another sensor positioned at 90° from the light source measures *side scatter*, which reflects cellular morphology and surface complexity. These measurements allow analysis of morphologically different sub-populations within a heterogeneous cell suspension. The use of fluorescent dyes and marker molecules allows the study of a wide variety of cellular functions, including (but by no means limited to) protein expression, alterations to cellular membranes, and enzyme activities (*Figure 12.1*).

Commercially available flow cytometers generally employ a 488 nm argon-ion laser source, allowing detection of fluorophores that emit in the green, orange and red spectra. Some cytometers still use mercury and xenon lamps (usually 100 W short-arc lamps) as a source of excitation, though these are increasingly considered obsolete. Selective wavelength filters and mirrors allow the exclusion of the excitation light and the separation and measurement of fluorescent intensity from multiple fluorescent dyes, permitting contemporary use of three or four fluorescent marker molecules.

The workhorse 488 nm laser, most common on single-laser machines, provides a range of emission values that allows contemporary use of up to three fluorescent markers, in the range from 530–670 nm; however, certain applications require simultaneous deployment of more than three dyes. A number of measures have been taken to extend the cytometer capabilities. Most simply, a UV laser can be employed in place of the 488 nm laser, enlarging the emission spectrum to 400–670 nm, thereby allowing use of up to five detection channels. Unfortunately, a UV laser often proves prohibitively expensive. A more cost-effective solution is the concurrent use of two lasers in the same machine. The range of emission wavelengths detectable in such machines is essentially identical to that employed in 488 nm cytometers, but multi-laser machines are capable of differentiating between the emission signals of a greater number of fluorophores. In such systems, cells are passed sequentially through multiple excitation lasers, with known lag times between excitation by each laser. The cytometer calculates which laser's excitation was responsible for a given emission event on the basis of that event's timing, thereby allowing differentiation between fluorophores with distinct excitation spectra but similar emission spectra. A common example of this system is the coupling of a low-cost red diode laser with the common 488 nm argon-ion laser, extending the number of possible data collection channels from three to four (Appendix A).

In addition, most new flow cytometers allow sorting of selected cell populations based on a variety of measured characteristics, yielding almost-pure subsets of cells that can be then analysed with other molecular techniques, or even cultured if particular precautions are taken to preserve sterility. Several different types of cell sorters are

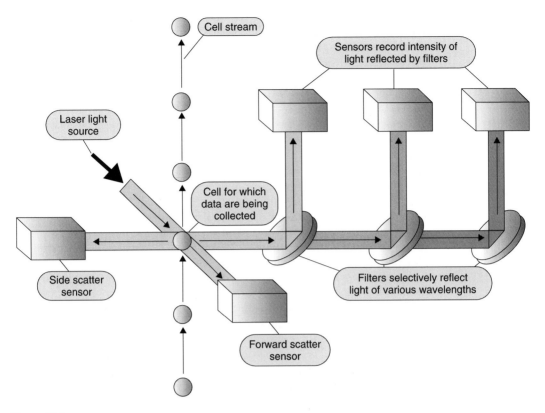

Figure 12.1

Schematic illustration of the flow cytometer. The cell stream passes through a laser light source; the resulting light scattering and fluorescence properties are recorded individually for each cell. The side scatter (left) and forward scatter (bottom) sensors record each cell's light scattering properties, which are related to cell size and morphology. Simultaneously, a series of reflective filters (right) separate the fluorescent emissions of each cell into specific wavelength categories, each of which is recorded by a sensor (upper right). This separation of the emission spectrum allows concurrent use of multiple fluorescent probes, as long as each probe emits at a different wavelength.

now available, with sorting speeds ranging from 300 to 10 000 particles second^{-1}. Cell sorting is usually achieved either by charging single-cell-containing droplets using high voltage plates and then deflecting them into different tubes using an electrostatic field (electrostatic sorters), or by employing a catcher tube that moves in and out of the sample stream to catch specific cells (mechanic sorters). The electrostatic sorter is much faster, but less precise; it is generally the method of choice for biochemical applications requiring millions of cells. The mechanical sorter has a maximum speed of only about 300 cells second^{-1}, but yields much higher purity. Its low speed renders it unsuitable for biochemical applications, but its superior precision is advantageous when culturing sorted cells [1–3].

12.1.2   The flow cytometer applied to apoptosis

Despite determined efforts, it remains a challenge to precisely predict the timing of the entry of a cell into the apoptotic programme; when widespread apoptosis is induced in a population of cells, morphologically they generally undergo apoptotic changes in an asynchronous manner. The population-wide nature of such phenomena presents challenges to traditional microscopy, which is capable of quantitative analysis of only a relatively small number of cells at a time. A primary advantage of flow cytometry is that it allows the cell-by-cell analysis of entire populations of cells. Because cells can be analysed at rates of many thousands per second, the flow cytometer can provide a 'snapshot' of tens of thousands of cells in a few seconds, a task that would take days or weeks with traditional microscopy. The percentage of cells exhibiting various apoptotic morphologies, from chromatin condensation to wholesale cellular shrinkage, can thus be quantified at specific points in time.

Unlike traditional microscopy, which provides images of entire cells, the flow cytometer only provides information on a cell's light scatter properties and fluorescence intensity. Therefore, the specific phenomena addressed by a given flow cytometric study depend heavily on the nature of the fluorescently labelled molecules employed. Nevertheless, labels and protocols have been developed for the study of a wide variety of apoptosis-related processes, including cell shrinkage and membrane blebbing, changes in membrane permeability, changes in membrane lipid distribution (i.e. phosphatidylserine exposure), chromatin condensation, DNA fragmentation and loss of mitochondrial intermembrane potential [4–7]. It is useful to keep in mind that these techniques are often employed in tandem, allowing more precise analysis of the chain of events that make up the apoptotic program. In the following sections, we'll take a closer look at these fundamental tools, and explore their uses in observing the myriad complexities of apoptosis.

12.2 Morphological assessment of apoptosis

Apoptosis was originally described on the basis of a typical morphology. While the changes may vary slightly between cell types, the general morphological hallmarks of apoptosis include cell shrinkage, chromatin condensation, and membrane blebbing [8, 9].

Microscopic analysis remains a simple and useful way to detect apoptosis in a cell population. Under the phase-contrast microscope, apoptotic cells often appear as dark bodies, smaller than the living cells, without the typical round appearance (for review see [10]). In the case of adherent cell lines, detachment is an early sign of apoptosis, though it cannot be regarded as diagnostic. Additionally, mitochondrial swelling and vacuolisation can be observed under the electron microscope. While microscopic analysis is essential to confirm that cells are dying with features characteristic of apoptosis, it is a time-consuming and generally qualitative technique, and is therefore unsuitable for analysis of large numbers of cells.

Fortunately, the same morphological hallmarks of apoptosis observed with microscopy can be used to differentiate apoptotic cells from healthy cells in the flow cytometer. The characteristic shrinkage, and associated increase in surface complexity (i.e. wrinkling), of apoptotic cells leads to a distinct change in their light scattering properties as compared to healthy cells. Specifically, the forward scatter value (a measure of cell size) decreases, while the side scatter value (a measure of surface complexity) increases. As shown in *Figure 12.2*, when a scatter plot of Forward Scatter (FSC) versus Side Scatter (SSC) is made, the apoptotic population appears above and to the left (low FSC and high SSC) of the healthy population [11].

While this method is both rapid and simple (no dyes or markers are required), it has certain limitations. Primarily, it is often difficult to clearly distinguish the apoptotic and healthy populations, as they frequently partially overlap. Additionally, characteristic morphological changes are hidden upon detachment in some adherent cell lines, precluding this type of analysis in such lines. In practice, this technique is most

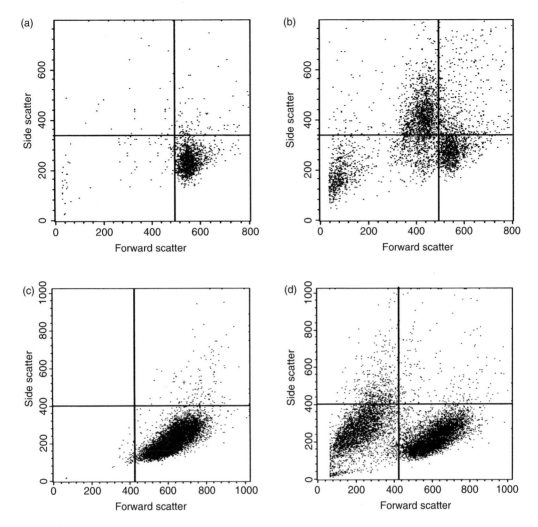

Figure 12.2

Changes in light scatter properties of Jurkat T cells and mouse thymocytes upon induction of apoptosis.
Jurkat T cells (a, b) and mouse thymocytes (c, d) were treated with α-CD95 antibody and UV radiation,
respectively. All panels show Forward Scatter on the X-axis, and Side Scatter on the Y-axis. Panels (a) and (c)
show control cells whereas panels (b) and (d) show the appearance of an apoptotic population with
increased side scatter and reduced forward scatter relative to the control population. Forward scatter is a
measure of cell size, while side scatter is a measure of cell complexity; the shrinkage and blebbing
characteristic of apoptosis reduces cell size and increases cell surface complexity, resulting in a population
with distinct light scattering properties.

useful for distinguishing apoptotic and healthy cells in thymocytes, freshly isolated
peripheral blood cells, and some lymphoid cell lines [12].

12.3   Methods based on membrane permeability changes occurring during apoptosis

In cells undergoing apoptosis, the permeability characteristics of the plasma membrane change significantly. In apoptotic cells, the cell membrane is more permeable

to a variety of molecules. This change can be utilised to distinguish apoptotic from healthy cells in the flow cytometer, and is particularly useful because it occurs early in the apoptotic process, allowing early detection of apoptotic cells.

There are several flow cytometric tests that distinguish between healthy and apoptotic cells on the basis of altered membrane permeability. PI and Trypan blue are common cell-staining fluorescent dyes. The cell membrane, normally impermeable to these dyes, becomes permeable upon cell death, causing dead cells to fluoresce with much more intensity in the presence of such markers.

The usefulness of these markers is limited by the fact that they do not distinguish between apoptotic and necrotic cells. In the early stages of apoptosis, there is a slight increase in membrane permeability, which allows the marker molecules to enter. However, this effect is transient and the apoptotic cell membrane soon regains its impermeability to the markers pending secondary necrosis. It is therefore impossible to distinguish between primary and secondary necrosis using these markers. Used alone, this method is primarily useful in distinguishing viable from non-viable cells. However, by analysing other parameters, such as DNA fragmentation, or by using these dyes in combination with other markers, such as fluorescein-diacetate (FDA) and Hoechst 33258, apoptosis and necrosis can be readily distinguished [13–15].

12.3.1 Dual staining of apoptotic cells

FDA/PI staining (Protocol 12.1) is very useful to discriminate between viable, necrotic and apoptotic cells. FDA is a low-cost lipase substrate that rapidly accumulates in viable cells, while PI is a DNA intercalating molecule able to stain nuclei in cells with a permeable plasma membrane. FDA will only stain living cells whereas PI will stain only necrotic cells (unless cells are previously fixed; see note on fixing, Section 12.5). As a consequence, when a population of cells is stained with FDA/PI, viable cells will be FDA positive and PI negative, necrotic cells will be PI positive but FDA negative, and apoptotic bodies will be negative for both dyes. This effect can be observed in *Figure 12.3* [13, 14].

Hoechst 33258/PI staining (Protocol 12.2) is conceptually similar to the FDA/PI method. Hoechst 33258 is a fluorescent dye whose rate of accumulation is strongly dependent on the state of the cell membrane. Living cells incorporate Hoechst 33258 very slowly, while apoptotic and necrotic cells take it up more quickly; apoptotic cells can further be separated from necrotic cells on the basis of PI staining, as previously mentioned. This method requires a cytometer equipped with an UV laser to excite the Hoechst 33258 (excitation 350 nm, emission 460 nm). The major advantage of this technique is that it does not require the use of the green detection channel, leaving it free for data collection. This is an important consideration, as the green channel is commonly used to measure a variety of phenomena, including free radical formation, calcium release, and the presence of specific proteins detected by immunostaining or transfected with a GFP-fusion protein [15].

Protocol 12.1: FDA/PI staining

EQUIPMENT, MATERIALS AND REAGENTS

Cells to be analysed.

Ice.

PBS pH 7.4.

FDA dye: dissolve FDA in DMSO to obtain a 1 µM solution. This solution should be stored at −20°C.

PI dye: dissolve PI in PBS to obtain a 100 µg ml^{-1} solution. If stored at −20°C in the dark, this solution is stable for several months.

Micro-centrifuge.

Flow cytometer.

METHOD

1. Detach cells as described in Appendix B.

2. Resuspend in 300 µl PBS. Between 0.5 and 2 million cells are required for the analysis.

3. Add FDA to a final concentration of 100 nM and PI to a final concentration of 2 µg ml^{-1}, and incubate for 5 minutes at room temperature.

4. Wash twice in 4 ml of cold PBS (4°C). Resuspend in 300 µl of PBS, keep on ice and immediately analyse the samples.

5. Samples must be analysed immediately and NOT fixed. Acquire 10 000 events using the green channel for FDA (maximum emission 530 nm) and the red channel for PI (maximum emission 640 nm). On an FDA versus PI scatter plot, apoptotic cells will be double-negative, appearing in the lower left corner. Necrotic cells will be positive for PI only, and will appear in the upper left quadrant, while living cells will be positive for FDA only, appearing in the lower right quadrant (*Figure 12.3*).

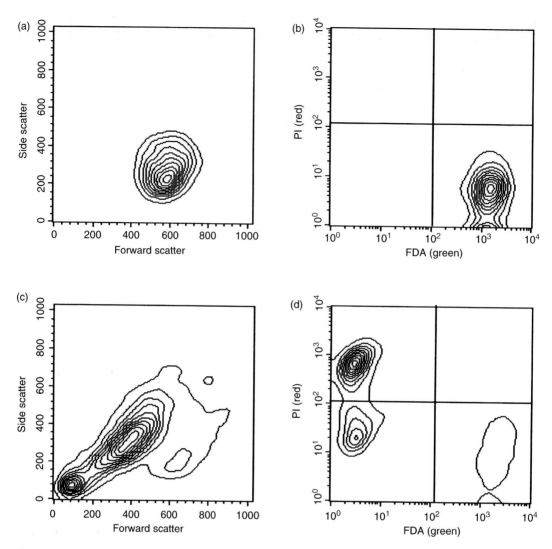

Figure 12.3

FDA/PI staining. Apoptosis and necrosis were induced in Jurkat T cells using α-CD95 antibody (c, d), detected by FDA/PI staining and compared to the control cells (a, b). FDA marks healthy cells, while PI marks necrotic (but not apoptotic) cells. Therefore, the PI positive and FDA negative population represents necrotic cells (d, upper-left quadrant); the PI negative and FDA negative population represents apoptotic cells (d, lower-left quadrant) and viable cells are FDA positive only (d, lower-right quadrant). Panels (a) and (c) show the concurrent changes in Forward- and Side-Scatter.

Protocol 12.2: Hoechst 33258/PI staining

EQUIPMENT, MATERIALS AND REAGENTS

Cells to be analysed.

Ice.

Hank's Balanced Salt Solution (HBSS) containing 1% BSA.

PBS pH 7.4.

Hoechst 33258 dye: dissolve Hoechst 33258 in H_2O to a final concentration of 0.1 mg ml^{-1}. Store this solution at $-20°C$.

PI dye: dissove PI in PBS to obtain a 100 μg ml^{-1} solution. If stored at $-20°C$ in the dark, this solution is stable for several months.

Micro-centrifuge.

Flow cytometer.

METHOD

1. Detach cells as described in Appendix B.

2. Wash in 2 ml HBSS containing 1% BSA: spin cells at 200 g for 10 minutes, remove supernatant, and thoroughly resuspend in 2 ml HBSS + BSA. Spin this suspension again at 200 g for 10 minutes, and remove supernatant, leaving cell pellet.

3. Resuspend 0.5–2 million cells in 1 ml HBSS containing 1% BSA.

4. Add 1 μg Hoechst 33258 (10 μl of the 0.1 mg ml^{-1} stock solution) and incubate for 2–10 minutes at 37°C.

5. Wash by adding 3 ml of ice-cold PBS and spinning for 10 minutes at 200 g at 4°C in a refrigerated centrifuge, then removing supernatant.

6. Resuspend cells in 1 ml PBS containing 2 μg ml^{-1} PI, incubate 10 minutes on ice, repeat washing step once, and analyse samples.

7. Both PI and Hoechst 33258 can be excited with a 340 nm wavelength laser (or xenon or mercury lamp), however PI (maximum emission ~640 nm) shows brighter fluorescence when excited with a 488 nm blue light. Healthy cells will be double negative; necrotic cells will be PI positive and Hoechst 33258 positive; apoptotic cells will be PI negative and Hoechst 33258 positive. If desired, populations can be gated, allowing a third parameter to be examined on a separate plot.

Note: PI can be substituted with 2 μg 7-AAD (maximum emission 650 nm) incubated for the same time and excited in the same way. 7-AAD presents an emission spectra shifted toward the far-red region, providing better signal separation from orange marker molecules in the case of multiple staining [16].

12.4 Annexin V staining to detect changes in membrane lipid distribution

In healthy cells, lipid translocases in the plasma membrane function to maintain lipid asymmetry. In such cells, some lipid species are found primarily on the outer leaflet of the plasma membrane, while others, most notably phosphatidylserine (PS), are found primarily on the inner leaflet. When a cell undergoes apoptosis, a bi-directional lipid scramblase protein is activated, allowing lipid species to freely diffuse between the membrane's two leaflets, an activity which results in the appearance of PS on the outer leaflet of cells [17–20]. PS exposure therefore represents a useful assay for apoptosis.

PS present on the outer leaflet can be detected using a fluorescent derivative of Annexin V, a 35–36 kDa human anticoagulant protein. In the presence of Ca^{2+}, annexin V shows phospholipid-binding activity and has a particularly high affinity for PS. Early apoptotic cells may be detected by FACS analysis (Protocol 12.3). The most common varieties of these derivatives are FITC, which fluoresces in the green spectrum at 488 nm, and ALEXA568, which fluoresces in the red spectrum at 568 nm [17–20]. Using a fluorescence microscope, cells exposing PS and labelled with fluorescent derivatives of Annexin V will appear stained on the surface with a slightly flecked pattern. Unfortunately not all cells expose PS during apoptosis in appreciable quantities, so this method cannot be used for all cell types.

Necrotic cells also expose PS, and will therefore also bind Annexin V. To differentiate between apoptotic and necrotic cells, PI or 7-AAD is often used in conjunction with Annexin V. As described above, these molecules will mark necrotic, but not apoptotic, cells. The effects of dual staining with Annexin V and PI may be observed in *Figure 12.4*: apoptotic cells are annexin positive but PI negative, while necrotic cells are positive for both markers.

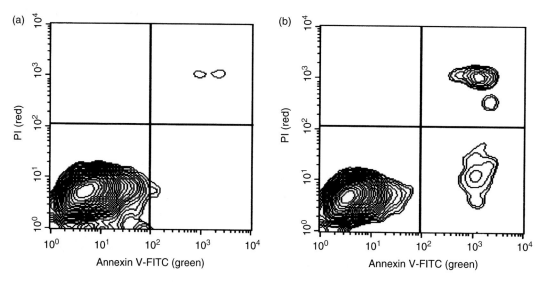

Figure 12.4

Annexin V/PI labeling of apoptotic cells. Jurkat T cells were treated with UV radiation (b), which induces both apoptosis and necrosis. Annexin V/PI fluorescence profiles are compared with control cells (a). Annexin V binds the phospholipid PS, marking apoptotic and necrotic cells, while PI binds DNA, marking only necrotic cells. Two new cell populations appear in panel (b): necrotic cells (PI and Annexin V positive, upper-right quadrant) and apoptotic cells (Annexin V positive but PI negative, lower-right).

Protocol 12.3: Annexin V binding

EQUIPMENT, MATERIALS AND REAGENTS

Cells to be analysed.

Ice.

5X Annexin V Binding Buffer (BB): 50 mM Hepes, 750 mM NaCl, 12.5 mM $CaCl_2$, 5 mM $MgCl_2$, 20% BSA. 5X sterile filtered BB can be stored at 4°C for 2 months.

Micro-centrifuge.

Flow cytometer.

Note: it is important that the buffer used in this protocol is not a Ca^{2+}-free buffer such as PBS, since Annexin requires Ca^{2+} to bind phosphatidylserine.

Annexin V preparation and storage:
Dissolve Annexin V in 1X BB to obtain a 1 μg ml^{-1} solution. If stored at −20°C in the dark, this solution is stable for several months.

METHOD

1. Detach cells as described in Appendix B.

2. Wash cells twice with 1X BB: spin cells at 200 g for 10 minutes, remove supernatant, and thoroughly resuspend in 2 ml 1X BB. Spin this suspension again at 200 g for 10 minutes, and remove supernatant, leaving cell pellet. Cells have now been washed once; repeat suspension, spinning, and supernatant removal for second wash.

3. Resuspend 0.5–2 million cells in 100 μl of 1X BB containing 1 μg ml^{-1} Annexin V conjugate.

4. Incubate at room temperature for 10–20 minutes.

5. Wash twice in 2 ml 1X BB (as described in step 2), and resuspend in 100 μl 1X BB.

6. (Optional) If PI staining is required: add PI to a final concentration of 2 μg ml^{-1} and incubate for 5 minutes at room temperature. If red fluorescent Annexin V is used, Sytox green at a final concentration of 100 nM can be added; in this case, incubate for 10 minutes at 4°C.

7. Sample analysis. Acquire at least 10 000 events; Annexin V positive cells will be in the lower right quadrant (*Figure 12.4*). When double staining with PI, electronic compensation might be required to remove interference from the Annexin V emission in the red channel. A similar compensation is required when using Sytox green to remove interference in the red channel where red Annexin V is measured. In an Annexin V/PI scatter plot (*Figure 12.4*), living cells will be double negative (near origin), apoptotic cells will be Annexin V positive (lower right quadrant), and necrotic cells will be PI positive (upper left quadrant).

12.5 Methods based on DNA fragmentation

One of the main features of the apoptotic process is the characteristic breakdown of DNA. Apoptotic DNA degradation proceeds via characteristic inter-nucleosomal double-stranded breaks, producing DNA fragments in multiples of 200 base pairs. This fragmentation can be visualised by agarose gel electrophoresis following total DNA extraction, producing a characteristic DNA ladder [21, 22]. In the cytometer, DNA fragmentation can be detected using fluorescent DNA-intercalating agents, such as ethidium bromide and PI, in conjunction with alcohol fixation or cell permeabilisation [23, 24]. These dyes produce cellular fluorescence proportional to cellular DNA content; since small DNA fragments are lost from cells upon alcohol fixation, cells with fragmented DNA will appear dimmer than those with intact genomes. DNA fragmentation can also be visualised using the TUNEL assay [25]. This reaction makes use of the ability of the TdT transferase enzyme to add fluorescently labelled nucleotides to DNA double-stranded breaks.

12.5.1 PI detection of DNA fragmentation

PI staining of ethanol-fixed cells is probably the most used test for DNA fragmentation in flow cytometry (Protocol 12.4). This method is based on the idea that small DNA fragments generated during apoptosis are lost after permeabilisation of the cell membrane with fixatives. Apoptotic cells will therefore have a decreased content of DNA when compared to living cells and can be detected using a fluorescent DNA-intercalating molecule such as PI or ethidium bromide. This method is normally used to analyse the distribution in the different cell cycle phases of a cell population. Cells in the G_1 phase will have a diploid DNA content whereas cells in G_2 or in mitosis will have a double DNA content. Cells in the S phase will have an intermediate DNA content. In this system, apoptotic cells are detectable as a G_1-like sub-population with a hypo-diploid DNA content (*Figure 12.5*). This method is therefore very useful because it allows contemporary study of cell death and cell cycle status in a given cell population [23, 24].

Fixation of a cell sample freezes cellular activity and allows storage of a cell sample for an extended period of time. There are a number of fixative agents commonly used. Among these, alcohol and PFA are probably the most common. Alcohol fixation causes membrane permeabilisation, and therefore can lead to leakage of cellular antigens and their associated fluorescent dyes. In contrast, PFA fixation leads to cross-linking at the cell surface, and therefore tenaciously conserves cellular content. Each of these techniques has advantages; for example, alcohol fixation is not ideal in many dye applications, as dye leakage can confuse the results, but is required for the DNA fragmentation assays, where fragmented DNA is allowed to leak from the cell [26, 27]. Fixing cells with PFA, however, results in an alteration of the cell cycle profile and should therefore be avoided when possible.

This method can also be used to study only a subset of cells if these are labelled with a fluorescent dye. Cells can either be stained by immunofluorescence with antibodies for a specific surface protein or can be transfected with a desired plasmid together with a GFP-expressing plasmid. In order to avoid leakage of cytoplasmic GFP from fixed cells that would require more delicate but more tedious fixing protocols, a GFP-fusion (such as GFP-spectrin) [28] protein, that localises in the cell membrane, can be used to label transfected cells. Transfected cells are then selected in a green channel/SSC or green channel/FSC dot plot, and only gated cells are analysed for cell cycle and apoptosis.

Protocol 12.4: PI detection of DNA fragmentation

EQUIPMENT, MATERIALS AND REAGENTS

Cells to be analysed.

Ice.

70% ethanol.

PBS pH 7.4.

PI dye: dissolve PI in PBS to obtain a 70 μg ml^{-1} solution. If stored at −20°C in the dark, this solution is stable for several months.

RNase (DNase free).

Micro-centrifuge.

37°C water bath.

Flow cytometer.

METHOD

1. Detach cells as described in Appendix B.

2. Wash in 4 ml PBS: spin cells at 200 g for 10 minutes, remove supernatant, and thoroughly resuspend in 4 ml PBS. Spin this suspension again at 200 g for 10 minutes, and remove PBS, leaving cell pellet.

3. Resuspend 0.5–2 million cells in 100 μl PBS then add 2 ml of −20°C 70% ethanol very slowly, while vortexing. Note: it is important that ethanol is added slowly and while vortexing to avoid unwanted cell aggregation.

4. Store at −20°C for at least 2 hours. Samples can be stored at −20°C for up to one month before analysis.

5. Wash once in 4 ml PBS (as described in step 2).

6. Resuspend in 50 μl of a 13 kunitz units solution of RNase (DNase free) in PBS and incubate for 15 min in a 37°C water bath.

7. Add 200 μl of a 70 μg ml^{-1} solution of PI in PBS and incubate for 20 minutes in a 37°C water bath.

Note: stained cells can be stored for up to 24 hours at 4°C; this usually results in a better staining of the cells.

Certain cell types may present some difficulty in DNA extraction. In this case, it is possible to treat the sample with a DNA extraction buffer after fixation. This buffer is composed of 192 ml of 0.2 M Na$_2$HPO$_4$ and 8 ml citric acid (adjust pH to 7.8).

An alternative protocol to improve DNA fragment extraction is nucleus extraction. This is a very aggressive fragment extraction protocol that results in the complete loss of cell morphology. However, the loss of the cytoplasm greatly improves the analysis of the cell cycle and apoptosis. Briefly, cells are harvested, washed in PBS and then resuspended directly in a solution of 0.1% sodium citrate containing 0.1% Triton X-100, 13 kunitz units RNase (DNase free) and 50 μg ml^{-1} PI, and incubated for 2 hours at room temperature or overnight at 4°C.

8. Sample analysis. Acquire 10 000 events. A one-dimensional plot of the red channel is used to analyse results (*Figure 12.5*). Cells in G_1 or G_0 will contain a single copy of the genome, and will appear as a sharp peak, while cells in G_2 or M phase will have a double content of DNA and will appear as a second peak with approximately double intensity in the red channel. Apoptotic cells will have a relatively low red fluorescence intensity, due to their decreased DNA content, and will appear as a sub-G_1 peak. Due to their lower fluorescence, the signal from apoptotic cells requires higher amplification, and should be analysed on a logarithmic scale plot. Cells in S phase will have variable DNA content and will be distributed in the area between the two peaks. A number of software packages are available to calculate the area of the different peaks thereby determining the number of cells in each cell cycle phase (*Figure 12.5*).

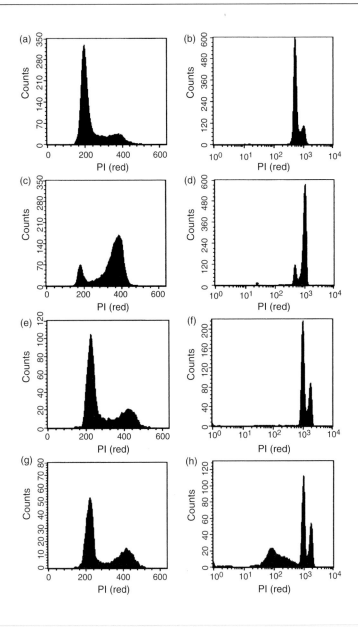

Figure 12.5

Cell cycle and DNA fragmentation. MCF-7 cells were treated (c, d) and untreated (a, b) with Taxol to induce accumulation in the G_2 phase. SaOs-2 cells stably transfected with an inducible p53 gene were treated (g, h) and untreated (e, f) with doxycycline, which induces p53 and therefore apoptosis. In all cases, cells were labelled with PI and then fixed. PI marks DNA, so a cell's position in the cell cycle can be studied based on its DNA content, which is directly proportional to its PI fluorescent intensity. The two peaks visible in each graph therefore represent cells in G_1 (low PI peak) and cells in G_2 (high PI peak), with S-phase cells in between. Cells are shown with two different x-axis scales; the left column is a linear view giving better resolution, while the right column is a log scale, giving a greater range of data and allowing clear viewing of apoptotic cells. The first four panels illustrate application of PI staining to the study of the cell cycle. Following Taxol treatment and subsequent G_2 arrest, the high PI peak can be observed to grow relative to the low PI peak; in panels (a) and (b) (control), there are more G_1 than G_2 cells, while in panels (c) and (d) (Taxol treated and therefore G_2 arrested), there are more G_2 than G_1 cells. The lower four panels illustrate concurrent observation of the cell cycle and apoptosis. Following apoptosis induction by p53, an apoptotic sub-population of cells with hypo-G_1 PI intensities appears (h, left side). Upon apoptosis, DNA fragmentation occurs, and when cells are fixed, these fragments leak out of the permeabilised cell membrane, giving apoptotic cells a sub-G_1 DNA complement.

12.5.2 **TUNEL assay**

Another method for detection of apoptotic cells based on DNA fragmentation is the TdT-mediated dUTP-biotin Nick End Labeling (TUNEL) method (Protocol 12.5). This approach is based on the observation that the DNA fragmentation that occurs during apoptosis produces double-stranded DNA breaks, which can easily be detected using mammalian TdT, which covalently adds nucleotides to the 3'-hydroxyl ends of these DNA fragments. If fluorescently labelled nucleotides are provided, this enzyme will also label the broken ends of DNA fragments. This method was originally developed for *in situ* staining of apoptotic cells and has since been adapted for flow cytometric use [25, 29].

Though long and somewhat complicated, the TUNEL technique performed in conjunction with PI staining provides an elegant and sensitive measure of the relationship between the cell cycle and apoptosis (i.e. whether apoptosis is preferentially induced at a certain point in the cycle). Additionally, this method allows very sensitive detection of even very small fractions of apoptotic cells in a given population.

12.5.3 **Combined cell cycle and DNA fragmentation analysis**

Though not directly related to apoptosis, this technique is particularly useful in marking the relative proportions of cells in each phase of the cell cycle. BrdU marks cells actively synthesising DNA, by replacing thymidine during DNA synthesis. BrdU is added to the medium of actively dividing cells; depending on the incubation time and rate of cell division, BrdU can mark only cells in S phase, or, if cells are given time to pass beyond S phase in the presence of BrdU, later cell cycle phases as well.

BrdU added to newly synthesised DNA can be detected using specific monoclonal antibodies. For such detection to take place, DNA has to be denatured to expose BrdU to the antibody; however, part of the DNA must remain double stranded to allow the PI intercalation required for DNA quantification. Several techniques have been developed to partially denature DNA based on the use of low pH solutions (0.1 M to 2 M HCl) and/or high temperatures (80–100°C). The choice of the best technique depends on the specific cellular model used and some trial-and-error is usually required to optimise the conditions for each situation. Further information on specific protocols for this technique can be obtained from the cited references [30–32].

Protocol 12.5: TUNEL assay

EQUIPMENT, MATERIALS AND REAGENTS

Cells to be analysed.

PBS pH 7.4.

Permeabilisation solution: 0.1% Triton X-100 dissolved in PBS.

Reaction buffer: 1 M sodium (or potassium) cacodylate, 12.5 mM Tris HCl (pH 6.6), 1.25 mg ml^{-1} BSA, 5 mM CoCl$_2$.

Washing buffer: 0.1% Triton X-100, 2% BSA in PBS.

Staining solution: 5 µg ml^{-1} PI in PBS, 13 kunitz units RNase (DNAse free)

TUNEL reaction mixture: 10 µl reaction buffer + 12.5 units TdT (25 units µl^{-1}), 1 µg f-dUTP (fluorescein 12-dUTP + 39 µl H$_2$O). Note: this reaction mixture must be prepared immediately before its use.

37°C incubator.

Micro-centrifuge.

Flow cytometer.

METHOD

1. Detach cells as described in Appendix B.

2. Wash in 4 ml PBS: spin cells at 200 g for 10 minutes, remove supernatant, and thoroughly resuspend in 4 ml PBS. Spin this suspension again at 200 g for 10 minutes, and remove PBS, leaving cell pellet.

3. Fix samples (maximum 1 million cells per tube) in a freshly prepared 4% PFA solution in PBS (pH 7.4) for 30 minutes at room temperature; cells should be directly resuspended in this solution following step 2.

4. Spin at 200 g for 10 minutes.

5. Permeabilise cells by resuspending the pellet in 500 µl permeabilisation solution and incubate for 2 minutes in ice.

6. Add 4 ml of PBS and centrifuge at 200 g at 4°C for 10 minutes.

7. Resuspend in 4 ml PBS, spin down and remove supernatant, leaving cell pellet.

8. Resuspend in 50 µl of TUNEL reaction mixture and incubate for 60 minutes at 37°C.

9. Wash cells twice with washing buffer: spin cells at 200 g for 10 minutes, remove supernatant, and thoroughly resuspend in 2 ml washing buffer. Spin this suspension again at 200 g for 10 minutes, and remove supernatant, leaving cell pellet. Cells have now been washed once; repeat suspension, spinning, and supernatant removal for second wash.

10. Stain for 30 minutes in 300 μl of staining solution at room temperature and proceed with cytometric analysis. This last step is optional but the contemporary use of PI gives additional information on the cell cycle status of the cell population.

11. Sample analysis. Acquire at least 10 000 events detecting green fluorescence for dUTP (fluorescein 12-dUTP, maximum emission 530 nm) and, if desired, red fluorescence for PI (maximum emission 640 nm; counter staining cells with PI allows contemporary analysis of cell cycle).

Note: apoptotic cells will fluoresce strongly in the green spectrum, due to dUTP staining. Both viable and necrotic cells will be either completely dUTP negative, or, in the case of DNA damage due to UV exposure or other irradiation, slightly dUTP positive, though still an order of magnitude weaker than apoptotic cells. A negative control prepared with the TUNEL reaction mixture without TdT should always be included. Several ready-to-use apoptosis detection kits for flow cytometry are now available from a number of companies.

12.6   Analysis of changes in mitochondrial potential

Mitochondrial dysfunction is a fundamental part of the apoptotic program. Indeed, permeabilisation of the outer and inner mitochondrial membrane represents a critical step in most pathways leading to apoptosis. This culminates in the release of molecules such as cytochrome c, inactive caspase precursors, DIABLO and AIF. Mitochondrial membrane permeabilisation is induced by pro-apoptotic molecules such as Bcl-2 family members, secondary messengers such as phosphatase and kinase and transcription factors such as p53 [33].

Mitochondrial alterations are almost always accompanied by changes in the membrane potential ($\Delta\Psi$), possibly due to the opening of the mitochondrial permeability transition pore. Several probes can be used to detect changes in mitochondrial potential, since the wavelength of their emission changes depending on the mitochondrial potential. The most reliable of these molecules seems to be JC-1 (5,5',6,6'-tetrachloro-1,1',3,3'-tetraethylbenzimidazolylcarbocyanine iodide) [34].

JC-1 selectively localises to mitochondria and exists as a monomer at low concentrations and at low membrane potentials. These monomers emit at 530 nm in the green region of the spectra. However, at higher concentrations or higher potentials, JC-1 forms red fluorescent aggregates: 'J-aggregates'. Measuring the ratio of the red to green emission allows detection of apoptotic cells (Protocol 12.6). Healthy cells will show a high ratio whereas apoptotic cells show lower ratios. JC-1 is excited at 488 nm and will emit at 527 nm or 590 nm in the monomeric or aggregated form, respectively (*Figure 12.6b*). The ratio of red to green JC-1 fluorescence is dependent only on the membrane potential and not on other factors, such as mitochondrial size, shape and density. While these factors may influence the strength of the JC-1 emission signals measured separately, they do not change the ratio of the two signals [35, 36].

Protocol 12.6: Evaluation of mitochondrial potential

EQUIPMENT, MATERIALS AND REAGENTS

Cells to be analysed.

Phenol red-free DMEM.

PBS pH 7.4.

JC-1 dye: dissolve JC-1 in anhydrous DMSO to obtain a 10 mM stock solution.

37°C incubator.

Micro-centrifuge.

Flow cytometer.

METHOD

1. Detach cells as described in Appendix B.

2. Wash cells twice with PBS: spin cells at 200 g for 10 minutes, remove supernatant, and thoroughly resuspend in 2 ml PBS. Spin this suspension again at 200 g for 10 minutes, and remove supernatant, leaving cell pellet. Cells have now been washed once; repeat suspension, spinning, and supernatant removal for second wash.

3. Resuspend 0.5–2 million cells in 200 μl of phenol red-free DMEM containing 1–20 μM JC-1. Incubate for 10–30 minutes at 37°C.

Note: correct volumes and incubation times vary between cell lines. A positive control of the technique can be obtained using FCCP or an equivalent uncoupler of oxidative phosphorylation to reduce mitochondrial potential to zero.

4. Resuspend in phenol red-free DMEM and proceed with data collection as soon as possible. Avoid long exposure to light, as this probe bleaches very quickly (especially the green monomeric form).

5. Sample analysis. Acquire 10 000 events from the green and the red channel (maximum emission wavelength ~530 nm for the monomeric form and 590 nm for the polymeric one).

Cells undergoing apoptosis often lose their $\Delta\Psi_m$ in the early phases, and will therefore have a lower red/green ratio compared to the healthy cells (*Figure 12.6*). It is important to note that, due to the rapid kinetics of the loss of membrane potential, these cells will not necessarily display the morphological hallmarks of apoptosis.

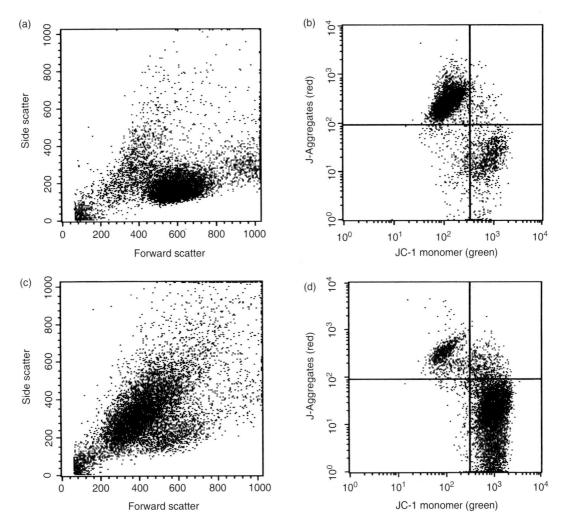

Figure 12.6

Mitochondrial potential evaluation. Apoptosis was induced in Jurkat T cells with α-CD95 antibody and mitochondrial potential was monitored using JC-1. The JC-1 probe has two possible emission spectra; it aggregates and emits in the red spectrum in cells with high mitochondrial potential, but emits as a monomer in the green spectrum in cells with lower mitochondrial potential. Apoptosis is accompanied by a loss of mitochondrial potential, so apoptosis leads to a shift in fluorescence from red (aggregated) to green (monomeric). This figure shows a forward scatter/side scatter plot along with a red/green plot. Untreated cells show a normal cell morphology (Forward vs Side Scatter, a) and a normal mitochondrial potential (JC-1 green vs red fluorescence, b) whereas treated cells are characterised by a change in cell morphology (c) together with a shift toward green fluorescence accompanying mitochondrial potential loss (d).

12.7   Other applications of flow cytometry to apoptosis

In addition to the above methods for studying apoptosis, a number of biochemical changes occurring during apoptosis can also be measured using cytofluorimetric techniques. This section describes some of these approaches.

12.7.1   Cytofluorimetric analysis of protease activity

A number of proteases have been implicated in apoptosis, including caspases, cathepsins and calpains. Caspases are a group of cysteine-dependent, aspartate-specific proteases that play a crucial role in the apoptotic process. These enzymes, initially described as the mammalian homologues of the *Caenorhabditis elegans* protease CED-3, are usually present in cells as inactive precursors and are activated, following apoptotic triggers, by proteolytic processing at Asp residues. Initial activation of caspases is mediated by molecules that generally promote caspase aggregation (e.g. death receptor molecules for caspase-8, and Apaf-1 for caspase-9). Once the first caspases are activated, they can activate other downstream caspases generating a cascade of events that leads to cell death. In particular, caspase-3 seems to play a central role in the apoptotic proteolytic cascade (for review see [37–42]).

Flow cytometry evaluation of caspase activation is possible using a caspase-3 fluorogenic substrate. The fluorochromes are coupled in these molecules in such a way that the intact complex shows relatively little fluorescence upon excitation. However, when the peptide is cleaved by one of caspase-3 family enzymes, the fluorochromes are released and became strongly fluorescent. Several fluorescent substrates are available for this purpose, though the only one used so far for flow cytometry is PhiPhiLux (OncoImmunine Inc.). This substrate is sold in an apoptosis detection kit and cannot be purchased separately. We will therefore not describe the protocol in this book. Among the other proteases, cathepsins are discussed elsewhere in this book (Chapter 8) and we will therefore focus on calpains.

Calpains (calcium-activated neutral proteases) are a large family of calcium-activated cysteine proteases that differ in structure and distribution. Calpains have been implicated in the proteolysis of a number of proteins during apoptosis and in a number of different models such as response to hypoxia in hepatocytes, neuronal degeneration, irradiation and dexamethasone treatment of thymocytes [43–47]. Although some overlap exists between the substrates for calpains and the caspases, the pattern of proteolytic cleavage is very different between the two families.

A cytofluorimetric assay for the evaluation of calpain activity has been developed (Protocol 12.7) and is based on the use of a CMAC-based t-BOC-Leu-Met-CMAC (tert-butyloxycarbonyl-Leu-Met-7-amino-4-chloromethylcoumarin) fluorogenic substrate (Molecular Probes). This probe passively diffuses into cells, where the thiol-reactive chloromethyl group is enzymatically conjugated to glutathione by intracellular glutathione S-transferase or reacts with protein thiols. These reactions transform the substrate into a membrane-impermeable probe that can no longer diffuse out of the cells. Subsequent cleavage by calpains results in a bright blue fluorescent glutathione conjugate.

12.7.2   Cytofluorimetric assessment of free radical formation

Cell survival requires multiple factors, including energy (ATP) and appropriate proportions of molecular oxygen or defined antioxidants. Although most oxidative insults can be overcome by the cell's normal defences, sustained perturbation of this balance can result in either apoptotic or necrotic cell death. Numerous recent studies have shown that the mode of cell death depends on the severity of insult and the effects of oxidative stress on components of the apoptotic machinery may mediate this decision (for review see [48, 49]). Cytofluorimetric evaluation of the formation of reactive oxidative species (ROS) may therefore prove very useful when studying apoptosis. Since different ROS may be involved in apoptosis, large spectrum probes can be used to evaluate the cell redox state, while selective probes directed to specific radical species are also available and can be used when a particular species is under investigation.

DCFDA (2′,7′,-dichlorodihydrofluorescein diacetate) is one of the most commonly used large spectrum probes (Protocol 12.8) and can freely diffuse through the plasma membrane because of its two acetate groups. Once in the cell, the acetate groups are cleaved by esterases and the compound is no longer capable of passing through the membrane and remains in the cell. Once oxidised, DCFDA becomes fluorescent in the green spectrum. This probe reacts with peroxynitrite, nitric oxide and hydrogen peroxide. Since hydrogen peroxide is generated from the action of many enzymes acting on ROS, its modification can be considered a measure of the general redox state of a cell [50, 51] (*Figure 12.7*).

12.7.3   Evaluation of cell calcium content

While neither ubiquitous nor diagnostic of apoptosis, changes in intracellular calcium levels are often associated with apoptotic signalling, and many researchers in the field find quantitation of calcium levels useful. Quantitative evaluation of cellular calcium levels can be performed by flow cytometry using probes that become fluorescent only after calcium binding (Protocol 12.9). One of the most frequently used calcium dyes in flow cytometry is Fluo-3 (from Molecular Probes) and its derivatives. This probe is characterised by a relatively high affinity for calcium, which allows detection of rapid, transient changes in calcium levels. Fluo-3 excited at 488 nm, emits at 530 nm following binding (*Figure 12.8*). Fura (Molecular Probes) is an alternative calcium-binding fluorescent probe, but requires UV excitation [52].

Protocol 12.7: Calpain activity assay

EQUIPMENT, MATERIALS AND REAGENTS

Cells to be analysed.

Ice.

PBS pH 7.4.

t-BOC-Leu-Met-CMAC: add directly in the culture medium to a final concentration of 10–30 μM.

37°C incubator.

Micro-centrifuge.

UV laser or filtered mercury lamp.

METHOD

1. Change culture media with the media containing t-BOC-Leu-Met-CMAC and incubate for approximately 15 to 30 minutes in a 37°C incubator with 5% CO_2.

2. Detach cells as described in Appendix B.

3. Wash cells once in PBS: spin cells at 200 g for 10 minutes, remove supernatant, and thoroughly resuspend in 2 ml PBS. Spin this suspension again at 200 g for 10 minutes, and remove supernatant, leaving cell pellet. Resuspend around 1 million cells in PBS, and acquire data as soon as possible.

4. Sample analysis. Acquire 10 000 events exciting at 360 nm with a UV laser or filtered mercury lamp; collect emission data at 530 nm. Calpain activity will be proportional to the amount of substrate cleaved, and therefore to the total blue fluorescence; cells with more blue fluorescence have more calpain activity.

Protocol 12.8: Cytofluorimetric assessment of free radical formation

EQUIPMENT, MATERIALS AND REAGENTS

Cells to be analysed.

PBS pH 7.4.

HBSS.

DCFDA dye: dissolve DCFDA in anhydrous DMSO to obtain a 10 mM stock solution.

Micro-centrifuge.

Flow cytometer.

METHOD

1. Detach cells as described in Appendix B.

2. Wash cells once in PBS: spin cells at 200 g for 10 minutes, remove supernatant, and thoroughly resuspend in 2 ml PBS. Spin this suspension again at 200 g for 10 minutes, and remove supernatant, leaving cell pellet. Resuspend 1–2 million cells in PBS.

3. Incubate with 200 μl of a 10 μM solution of DCFDA in HBSS for 15–20 minutes, depending on the cell line under investigation. To determine proper dye concentration, 10 mM hydrogen peroxide can be used as a positive control for dye activity. Upon hydrogen peroxide addition, green fluorescence should increase at least 5-fold.

4. Wash cells once in PBS: spin cells at 200 g for 10 minutes, remove supernatant, and thoroughly resuspend in 2 ml PBS.

5. Sample analysis. Acquire 10 000 events using the green channel (maximum emission ~530 nm). ROS concentration will be proportional to the amount of DCFDA oxidized, and therefore to green fluorescence intensity. The mean green intensity is proportional to ROS concentration (*Figure 12.7*).

Green fluorescent DCFDA may be counterstained with PI (2 μg ml^{-1} for 5 minutes, maximum emission ~640 nm) to assess cell viability at the same time, but is not necessary for ROS measurement.

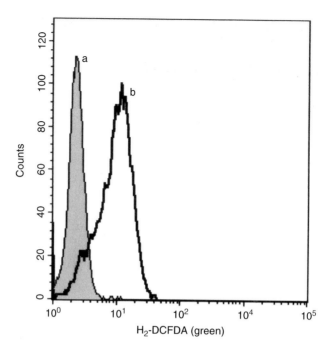

Figure 12.7

Reactive oxygen species (ROS) generation assayed by DCFDA. ROS generated in SH-5YSY cells treated with 4-hydroxyphenylretinamide (b), and in untreated control cells (a) were detected using the H_2-DCFDA probe. H_2-DCFDA fluorescence is activated by ROS, so 4-hydroxyphenylretinamide treatment results in increased fluorescence.

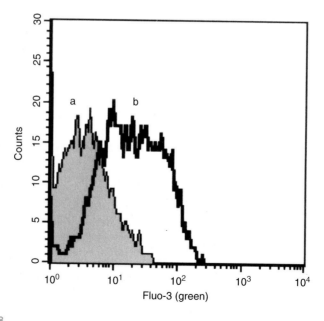

Figure 12.8

Calcium intracellular content assayed by Fluo-3. Ionomycin-treated CHP-100 cells show a fluorescence profile shift (b) due to intracellular Ca^{2+} increase with respect to the control (a). The Fluo-3 probe becomes fluorescent in the presence of ionic calcium, resulting in increased fluorescence in ionomycin-treated cells.

Protocol 12.9: Evaluation of cell calcium content

EQUIPMENT, MATERIALS AND REAGENTS

Cells to be analysed.

Ca^{2+}- and Mg^{2+}-free HBSS.

HBSS with Ca^{2+} and Mg^{2+}

Fluo-3 dye: dissolve Fluo-3 powder in DMSO to a final concentration of 10 mM; this concentration can be used as a stock solution.

Micro-centrifuge.

37°C incubator.

Flow cytometer.

METHOD

1. Detach cells as described in Appendix B.

2. Resuspend 0.5–2 million cells in Ca^{2+}- and Mg^{2+}- free HBSS, add the Fluo-3 stock solution to obtain a final probe concentration of 10 µM and incubate for 20 minutes at 37°C.

3. Wash in Ca^{2+}- and Mg^{2+}-free HBSS: spin cells at 200 g for 10 minutes, remove supernatant, and thoroughly resuspend in 2 ml HBSS. Spin this suspension again at 200 g for 10 minutes, and remove supernatant, leaving cell pellet. Resuspend in 2 ml HBSS with Ca^{2+} and Mg^{2+} and analyse cells as soon as possible.

Note: since calcium levels change very quickly following a particular stimulus it is possible to add the stimulus directly to the cytometry tube, while acquiring cells for analysis, and follow the changes in fluorescent emission. In order to do this, cells have to be resuspended in media containing Ca^{2+} (2 mM).

4. Sample analysis. Acquire 10 000 events and analyse the green fluorescence (maximum emission ~540 nm); high levels of green fluorescence correspond to high calcium levels (*Figure 12.8*).

A positive control can be obtained by adding ionomycin, to a final concentration of 3 µM, to the cell solution. This compound induces massive calcium uptake by the cell, saturating the fluorescent dye. Quantitative measurement of Ca^{2+} can be calculated with the following formula:

$$[Ca^{2+}] = Kd \frac{[F - F_{min}]}{[F_{max} - F]}$$

where 'F' is the fluorescence emitted by our sample, and maximum fluorescence 'F_{max}' is the fluorescence registered by adding to the same sample 3 μM ionomycin in the presence of 2 mM Ca^{2+}. Baseline fluorescence (F_{MnCl_2}) is measured by adding 3 μM ionomycin in the presence of 2 mM $MnCl_2$. In this way Mn^{2+} enters the cell but does not bind to the dye.

Minimal fluorescence can now be calculated (F_{min}) by the following formula:

$$F_{min} = 1.25 \times F_{MnCl_2} - 0.25 \times F_{max}$$

12.8   References

1. Shapiro, H.M. (1995) *Practical Flow Cytometry*. Wiley-Liss, New York.
2. Darzynkiewicz, Z., Robinson, J.P. and Crissman, H. (1994) *Flow Cytometry*. Academic Press, San Diego.
3. Ormerod, M.G. (1998) The study of apoptotic cells by flow cytometry. *Leukemia* **12**: 1013–1025.
4. Blackstone, N.W. and Green D.R. (1999) The evolution of a mechanism of cell suicide. *Bioassays* **21**: 84–88.
5. Evan, G.I. and Vousden, K.H. (2001) Proliferation, cell cycle and apoptosis in cancer. *Nature* **411**: 342–348.
6. Ferri, K.F. and Kroemer G. (2001) Organelle-specific initiation of cell death pathways. *Nat. Cell Biol.* **3**: E255–E263.
7. Lockshin, R.A. and Zakeri, Z. (2001) Timeline: programmed cell death and apoptosis: origin of the theory. *Nat. Rev. Mol. Cell Biol.* **2**: 545–550.
8. Kerr, J.F.R., Wyllie, A.H. and Currie, A.R. (1972) Apoptosis: a basic biological phenomenon with wide-ranging implication in tissue kinetics. *Br. J. Cancer* **26**: 239–257.
9. Kerr, J.F.R., Harmon, B.V. and Searle, J. (1974) An electron-microscope study of cell deletion in the anuran tadpole tail during spontaneous metamorphosis with special reference to apoptosis of striated muscle fibres. *J. Cell. Sci.* **14**: 571–585.
10. Martelli, A.M., Zweyer, M., Ochs, R.L., Tazzari, P.L, Tabellini, G., Narducci, P. and Bortul, R. (2001) Nuclear apoptotic changes: an overview. *J. Cell Biochem.* **82**: 634–646.
11. Ormerod, M.G., Paul, F., Cheetham, M. and Sun, X.M. (1995) Discrimination of apoptotic thymocytes by forward light scatter. *Cytometry* **21**: 300–304.
12. Mentz, F., Baudet, S., Blanc, C., Issaly, F., Binet, J.L. and Merle-Beral, H. (1998) Simple, fast method of detection apoptosis in lymphoid cells. *Cytometry* **32**: 95–101.
13. Sasaki, D.T., Dumas, S.E. and Engleman, E.G. (1987) Discrimination of viable and non-viable cells using propidium iodide in two color immunofluorescence. *Cytometry* **8**: 413–420.
14. Ormerod, M.G., Sun, X.M., Snowden, R.T., Davies, R., Fearnhead, H. and Cohen, G.M. (1993) Increased membrane permeability of apoptotic thymocytes: a flow cytometric study. *Cytometry* **14**: 595–602.
15. Schmid, I., Uittenbogaart, C.H., Keld, B. and Giorgi, J.V. (1994) A rapid method for measuring apoptosis and dual-color immunofluorescence by single laser flow cytometry. *J. Immunol. Methods* **170**: 145–157.
16. Schmid, I., Krall, W.J., Uittenbogaart, C.H., Braun, J. and Giorgi, J.V. (1992) Dead cell discrimination with 7-amino-actinomycin D in combination with dual color immuno-fluorescence in single laser flow cytometry. *Cytometry* **13**: 204–208.
17. Fadok, V.A., Voelker, D.R., Campbell, P.A., Cohen, J.J., Bratton, D.L. and Henson, P.M. (1992) Exposure of phosphatidylserine on the surface of apoptotic lymphocytes triggers specific recognition and removal by macrophages. *J. Immunol.* **148**: 2207–2216.
18. Vermes, I., Haanen, C., Steffens-Nakken, H. and Reutelingsperger, C. (1995) A novel assay for apoptosis. Flow cytometric detection of phosphatidylserine expression on early apoptotic cells using fluorescein labelled Annexin V. *J. Immunol. Methods* **184**: 39–51.
19. Martin, S.J., Reutelingsperger, C.P., McGahon, A.J., Rader, J.A., van Schie, R.C., LaFace, D.M. and Green, D.R. (1995) Early redistribution of plasma membrane phos-phatidylserine is a general feature of apoptosis regardless of the initiating stimulus: inhibition by overexpression of Bcl-2 and Abl. *J. Exp. Med.* **182**: 1545–1556.
20. van Engeland, M., Ramaekers, F.C., Schutte, B. and Reutelingsperger, C.P. (1996) A novel assay to measure loss of plasma membrane asymmetry during apoptosis of adherent cells in culture. *Cytometry* **24**: 131–139.

21. Wyllie, A.H. (1980) Glucocorticoid-induced thymocyte apoptosis is associated with endogenous endonuclease activation. *Nature* **284**: 555–556.

22. Montague, J.W. and Cidlowski, J.A. (1996) Cellular catabolism in apoptosis: DNA degradation and endonuclease activation. *Experientia* **52**: 957–962.

23. Nicoletti, I., Migliorati, G., Pagliacci, M.C., Grignani, F. and Riccardi, C. (1991) A rapid and simple method for measuring thymocyte apoptosis by propidium iodide staining and flow cytometry. *J. Immunol. Methods* **139**: 271–279.

24. Tounekti, O., Belehradek, J. Jr. and Mir, L.M. (1995) Relationships between DNA fragmentation, chromatin condensation, and changes in flow cytometry profiles detected during apoptosis. *Exp. Cell Res.* **217**: 506–516.

25. Gavrieli, Y., Sherman, Y. and Ben-Sassonm, S.A. (1992) Identification of programmed cell death in situ via specific labeling of nuclear DNA fragmentation. *J. Cell Biol.* **119**: 493–501.

26. Maciorowski, Z., Veilleux, C., Gibaud, A., Bourgeois, C.A., Klijanienko, J., Boenders, J. and Vielh, P. (1997) Comparison of fixation procedure for fluorescent quantitation of DNA content using image cytometry. *Cytometry* **28**: 123–129.

27. Schimenti, K.J. and Jacobberger, J.W. (1992) Fixation of mammalian cells for flow cytometric evaluation of DNA content and nuclear immunofluorescence. *Cytometry* **13**: 48–59.

28. Kalejta R.F., Shenk, T. and Beavis, A.J. (1997) Use of a membrane-localized green fluorescent protein allows simultaneous identification of transfected cells and cell cycle analysis by flow cytometry. *Cytometry* **29**: 286–291.

29. McKenna, S.L., Hoy, T., Holmes, J.A., Whittaker, J.A., Jackson, H. and Padua, R.A. (1998) Flow cytometric apoptosis assays indicate different types of endonuclease activity in haematopoietic cells and suggest a cautionary approach to their quantitative use. *Cytometry* **31**: 130–136.

30. Hedenfalk, I.A., Baldetorp, B., Borg, A. and Oredsson, S.M. (1997) Activated cell cycle checkpoints in epirubicin-treated breast cancer cells studied by BrdUrd-flow cytometry. *Cytometry* **29**: 321–327.

31. Ninomiya, Y., Adams, R., Morriss-Kay, G.M. and Eto, K. (1997) Apoptotic cell death in neuronal differentiation of P19 EC cells: cell death follows reentry into S phase. *J. Cell Physiol.* **172**: 25–35.

32. Faretta, M., Bergamaschi, D., Taverna, S., Ronzoni, S., Pantarotto, M., Mascellani, E., Cappella, P., Ubezio, P. and Erba, E. (1998) Characterization of cyclin B1 expression in human cancer cell lines by a new three-parameter BrdUrd/cyclin B1/DNA analysis. *Cytometry* **31**: 53–59.

33. Kroemer, G. and Reed, J.C. (2000) Mitochondrial control of cell death. *Nat. Med.* **6**: 513–519.

34. Cossarizza, A., Baccarani-Contri, M., Kalashnikova, G. and Franceschi, C. (1993) A new method for the cytofluorimetric analysis of mitochondrial membrane potential using the J-aggregate forming lipophilic cation 5,5′,6,6′-tetrachloro-1,1′,3,3′-tetraethyl-benzimidazolcarbocyanine iodide (JC-1). *Biochem. Biophys. Res. Comm.* **197**: 40–45.

35. Salvioli, S., Ardizzoni, A., Franceschi, C. and Cossarizza, A. (1997) JC-1, but not DiOC6(3) or rhodamine 123, is a reliable fluorescent probe to assess delta psi changes in intact cells: implications for studies on mitochondrial functionality during apoptosis. *FEBS Lett.* **411**: 77–82.

36. Bkaily, G., Jacques, D. and Pothier, P. (1999) Use of confocal microscopy to investigate cell structure and function. *Meth. Enzymol.* **307**: 119–135.

37. Nicholson, D.W. (1999) Caspase structure, proteolytic substrates, and function during apoptotic cell death. *Cell Death Differ.* **6**: 1028–1042.

38. Zheng, T.S., Hunot, S., Kuida, K. and Flavell, R.A. (1999) Caspase knockouts: matters of life and death. *Cell Death Differ.* **6**: 1043–1053.

39. Stennicke, H.R. and Salvesen G.S. (1999) Catalytic properties of the caspases. *Cell Death Differ.* **6**: 1054–1059.

40. Kumar, S. (1999) Mechanisms mediating caspase activation in cell death. *Cell Death Differ.* **6**: 1060–1066.

41. Slee, E.A, Adrain, C. and Martin, S.J. (1999) Serial killers: ordering caspase activation events in apoptosis. *Cell Death Differ.* **6**: 1067–1074.

42. Zeuner, A., Eramo, A., Peschle, C. and De Maria, R. (1999) Caspase activation without death. *Cell Death Differ.* **6**: 1075–1080.

43. Bronk, S.F. and Gores, G.J. (1993) pH-dependent nonlysosomal proteolysis contributes to lethal anoxic injury of rat hepatocytes. *Am. J. Physiol.* **264**: G744–G751.

44. Saito, K., Elce, J.S., Hamos, J.E. and Nixon, R.A. (1993) Widespread activation of calcium-activated neutral proteinase (calpain) in the brain in Alzheimer disease: a potential molecular basis for neuronal degeneration. *Proc. Natl Acad. Sci. U. S. A.* **90**: 2628–2632.

45. Vanags, D.M, Porn-Ares, M.I., Coppola, S., Burgess, D.H. and Orrenius, S. (1996) Protease involvement in fodrin cleavage and phosphatidylserine exposure in apoptosis. *J. Biol. Chem.* **271**: 31075–31085.

46. Squirer, M.K., Miller, A.C., Malkinson, A.M. and Cohen, J.J. (1994) Calpain activation in apoptosis. *J. Cell Physiol.* **159**: 229–237.

47. Squirer, M.K. and Cohen, J.J. (1997) Calpain, an upstream regulator of thymocyte apoptosis. *J. Immunol.* **158**: 3690–3697.

48. Simon, H.U., Haj-Yehia A. and Levi-Schaffer F. (2000) Role of reactive oxygen species (ROS) in apoptosis induction. *Apoptosis* **5**: 415–418.

49. Datta, K., Sinha S. and Chattopadhyay, P. (2000) Reactive oxygen species in health and disease. *Natl Med. J. India* **12**: 304–310.

50. Possel, H., Noack, H., Augustin, W., Keilhoff, G. and Wolf, G. (1997) 2,7-Dihydrodichlorofluorescein diacetate as a fluorescent marker for peroxynitrite formation. *FEBS Lett.* **416**: 175–178.

51. Xie, Z., Kometiani, P., Liu, J., Li, J., Shapiro, J.I. and Askari, A. (1999) Intracellular reactive oxygen species mediate the linkage of Na^+/K^+-ATPase to hypertrophy and its marker genes in cardiac myocytes. *J. Biol. Chem.* **274**: 19323–19328.

52. Struk, A., Szucs, G., Kemmer, H. and Melzer, W. (1998) Fura-2 calcium signals in skeletal muscle fibres loaded with high concentrations of EGTA. *Cell Calcium* **23**: 23–32.

Appendix A: Typical emission spectra of different dyes

450–500 nm: blue emission (DAPI, Hoechst 33258, blue GFP) only for UV or multi-photon lasers.

500–550 nm: green-FL1 for Becton Dickinson cytometers (FITC, e-GFP).

550–600 nm: yellow; rarely used in cytometry due to strong overlap of emission with green and red channel (yGFP).

600–650 nm: orange-FL2 for Becton Dickinson cytometers (PI, PE, RFP (red fluorescent protein)).

650–750 nm: red-FL3 Becton Dickinson cytometers (PI, 7-AAD, APC). APC can be detected using a 635 nm solid state laser (4-channel configuration in far-red double laser Becton Dickinson system).

Appendix B: Cell collection protocol

For non-adherent cells, spin at $200\,g$ for 10 minutes and collect pellet. For adherent cells, first collect supernatant and spin at $200\,g$ to collect detached cells, that may contain a considerable number of apoptotic cells, then detach the remaining cells using a 0.025%/0.02% trypsin/EDTA solution for approximately 10 minutes. Trypsinisation conditions vary between cell lines, so some experimentation is required to find the time and concentration most suitable for each. Trypsinisation is recommended and mechanical detachment of cells should be avoided since it does not yield the single cell suspension required for cytofluorimetric analysis.

Appendix I. Suppliers of Reagents and Equipment as Recommended by Authors in this Book

General reagent and equipment suppliers

BDH Merck
See VWR International

Merck
See VWR International

Sigma-Aldrich Co Ltd (biochemicals, reagents and some equipment)
Fulka, Fancy Road, Poole, Dorset BH12 4QH, UK.
Tel: +44 (0)1202-733114 Fax: +44 (0) 1202 740171 http://www.sigma-aldrich.com

VWR International Ltd
Hunter Boulevard, Magna Park, Lutterworth, Leics LE17 4XN, UK.
UK Freephone 0800-223344 Fax: +44 (0)1455-558586 http://www.vwr.com

Molecular biology reagents

Advanced Biotechnologies (molecular biology kits and reagents, rapid chromosome identification)
ABgene House, Blenheim Road, Epsom, Surrey KT19 9AP, UK.
Tel: +44(0)1372-723456   Fax: +44(0)1372-741414   http://www.adbio.co.uk   or www.abgene.com/

Affiniti Research Products Ltd
Mamhead Castle, Mamhead, Exeter, Devon EX6 8HD, UK.
Tel: +44 (0)1626-891010 Fax: +44 (0)1626-891090 http://www.affiniti-res.com/

Alexis Biochemicals
QBiogene-Alexis Ltd, P.O. Box 6757, Bingham, Nottingham NG13 8LS, UK.
Tel: +44 (0)1949-836111 Fax: +44 (0)1949-836222 http://www.qbiogene.com

Boehringer Mannheim
See Roche Molecular Biochemicals

Cell Proliferation and Apoptosis, David Hughes and Huseyin Mehmet (Eds)
© 2003 BIOS Scientific Publishers Ltd, Oxford

Cambridge Bioscience (distributes products from Molecular Probes, Clonetech)
24–25 Signet Court, Newmarket Road, Cambridge CB5 8LA, UK.
Tel: +44 (0)1223-316855 http://www.bioscience.co.uk

Molecular Probes
For UK see Cambridge BioScience

New England Biolabs (UK) Ltd
73 Knowl Piece, Wilbury Way, Hitchin, Hertfordshire SG4 0TY, UK.
Tel: +44 (0)1462-420616 Fax: +44 (0)1462-421057 http://www.neb.com/neb/

NeXins Research BV
Monnikendijk 4, 4474 ND Kattendijke, The Netherlands.
Tel: +31-113-626062 Fax: +31-113-626035 http://www.nexins.com/

Nutacon
Tuinderij 25, Postbus 94, 2450 AB, Leimuiden, The Netherlands.
Tel: +31-172-506214 Fax: +31-172-506515 http://www.nutacon.nl/uk/

Pharmingen
See BD Biosciences

Pierce Biotechnology
Perbio Science UK Ltd, Century House, High Street, Tattenhall, Cheshire CH3 9RJ, UK.
Tel: +44 (0)1829-771744 Fax: +44 (0)1829-771644 http://www.piercenet.com/

Promega Ltd
Delta House, Chilworth Research Centre, Southampton, UK.
UK Freephone: 0800 378994 UK Freefax: 0800 181037 http://www.promega.com/uk/

Qiagen Ltd
Boundary Court, Gatwick Road, Crawley, West Sussex RH10 2AX, UK.
Tel: +44 (0)1293-422911 Fax: +44 (0)1293-422922 http://www.qiagen.com/uk

Roche Molecular Biochemicals
Roche Diagnostics Ltd, Bell Lane, Lewes, East Sussex BN7 1LG, UK.
Tel: +44 (0)1273 480444

Schleicher & Schuell MicroScience GmbH
Hahnestrasse 3, D-37586 Dassel, Germany.
Tel: +49 (0)5561-791580 Fax: +49 (0)5561-791536 http://www.s-and-s.com/

Upstate Biotechnology
Unit 3, Mill Square, Featherstone Road, Wolverton Mill South, Milton Keynes MK12 5YU, UK.
Tel: +44 (0)1908-552820 Fax: +44 (0)1908-552821 http://www.upstate.com

Transduction Laboratories
See BD Biosciences under *immunocytochemistry*

Immunocytochemistry reagents and equipment; flow cytometry

Aurion
Costerweg 5, 6702 AA Wageningen, The Netherlands.
Tel: +31-317-497676 Fax: +31-317-415955 http://www.aurion.nl/

Beckman Coulter UK Ltd
Oakley Court, Kingsmead Business Park, London Road, High Wycombe, Bucks HP11 1JU, UK.
Tel: +44 (0)1494-442233 Fax: +44 (0)1494-447558 http://www.beckmancoulter.com

BD Biosciences (Becton Dickinson)
Becton Dickinson UK Ltd, Between Towns Road, Cowley, Oxford OX4 3LY, UK.
Tel: +44 (0)1865-748844 Fax: +44 (0)1865-717313 email: customerservice@europe.bd.com

British BioCell International
BBInternational Ltd, Golden Gate, Ty Glas Avenue, Cardiff CF14 5DX, UK.
Tel: +44 (0)2920-747232 Fax: +44 (0)2920-747242 http://www.british-biocell.co.uk/

Calbiochem
CN Biosciences (UK) Ltd, Boulevard Ind'l Park, Padge Road, Beeston, Nottingham NG9 2JR, UK.
Tel: +44 (0)115-943-0840 Fax: +44 (0)115-943-0951 http://www.calbiochem.com/

DakoCytomation
Denmark House, Cambridgeshire Business Park, Angel Drive, Ely, Cambs CB7 4ET, UK.
Tel: +44 (0)1353-669911 Fax: +44 (0)1353-668989 www.dakocytomation. com

MBL (Medical and Biological Laboratories) Ltd
See Stratech Scientific Ltd.

Stratech Scientific Ltd
61–63 Dudley Street, Luton, Bedfordshire LU2 0NP, UK.
Tel: +44 (0)1582-529000 Fax: +44 (0)481895 http://stratech.co.uk

Vector Laboratories Ltd (immunology and antibody technology; distributors for Novocastra)
3 Accent Park, Bakewell Rd, Orton Southgate, Peterborough PE2 6XS, UK.
Tel: +44 (0)1733-237999 Fax: +44 (0)1733-237119 http://www.vectorlabs.com

Miscellaneous and microscopy

Amersham Pharmacia Biotech
Amersham Biosciences UK Limited, Amersham Place, Little Chalfont, Bucks HP7 9NA UK.
Tel: +44 (0)870-6061921 Fax: +44 (0)1494-544350 http://www.amershamhealth. com/

BioRad
Bio-Rad Laboratories Ltd, Bio-Rad House, Maylands Avenue, Hemel Hempstead HP2 7TD, UK.
Tel: +44 (0)20-8328-2000 Fax: +44 (0)20-8328-2550 http://www.bio-rad.com/

Gibco-BRL

Life Technologies Ltd, 3 Fountain Drive, Inchinnan Business Park, Paisley PA4 9RF, UK.
Tel: +44 (0)141-8146100 Fax: +44 (0)141-8146287 www.lifetech.com

Halocarbon Products Corp

P.O. Box 661 River Edge, New Jersey 07661, USA.
Tel: +1-011-201-262-8899 Fax: +1-011-201-262-0019 email: info@kmzchemicals.ltd.uk

Kodak Ltd

Kodak House, Station Road, Hemel Hempstead, Herts HP1 1JU, UK.
Tel: +44 (0)1442-261122 http://www.kodak.co.uk/

Sysmex Corporation UK Ltd

Sunrise Parkway, Linford Wood (East), Milton Keynes, Buckinghamshire MK14 6QF, UK.
Tel: +44 (0)1908-669555 Fax: +44 (0)1908-669409 http://www.sysmex.co.uk/

Theseus Imaging

North American Scientific, Inc., 20200 Sunburst Street, Chatsworth, CA 91311, USA.
Tel: +1-818-734-8600 Fax: +1-818-734-5223 http://www.nasi.net/index.html

Zeiss (microscopes, image analysis, confocal microscopes)

Carl Zeiss Ltd, PO Box 78, Woodfield Road, Welwyn Garden City, Herts AL7 1LU, UK.
Tel: +44 (0)1707-871350 Fax: +44 (0)1707-871287 http://www.zeiss.co.uk

Appendix II. Safety Notes Regarding the Use of Procedures Described in this Book

Many of the chemicals employed in the protocols described in this book are potentially hazardous. Always perform your own risk assessments before following the methods, paying attention to the guidelines issued by your own establishment. Always read the data sheets provided by the suppliers and use appropriate safety measures to avoid personal and third-party exposure, and to comply with local disposal regulations.

Particularly great care should be taken when using DIFP, NEM and other similar inhibitors.

TAT fusion proteins are capable of translocating into cells of all known tissues, and so are capable of causing damage to those tissues if accidentally introduced into the body. The generation and use of these proteins should be treated with caution. This is particularly important when fusion proteins are designed to modulate the cell cycle and may induce apoptosis. You should adhere to locally approved safety procedures specified for the particular TAT-fusion protein being generated.

Appropriate care should be taken when using all electrical equipment, and particularly with respect to the high voltages involved in the use of electrophoresis apparatus.

Before starting experiments involving radioactivity you should discuss the health and safety issues with the appropriate Local Safety Officer.

Cell Proliferation and Apoptosis, David Hughes and Huseyin Mehmet (Eds)
© 2003 BIOS Scientific Publishers Ltd, Oxford

Appendix III.
Acknowledgements

The editors are most grateful to the chapter contributors for their prompt submission of manuscripts and their forbearance during the editorial process. Especial thanks are due to the editorial staff at BIOS Scientific Publishers for their professional guidance during the formulation of this volume. The editors also wish to record their gratitude to *DakoCytomation* for their generous sponsorship of the colour plates.

Chapter 4

The authors thank Steve Orr for some of the data shown in *Figures 4.1* and *4.2*, and Alison John for critical comments. Research in the Cell Cycle laboratory at the Rayne Institute is funded by the Charles Wolfson Charitable Trust and the Leukaemia Research Fund.

Chapter 5

The work in the author's laboratory is supported by grants from the United Kingdom Co-ordinating Committee for Cancer Research (radiation program) and EUROATOM Grants, FIGH-CT1999-00011 and FIGH-CT1999-00009 from the European Union. The author is grateful to Drs. Paul Finnon, Andrew Silver and Simon Bouffler for providing the protocol for TRF measurement and *Figure 5.3*.

Chapter 10

NJW is a Peter Doherty Fellow of the NH&MRC, Australia. The work was supported by NIH grants AI40646 and CA69381.

Index